Advanced Landslide Susceptibility Modeling

R. M. Yuvaraj · M. V. Karunambigai · V. Gopal
Editors

Advanced Landslide Susceptibility Modeling

Effects on Sustainability and Human and Economic Welfare

Springer

Editors
R. M. Yuvaraj
Department of Geography
Institute of Distance Education
University of Madras
Chennai, Tamil Nadu, India

M. V. Karunambigai
Department of Tourism and Travel Management
J. B. A. S. College for Women
Chennai, Tamil Nadu, India

V. Gopal
Centre for Earth and Atmospheric Sciences (CEAS)
Sathyabama Institute of Science and Technology
Chennai, Tamil Nadu, India

ISBN 978-3-032-14484-3 ISBN 978-3-032-14485-0 (eBook)
https://doi.org/10.1007/978-3-032-14485-0

This Springer imprint is published by the registered company Springer Nature Switzerland AG
The registered company address is: Gewerbestrasse 11, 6330 Cham, Switzerland

Foreword

Landslides are among the most destructive and unpredictable natural hazards, posing severe threats to human life, infrastructure, and ecosystems worldwide. Their occurrence is intricately linked to complex interactions between geology, geomorphology, climate dynamics, and human activities. In recent years, accelerated urbanization, deforestation, and the intensifying effects of climate change have significantly increased the vulnerability of fragile hill and mountain ecosystems to landslide disasters. These events not only cause physical damage but also have far-reaching implications for economic stability, community livelihoods, and sustainable development.

In this context, adopting advanced scientific tools and holistic approaches for landslide prediction and management is no longer an option but a necessity. The integration of **geospatial technologies, machine learning models, and sustainability-driven frameworks** has opened new horizons for understanding the spatial-temporal dynamics of landslide susceptibility. This book, *Advanced Landslide Susceptibility Modeling: Effects on Sustainability and Human and Economic Welfare*, arrives at a critical juncture, offering a comprehensive and multidisciplinary exploration of this pressing issue.

The volume brings together contributions from leading experts who examine landslide risk through multiple lenses—ranging from terrain-based and geostatistical modeling to the socio-economic dimensions of hazard impacts and the role of indigenous knowledge in building resilience. Each chapter not only presents scientific rigor but also emphasizes practical strategies for policy-making, risk reduction, and sustainable land-use planning.

I commend the editors and contributors for curating this insightful work, which bridges the gap between academic research and real-world applications. The book is an essential resource for researchers, environmental scientists, planners, and disaster management professionals seeking innovative, evidence-based solutions to mitigate landslide hazards and enhance community resilience.

R. Jaganathan
Professor and Head
Department of Geography
School of Earth and Atmospheric Science
University of Madras
Chennai, India

Preface

The genesis of this book lies in the growing concern over the increasing frequency and intensity of landslides worldwide, particularly in fragile hill ecosystems. Landslides are not merely geological phenomena; they have profound implications for human lives, economic stability, and environmental sustainability. With the advent of advanced geospatial tools, artificial intelligence, and multi-criteria decision-making techniques, researchers now have unprecedented opportunities to model, predict, and mitigate landslide hazards with greater precision.

This book, *Advanced Landslide Susceptibility Modeling: Effects on Sustainability and Human and Economic Welfare*, brings together contributions from distinguished researchers and practitioners across various disciplines. The chapters encompass a wide spectrum of themes—geographical and geological determinants of landslides, innovative modeling techniques like AHP, Maxent, MGWR, and ANN, climate change-induced vulnerabilities, socio-economic consequences, and community-based risk reduction strategies.

Our aim is to provide a comprehensive resource that not only advances academic knowledge but also informs policy, planning, and practical interventions. We strongly believe that sustainable landslide risk management requires an integrated approach that combines scientific innovation with local wisdom and participatory strategies.

We express our sincere gratitude to all the authors for their valuable contributions, the reviewers for their insightful feedback, and Springer Nature for their continued commitment to publishing high-quality scientific works. We hope this book will inspire further research and practical solutions toward achieving resilience and sustainability in landslide-prone regions.

Chennai, India

R. M. Yuvaraj
M. V. Karunambigai
V. Gopal

Acknowledgments

The successful completion of this edited volume, *Advanced Landslide Susceptibility Modeling: Effects on Sustainability and Human and Economic Welfare*, would not have been possible without the valuable contributions, guidance, and support of many individuals and institutions.

First and foremost, we express our profound gratitude to **Springer Nature** for providing this distinguished platform to publish and disseminate cutting-edge research on landslide susceptibility, environmental sustainability, and geospatial innovation. Their continued commitment to academic excellence and scientific advancement has been a great source of encouragement.

We sincerely thank all the **chapter contributors** for their scholarly inputs, rigorous research efforts, and timely cooperation, which have enriched this book with diverse perspectives and scientific depth. Each chapter reflects the dedication and collaborative spirit of researchers who share a common goal of mitigating landslide risks through scientific innovation and sustainable approaches.

We extend our heartfelt appreciation to the **reviewers** for their insightful comments, constructive suggestions, and meticulous efforts that ensured the high academic quality of this publication.

This book is dedicated to all researchers, scholars, and practitioners striving toward a safer, sustainable, and resilient world through scientific knowledge, innovation, and the continued advancement of the discipline of geography at the international level.

Contents

Editors and Contributors

About the Editors

Dr. R. M. Yuvaraj is an accomplished academician and researcher in the field of geography, with a specialization in environmental geography, geo-informatics, and disaster risk assessment. He currently serves as an assistant professor in the Department of Geography, Institute of Distance Education, University of Madras, where he is dedicated to advancing research and education in geospatial technologies and environmental sustainability.

Dr. Yuvaraj earned his Ph.D. in Geography from the University of Madras, focusing on agricultural land transformation, productivity patterns, and land degradation. He has cleared the UGC-NET twice and qualified as both Junior Research Fellow (JRF) and Senior Research Fellow (SRF) in Geography. His research expertise encompasses GIS and remote sensing applications, environmental modeling, landslide susceptibility mapping, air and water quality assessment, and sustainable tourism development.

With a prolific academic record, Dr. Yuvaraj has authored two books, including *Eco-Tourism: A Geographical Perspective and Physical Geography*, and contributed several book chapters with Springer Nature. He has published extensively in Scopus and Web of Science indexed journals, such as *Journal of Coastal Conservation*, *Quaternary Science Advances*, and *Applied Water Science*, covering cutting-edge topics like dynamic landslide risk forecasting, multi-criteria decision analysis for tourism, and environmental monitoring using geospatial tools.

Dr. Yuvaraj has actively contributed to the academic community as a reviewer for reputed journals including *Applied Water Science* (Springer) and as an editor for international journals in *Earth and Environmental Sciences*. He has organized faculty development programs and academic writing workshops to promote quality research and knowledge dissemination. His academic achievements have earned him recognition, including the Young Scientist Award.

A strong advocate of geospatial technology for sustainable development, Dr. Yuvaraj's research integrates AI, GIS, and statistical modeling to address pressing issues in environmental management and disaster resilience. His contributions aim to bridge the gap between research and practical applications, fostering innovations that support sustainability and resilience in the face of global challenges.

Dr. M. V. Karunambigai is a distinguished academician and researcher with over 14 years of teaching and academic leadership in the field of tourism and travel management, coupled with several years of industry experience. Currently serving as the associate professor and head of the Department of Tourism and Travel Management at J. B. A. S. College for Women, Chennai, she has played a pivotal role in curriculum development, academic research, and institutional quality enhancement.

Her multidisciplinary academic background spans geography, tourism administration, human resource management, and history. She holds a Ph.D. in Geography with a specialization in Tourism, an M.Phil. in Geography, an MBA in Human Resources, and two Master's degrees—M.Sc. in Geography and Tourism Administration and M.A. in History. She is also a qualified UGC-NET (Tourism Administration and Management) academic.

Dr. Karunambigai's research interests include sustainable tourism, religious tourism, heritage management, cultural geography, and tourism education. She has presented papers at national and international conferences, with notable presentations such as "Temples as a Religious Resource of Tourism" at Queen Mary's College, Chennai. Her scholarly contributions extend to editing academic volumes and authoring book chapters focusing on India's tourism resources, culinary heritage, and UNESCO World Heritage Sites.

Beyond academia, Dr. Karunambigai brings rich professional experience from the tourism industry, having worked with leading travel agencies such as Sugal Tours and Pegasus Travels, as well as in the corporate sector with Perot Systems and Wipro Technologies. Her corporate exposure adds practical depth to her teaching and research.

She has actively participated in various national and international workshops and seminars on topics including higher education quality, disability sensitization, human rights, and scientific computing. Her commitment to academic excellence is reflected in her ongoing contributions to faculty development and student mentoring.

Dr. Karunambigai continues to inspire students and peers alike through her interdisciplinary approach, research productivity, and unwavering dedication to the advancement of tourism studies in India and beyond.

Dr. V. Gopal is an environmental geoscientist with extensive research expertise across various domains, including geochemistry, environmental pollution, sedimentology, and coastal ecosystem studies. He presently holds the position of an assistant professor (research) in the Centre for Earth and Atmospheric Sciences at Sathyabama Institute of Science and Technology in Chennai, India. He has contributed significantly to enhancing comprehension of coastal and estuarine systems, particularly throughout the Indian coastline.

Dr. Gopal possesses a Ph.D. in Applied Geology from the University of Madras and a UGC-DS Kothari Postdoctoral Fellowship from Anna University. This indicates he possesses extensive foundational and advanced knowledge on the interactions between land and sea, the introduction of trace elements into the environment, and the dynamics of sediment transport. He has authored over 39 peer-reviewed publications and book chapters concerning heavy metal contamination, ecological risk evaluation, and paleoclimate reconstructions derived from coastal sediments.

His extensive fieldwork and laboratory research, encompassing multi-proxy geochemical studies, magnetic susceptibility analyses, stable isotope methodologies, and pollution indices, is directly pertinent to the primary subject of this edited collection. Dr. Gopal has already authored numerous topics encompassed in the proposed chapters, including pollution pathways, sediment-water interactions, ecological concerns, and integrated management frameworks.

Dr. Gopal has mentored over 30 postgraduate research students and collaborated with national institutes on environmental research. He is the ideal lead editor for this significant and timely book, with both subject matter expertise and editorial acumen, as well as an awareness of regional case studies and a global perspective on coastal and estuary sustainability.

Contributors

Kashif Abbas Department of Zoology, Aligarh Muslim University, Aligarh, India

Mudassir Alam Department of Zoology, Aligarh Muslim University, Aligarh, India

Sawant Sushant Anil Division of Geoinformatics, School of Life Sciences, JSS Academy of Higher Education and Research, Mysuru, Karnataka, India

S. Aravind Raj Centre for Water Resources Management (CWRM), University of Madras, Chennai, Tamil Nadu, India

Muruganantham Arumugam Department of Geology, Faculty of Science, Alagappa University, Karaikudi, Tamil Nadu, India

P. Baby Department of Geography, Bharathi Women's College, Chennai, Tamil Nadu, India

Asfiya Banu Department of Tourism and Travel Management, JBAS College for Women, Chennai, India

K. S. Banu Department of Geography, Queen Mary's College (A), Chennai, Tamil Nadu, India

G. Bhaskaran Centre for Water Resources Management (CWRM), University of Madras, Chennai, Tamil Nadu, India

Bhagyasree Dolui Department of Geography, Queen Mary's College, Chennai, Tamil Nadu, India

Jayasree Dolvi Department of History, JNRM College, Port Blair, Andaman and Nicobar Island, India

J. Eugene Institute of Hotel Management Catering Technology and Applied Nutrition, CIT Campus, Taramani, Chennai, India

V. Gopal Centre for Earth and Atmospheric Sciences, Sathyabama Institute of Science and Technology, Chennai, Tamil Nadu, India

Sivakumar Karthikeyan Department of Geology, Faculty of Science, Alagappa University, Karaikudi, Tamil Nadu, India

M. V. Karunambigai Department of Tourism and Travel Management, Justice Basheer Ahmed Sayeed College for Women, Chennai, Tamil Nadu, India

Prabakaran Kulandaisamy Department of Geology, Faculty of Science, Alagappa University, Karaikudi, Tamil Nadu, India

Manoj Kumar Department of Geography, Kumaun University, Nainital, Uttarakhand, India

M. Nithya Kerala State Remote Sensing and Environment Centre, Thiruvananthapuram, India

R. K. Pande Department of Geography, Kumaun University, Nainital, Uttarakhand, India

Muthuramalingam Rajendran Department of Geology, Faculty of Science, Alagappa University, Karaikudi, Tamil Nadu, India

M. Sasi Division of Geoinformatics, School of Life Sciences, JSS Academy of Higher Education and Research, Mysuru, Karnataka, India

Venkatramanan Senapathi PG and Research Department of Geology, National College (Autonomous), Tiruchirappalli, Tamil Nadu, India

Rushda Sharf Department of Environmental Science, Integral University, Lucknow, India

Yusra Sharf Department of Zoology, Aligarh Muslim University, Aligarh, India

Vandana Sharma Department of Geography, Kumaun University, Nainital, Uttarakhand, India

H. Sivasankari Department of Geography, Ethiraj College for Women (A), Chennai, Tamil Nadu, India

S. Bala Subramaniyam CartoXR Research Center and Consultancy LLP, Chennai, India

Bangaru Priyanga Sundaram Department of Geology, Faculty of Science, Alagappa University, Karaikudi, Tamil Nadu, India

Kongeswaran Thangaraj Department of Geology, Faculty of Science, Alagappa University, Karaikudi, Tamil Nadu, India

Perumal Velmayil Department of Geology, Faculty of Science, Alagappa University, Karaikudi, Tamil Nadu, India

B. Vijayakumari Department of Geography, Queen Mary's College (A), Chennai, Tamil Nadu, India

R. M. Yuvaraj Department of Geography, Institute of Distance Education (IDE), University of Madras, Chennai, Tamil Nadu, India

Part I
Introduction

1 Environment, Geography, Landslide, and Sustainability: An Overview of Geography–Landslide Interface

Rushda Sharf, Yusra Sharf, Kashif Abbas, and Mudassir Alam

Abstract

This chapter provides a comprehensive overview of the intricate relationships between environment, geography, landslides, and sustainability. Landslides, as complex geomorphological phenomena, play a crucial role in shaping landscapes and pose significant risks to human societies worldwide. We explore the geographical aspects of landslides, examining their spatial distribution using high-resolution DEMs and the geomorphological factors influencing their occurrence. The chapter probes into the environmental determinants of landslide susceptibility, including lithology, and hydrological processes. The impacts of climate change on landslide dynamics are critically assessed using the latest IPCC AR6 projections. The chapter examines cutting-edge approaches in landslide hazard assessment, including the integration of InSAR for ground deformation monitoring, LiDAR for high-resolution topographic mapping, and deep learning models such as convolutional neural networks for automated landslide detection. The chapter concludes by identifying emerging research directions, such as the application of UAV-based photogrammetry for rapid landslide inventory mapping and the integration of landslide risk assessment into sustainable development planning using multi-criteria decision analysis techniques. This holistic approach aims to bridge the gap between scientific understanding and practical implementation, fostering resilience and sustainability in landslide-prone regions.

R. Sharf (✉)
Department of Environmental Science, Integral University, Lucknow, India
e-mail: rush.khan09@gmail.com; rsharf@iul.ac.in

Y. Sharf · K. Abbas · M. Alam
Department of Zoology, Aligarh Muslim University, Aligarh, India

R. M. Yuvaraj et al. (eds.), *Advanced Landslide Susceptibility Modeling*,
https://doi.org/10.1007/978-3-032-14485-0_1

Keywords

Landslides • Environmental factors • Climate change • Geohazards • Sustainability • Risk management

1.1 Introduction

Extreme events that deviate significantly from the norm inside the Earth's system and cause death or serious wound to people as well as loss of "goods" like agricultural land, forests, and the buildings, natural environment are known as natural disasters. Natural disasters may have slow-moving effects, like drought, or quick-moving effects, like earthquakes. It is critical to understand the difference between hazard and disaster. When an earthquake occurs in an area that is not inhabited, it is not regarded as a disaster even though it is a potentially harmful phenomenon (hazard). When something happens in a populated area and causes harm, or damage to the socioeconomic system, it is referred to as a disaster (Alexander, 1993). Although every type of disaster is limited to specific regions, natural disasters happen all over the world. The primary determinants that contribute to a disaster are the subject of another subdivision. These can be classified as extraterrestrial, global environmental, ecological (pertaining to flora and fauna), technological (man-made), geomorphological/geological, and meteorological. Natural disasters' effects on the environment around the world are getting worse over time. Both the number of people impacted and the cost to the world economy have significantly increased, as has the reported frequency of disasters. Disaster relief, treatment, and renovation are the three phases that occur after a disaster occurs, while disaster prevention and preparedness are the two phases that occur prior to a disaster (UNDRO, 1991). The deliberate lowering of risks to human health and safety is known as disaster prevention. This could entail altering the hazard's causes or effects, the population's susceptibility, or how losses are distributed.

The term "landslide" in geology refers to a wide range of processes that cause the materials that form slopes to fall, topple, slide, spread, or flow down the mountainside (Kirschbaum et al., 2010). Smaller seasonal landslides also have long-term negative effects on the residents of the risk zones, although large and acute landslides typically attract public attention because of their immediate destructive nature (Sidle et al., 2017; Sudmeier-Rieux et al., 2012). In addition to posing a direct threat to human life and health, seasonal landslides cause regular property and land loss, which can pose a serious threat to mountain communities' livelihoods and food security (Arouri et al., 2015). However, chronically occurring hazards' recurrent and unrelenting nature offers an unmatched chance for community-level experimentation and learning. In contrast to acute landslides, which frequently lead to extensive resettlement plans (Vlaeminck et al., 2016), chronic landslide hazard populations frequently continue to reside in risk areas while implementing various adaptation and mitigation techniques (Cieslik et al.,

2019; Vaciago, 2013). A natural occurrence that occurs everywhere in the world is landslides. When they occur in cities, they turn into catastrophes that upend a community's or society's way of life. Among the consequences of the landslide are structural damage, social repercussions, and human losses. Extreme weather conditions, such as protracted droughts or heavy, prolonged rainfall, are becoming more common due to the current climate variability. One of the deadliest natural disasters in Brazil is landslides, which have already impacted over four million people and are typically caused by intense rainfall. Furthermore, areas with high declivities have been occupied and transformed into urban areas due to population growth. Residents there run the risk of experiencing landslides, losing their homes, and in the worst situations, losing their lives. To prevent catastrophes, landslide-prone areas must be identified and monitored. To determine the landslide-prone areas, a number of sophisticated models using various methodologies were created. Natural disasters occur all over the world, and in recent decades, their frequency has been on the rise. The vulnerability of society has changed due to population growth, and changes in the global climate may also have some impact (Marcelino et al., 2006; Peilke, 2005). Extreme weather conditions, such as protracted drought or heavy, prolonged rainfall, are becoming more common due to the current climate variability. Heavy rainfall typically precedes and causes landslides. In urban areas, landslides have an impact on the community, destroying homes, damaging structures, and occasionally resulting in fatalities (Mendes et al., 2018). By highlighting and identifying the most important geotechnical parameters (soil cohesion, internal friction angle, and specific weight) that contributed to the landslides, the combined use of various models enhances the quality and dependability of each method (Konig et al., 2022; Zizioli et al., 2013). Since each model requires a different set of inputs, the geological, geomorphological, and geotechnical data available determine which model is best for mapping landslide-prone areas.

Decision makers and the scientific and technical literature are becoming more interested in the management, implementation, design, and authentication of "landslide early warning systems" (LEWSs). An analysis shows a definite focus on "local" systems that are intended to forecast the temporary behavior of individual slopes, or landslides. According to this, an "early warning system" (EWS) is a tool, system, or collection of capabilities that produces and distributes pertinent information in a timely manner so that people, communities, and organizations that are at risk can take prompt, appropriate action to prevent or lessen the threat's impact. (Alfieri et al., 2012; United Nations, 2016). While a "national" LEWS covers the entire country or state, a "regional" LEWS covers a metropolitan area, huge municipality, an administrative district, or a region. For LEWSs that cover a greater percentage of the world, we save the term "global."

1.2 Landslide Susceptibility

Because landslides are expected to occur more frequently and in greater quantities in areas with a high susceptibility, LEWSs must be aware of which areas are at risk of experiencing landslides (Stähli et al., 2015). Furthermore, landslide susceptibility data is used by many LEWSs (Guzzetti, 2006; Guzzetti et al., 2020). The idea of "non-susceptibility," or the identification of regions where landslide vulnerability is zero or very low and where landslides are not anticipated, has been studied by Marchesini et al. (2014). LEWSs can also make use of this geographic data. Nevertheless, there are still a number of problems with incorporating susceptibility data into LEWSs for operational landslide prediction. First of all, landslide vulnerability assessments and rainfall data usually differ greatly in terms of scale, resolution, and spatial domains. It is still unclear how to integrate threshold and susceptibility models created at drastically different scales. Making susceptibility assessments especially for operational landslide forecasting at spatial resolutions that match those needed or anticipated by landslide forecasts is one option. Second, according to Guzzetti, (2006), susceptibility is time invariant by definition, meaning that it should not alter when a particular trigger forces a change in the landscape.

1.3 Rationale for Landslide Forecasting

Assessing how climate change is affecting the environment presents a challenge to the scientific community as well as thought-provoking issues for politicians and decision makers. Global warming is undeniable (IPCC, 2014; LoPresti et al., 2015), but it is still challenging to identify and forecast how it will affect geo-hydrological hazards like droughts, floods, and landslides. On both naturally occurring and artificially created slopes, landslides are a form of mass wasting process. Under the influence of gravity, it is the movement of debris and rock mass (Hungr et al., 2013). Numerous landslides exhibit a combination of various movement types, either simultaneously or over the course of their lifetime. Landslides can involve flowing, toppling, sliding, falling, or spreading. Every continent has landslides, which are crucial to the changing of the landscape. They are also a major threat to the populace in many places (Petley, 2012). A variety of phenomena, such as precipitation, snowmelt, temperature fluctuations, seismic activity, volcanic eruptions, and human activity, affect the slopes stability and result in landslides. Some of these phenomena, primarily temperature and precipitation, are controlled or influenced by climate and its fluctuations. Accordingly, it is anticipated that slope stability is influenced by climate (and its variations) at various geographical and temporal scales (Gariano & Guzzetti, 2016; Seneviratne et al., 2012). Slope stability, landslide hazards, and the associated risk are all poorly understood in relation to climate and its variability. The definition of "early" varies according to the kind of hazard, the viewpoint and duties of the person issuing the warning, and the person who receives and applies it (Capparelli & Versace, 2011;

Guzzetti et al., 2020). Landslides are weather-induced, including those brought on by rainfall and snowmelt. Systematic attempts were undertaken to look into ideas and frameworks for the EWS of natural hazards as affordable instruments for risk reduction and disaster prevention under the aegis of the United Nations International Decade for Natural Disaster Reduction Early Warning Programme. First in the Hyogo Framework for Action 2005–2015 (UNISDR, 2005) and later in the Sendai Framework for Disaster Risk Reduction 2015–2030 (UNISDR, 2015), these initiatives persisted under the United Nations secretariat for disaster risk reduction.

Campbell (1975) was likely the first to think about the possibility of predicting landslides caused by rainfall and to suggest a framework for an LEWS, specifically for debris flows and shallow landslides in southern California. A warning system could be built with three main components, each of which is currently, partially, or fully operational. (1) A rain gauge system that records rainfall on basis of hour, (2) a weather mapping system that can identify higher-intensity rainfall centers in the storm area and plot their locations frequently in relation to gauge locations with sufficient registries for precise transfer to topographic, and (3) an communications and administrative network that can compile the data, identify when serious factors have been exceeded in a specific area, and notify the local residents. Although a system is most likely fine in current technological capabilities. In order to reduce the complex and still unsolved challenge of predicting the potential landslides occurrence, as well as the operational and technological challenges associated with the timely distribution in order to organizations, communities, administrations, and individuals, a number of simplifying assumptions have been implicitly or explicitly adopted in all attempts at operational landslide forecasting (United Nations, 2016).

1.4 Material and Method

We started our investigation by thoroughly examining the secondary sources that were already available. These incorporated research papers about the study areas" susceptibility to landslides as well as demographic information gleaned from literature reviews. In order to understand the geomorphology of the locations and the common traits of chronic and acute landslides, as well as the potential effects they may have on local livelihoods, we also reviewed the geological reports. In the WOS document data platform, we choose "Search in: Web of Science Core Collection" as the data source. "Editions: All" was chosen for the citation index, "Topic" for the document fields, and "Publication: All years (1900–2022)" for the retrieval time. The WOS document data platform yielded 1328 records between 1995 and 2022. This allowed the retrieval results to be quantitatively analyzed using WOS's bibliometric analysis function. The two researchers filtered and compiled the documents using the inclusion and exclusion criteria including titles, abstracts, and keywords, after they had been de-duplicated based on title and DOI. The research object must be landslides caused by rainfall, and the research materials must use GIS and RS technologies in geoinformatics. These were the inclusion criteria.

1.5 Impact of Landslides on Environment

Landslide is the movement of a rock, debris, soil, or mass of soil downward. When gravity or other forms of shear stress in the slope surpass the shear strength of the material that forms the slope, landslides happen. Landslides may also be triggered by transient loads like earthquakes and storms. The deterioration of slope materials' shear strength can also result in landslides. Shear strength is influenced by two primary factors. The spatial arrangement of the sedimentary structures the particles that make up embankment materials also influences the materials' shear strength. When mechanically disturbed or submerged in water, certain materials with loosely open sedimentary structures become weaker. Increases in water content, whether from natural or man-made sources, usually weaken clays by dissolving inter-particle friction, hydrating clay minerals, and removing inter-particle tension, while weakening sandy materials by decreasing inter-particle friction. Landslides are a constant threat to human life and livelihood in most parts of the world, but they are especially dangerous in a few places with rapidly growing populations and economies (Sams, 2022). Precautionary measures, such as limiting or possibly casting off populations, are particularly effective in reducing hazards; from areas with a history of landslides, by restricting specific land uses where slope balance is a concern, and by installing early warning systems that are mainly focused on monitoring floor conditions such as groundwater levels, slope displacement, and rock and soil pressure. Overflowing landslides have the potential to contaminate streams and other bodies of water with surplus silt. Large areas of forest, wildlife habitat, and productive soils on slopes can all be destroyed by landslides. Ecosystems, agriculture, built infrastructure, economic losses, fatalities, social unrest, and other risks are all harmed or destroyed by them. Landslides have an effect on the morphology of the Earth's subaerial and submarine surfaces, forests and grasslands, and the habitats of native plants and animals. The general trend of surface degradation due to erosion and mass wasting includes morphologic effects (Fig. 1.1).

1.6 Destruction of Seasonal Landslides

According to the classification of deep-seated landslides, slope failures, majority of the slide surface is beneath the tree roots (Barik et al., 2017). Changes in local hydrologic and geologic processes, like earthquakes and shifting groundwater levels, are frequently associated with the frequency and speed of these slides (Van Den Eeckhaut et al., 2009). Conversely, land mass flows that take place inside the forest rooting zone are what define shallow landslides (Cohen & Schwarz, 2017). Most often, shallow landslides are seasonal in nature, triggered by heavy monsoon rainfall that loosens the soil structure due to ground saturation. Mudflows and/or debris flows frequently accompany them. Shallow landslides' seasonal and relentless nature gradually depletes human settlements' capacity for adaptation. As agricultural land continues to shrink, significant structural changes in livelihood

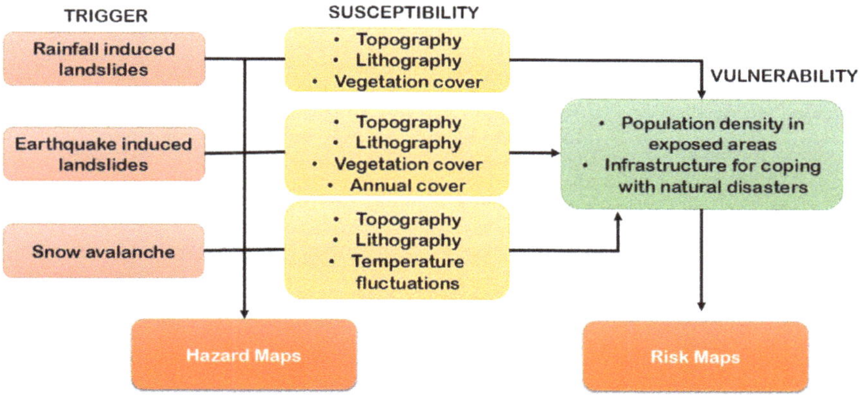

Fig. 1.1 Social and environmental impact of landslide

strategies are imposed, which have a profound impact on the social institutions of mountain communities (Pisano et al., 2017). The distinction between shallow and seasonal landslides that cause one-time, major events has significant policy and practice implications from the perspective of disaster management. The advancement of shallow landslides can occasionally be slowed down with the proper mix of land engineering, drainage, reinforcement, and extensive afforestation projects, but there is very little that may be done to stop deep-seated landslides (Stokes et al., 2014). Specifically, alterations in land use have the potential to significantly lessen the effects of landslides on nearby communities (Cieslik et al., 2019; Persichillo et al., 2017).

1.7 Impact of Climate Change on Land Slide

In assessing the factors and mechanisms influencing landslide phenomena, Sidle and Ochiai (2006) examined the climate change effects on landslides and came to the conclusion that the most significant climate variations that could have an impact on landslides were changes in seasonal, annual, and regional precipitation and a rise in mean air temperature (Table 1.1). They also took into account how vegetation, soil, and land cover are affected by climate change, but they admitted that the associated feedback processes added more intricate interactions to a previously challenging assessment (Fig. 1.2). Coe and Godt (2012) categorized 14 distinct methods for determining how climate change is affecting landslides into three main groups: (i) long-standing monitoring, (ii) retrospective methods, and (iii) prospective methods. The study's relevant conclusion was that all of the studies that were taken into consideration and that tried to forecast the activity of debris flows and shallow landslides had higher doubts because it is difficult to forecast short-term extreme storms. The modeling approach uses long-lasting rainfall variations derived from future rainfall projections produced by downscaled global

Table 1.1 Potential stability responses to changes in climatic factors (modified after Crozier, 2010)

Change in climatic factor	Process affected	Effect on landslide
Increased precipitation total	Wetter antecedent conditions Increased weight Higher water table for longer periods Increased river discharge	Less rainfall required to attain critical water content Reduction in soil suction and cohesion
Increased rainfall intensity	Infiltration exceeds subsurface drainage Increased through flow Increased surface runoff	Build of perched water tables Reduction of effective normal stress
Increase in air temperature	Higher evapotranspiration More abundant vegetation Higher hydraulic conductivity Rapid snowmelt Reduction in interstitial ice and permafrost	Reduction in antecedent water conditions More rainfall required to trigger landslides
Change in wind speed and duration	Enhanced evapotranspiration Enhanced root levering by trees Increased wave action on shorelines	Reduction in soil moisture Enhanced cracking, reduction of cohesion and soil strength
Change in weather systems	Areas previously unaffected (affected) subject to higher (lower) rainfall	Adjustment of slopes to changed weather conditions
Larger meteorological variability	More (less) frequent wetting and drying cycles	Increased fissuring Widening of joint systems Reduction of cohesion and rock mass friction

climate models to analyze changes in the constancy conditions of individual slopes or landslides as an input to statistical models (Fowler et al., 2007). The empirical method examines landslide records in an effort to identify temporal and geographic differences in the frequency or reactivation of landslides rates. Depending on the time frame of the study, two categories of empirical approaches can be distinguished, which influences the techniques employed to reconstruct the climate and landslide records.

1.8 Geographical Analysis

When taken as a whole, these LEWSs only make up a small portion of the world's potentially landslide-prone terrain (Nadim et al., 2006). Kirschbaum and Stanley's (2018) global system fills in the gaps left by the lack of an alternative, but it may not replace regional or national LEWSs due to inbuilt limitations in the variables taken into account, the difficulty of capturing local extreme rainfall

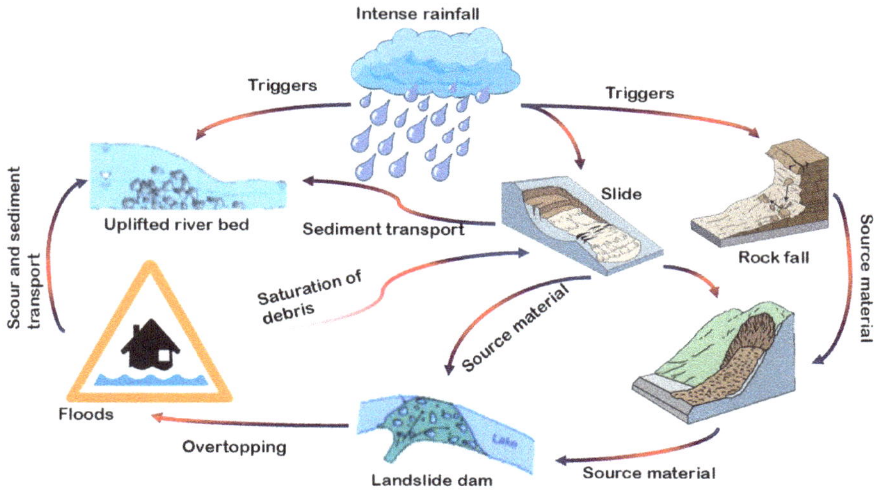

Fig. 1.2 Impact of climate change on landslide hazard and risk

events that cause landslides using IMERG rainfall estimates, and the incapacity of an easy model to explain the broad diversity of hillslope processes. The distribution and frequency of landslides that are not caused by seismic activity are known to be influenced by geology and seismicity (Guzetti et al., 2020; Vanmaercke et al., 2017). We assigned a common geological domain to every LEWS using the general geological map developed by Chorlton (2007), which considered five major rock type domains: tectonic, metamorphic, sedimentary, volcanic, and igneous rocks. We discovered that the national and regional LEWSs were applied across all geological domains, with sedimentary rocks accounting for the majority (55.0%) and volcanic and metamorphic rocks coming in second and third, respectively, at 20.0% and 17.5%. In a similar vein, we discovered that LEWSs were used in every seismic class, with the high (28.0%) and medium (36.0%) seismicity classes having the highest numbers, followed by the very high (16.0%) and low (20.0%) seismicity classes.

1.9 Temporal Analysis

In 2015, the Sendai conference, for a rehabilitated endeavor to use EWSs as cost-effective tools to alleviate the consequences of natural hazards there were 17 LEWSs operating, pre-operational, or in the design phase worldwide. These included the metropolitan areas of Rio de Janeiro, and Hong Kong as well as six countries (Indonesia, Norway, Taiwan, and USA etc.) (Alfieri et al., 2012; UNISDR, 2015). Since 2005, LEWSs number have grown at a roughly rate one innovative system annually, with that number rising to 1.5 original systems annually in the lastly five years due to improved technological readiness, data

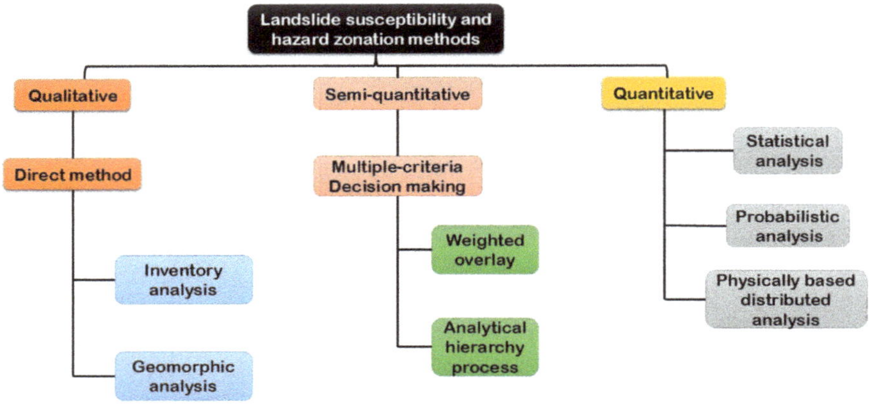

Fig. 1.3 Spatial temporal magnitude prediction of landslide hazard

availability, and increased interest among landslide scientists and decision makers (Fig. 1.3). We also observe that during the 42.5-year period under review, three LEWSs ceased. Due to a lack of funding, human resources, and motivation, only the USGS–NWS LEWS for the SFBA was discontinued (Guzzetti et al., 2020; Wilson, 2012). The other LEWSs that were turned down were either prototype, design, or proof of concept systems, or they were managed by university teams that lacked the authority and motivation to oversee the LEWSs in the extended run.

1.10 Landslide, Rainfall, and Threshold Models

Determining a rainfall threshold requires knowing when landslides occur, which is frequently challenging because there is insufficient information on when or how long landslides occur (Peres et al., 2018; Peruccacci et al., 2017). Inaccurate thresholds, inaccurate forecasts, and flawed verifications can arise from the use of ambiguous or inaccurate temporal landslide data (Fig. 1.4). The number of observations matters for thresholds that are established statistically or heuristically (i.e., visually). A very few number of landslides served as the basis for some of the thresholds, or once did. A total of 14 landslides were used to determine the threshold for the city of Chittagong (Ahmed et al., 2018), 65 landslides for Rio de Janeiro (D'Orsi et al., 1997), 83 landslides for Indonesia (Mulyana et al., 2019), and 70 landslides worldwide for Hong and Adler's (2007) global LEWS. Large uncertainties and possibly incorrect thresholds are produced by a small landslide catalogue (Kirschbaum et al., 2012). Researchers are forced to gather historical landslide records and spatially and temporally immediate rainfall data in every novel area due to the persistent problem of rainfall thresholds being difficult to export. The challenge is lessened by tools that make it easier and faster

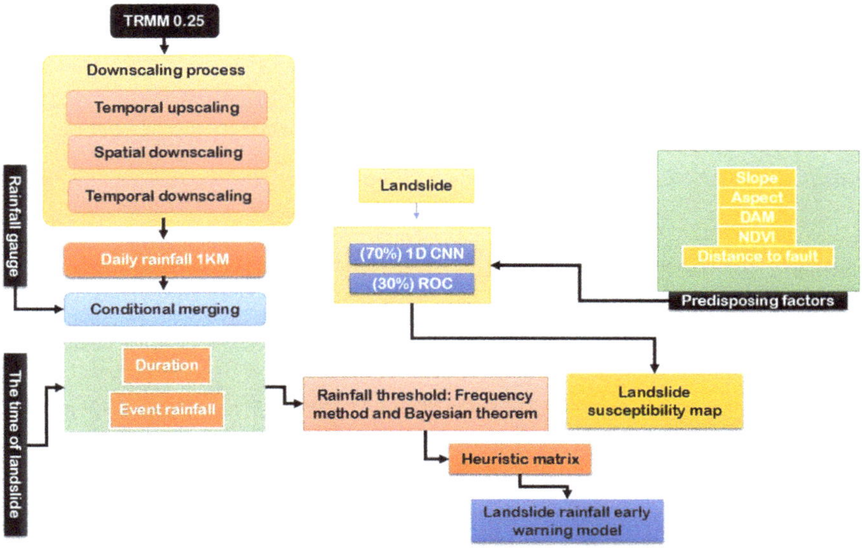

Fig. 1.4 Rainfall induced rainfall threshold model in mountain areas

to gather landslide data, organize and manage it (Napolitano et al., 2018), and use software and algorithms to determine rainfall thresholds and reconstruct rainfall events (Baum et al., 2018; Guzzetti et al., 2020).

1.11 Landslide Recognition

Any activity intended to identify previous landslide incidents that took place in a particular area is referred to as "landslide recognition." These methods enable the creation of landslide inventories and feeding databases. Landslide inventory maps, according to (Malamud et al., 2004) are maps that show the landslides special distribution at a specific cartographic scale (Table 1.2). Such inventories should be produced using optical images with a scale of at least 1:10,000, according to (Mantovani et al., 1996). In actuality, landslide detection necessitates extensive fieldwork. Landslide recognition uses four primary types of RS data: laser scanning data, thermal and microwave radar images, and optical data (Fig. 1.5). In these types of applications, optical and thermal RS are typically carried out from space and aircraft and infrequently from platforms on the ground. Microwave sensors are mounted on platforms that are grounded, in space, and in the air. Both ground-based and aerial platforms use laser scanning. Aerial photo interpretation and field surveys are the traditional approaches for creating a landslide inventory. The benefits and drawbacks of these two approaches were thoroughly discussed in earlier reviews and are not reiterated here (Guzzetti et al., 2012). While they used

Table 1.2 Analytical/physical model-based methods for assessing vulnerability (Corominas et al., 2014)

Exposed element	Landslide mechanism	Application scale	Methodologies
Buildings (RC frames)	Rock falls	Site-specific, Local	Evaluation of column resistance. Application of the finite element method for the progressive collapse potential. Yields a vulnerability matrix and v curves associated with the impact energy and the uncertainty of the impact location
Buildings (RC frames)	Slow moving	Site-specific	Calculation of damage from earthquake-induced landslides using a finite slope model and the finite difference method. Parametric analysis and statistical evaluation for the construction of fragility curves
Buildings (RC frames)	Slow moving	Site-specific	Use of numerical simulations and earthquake analysis concepts for the calculation of vulnerability curves associated with differential settlements
Buildings	Debris flow	Site-specific, local, regional	Uses the principles of the dynamic responses of simple structures to earthquake excitation. Vulnerability curves associated with the impact force
Buildings (RC frames and masonry)	Debris flow	Site-specific, local, regional	Probabilistic evaluation of damage by calculating element resistance, using the Monte Carlo method for various structural typologies Calculation of fragility curves

to be the main burden for landslide inventory mapping, operations on the ground are now primarily approved as support and incorporation of RS techniques. Three primary areas of research have emerged in the field of landslide recognition in recent years: (i) using high-resolution digital elevation models (HR-DEMs); (ii) automating feature extraction and images analysis, like synthetic aperture radar (SAR), multispectral, hyperspectral, and panchromatic data; and (iii) integrating various types of RS data. The upcoming subsections will cover these categories.

1.11.1 Optical Passive Sensors

The shortwave, visible, near infrared portions of the electromagnetic spectrum are all used by optical sensors. Platforms on the ground, in the air, and space are frequently utilized. Unmanned aerial vehicles (UAV), also known as unmanned aerial systems (UAS) are promising tool for landslide detection in the context of aerial platforms because they can be operated at very low heights, which improves precision and geometric resolution. The majority generally used optical RS tools are high-resolution images (HR) from satellite sensors, hyper and multispectral sensors, different aerial image types, and video cameras.

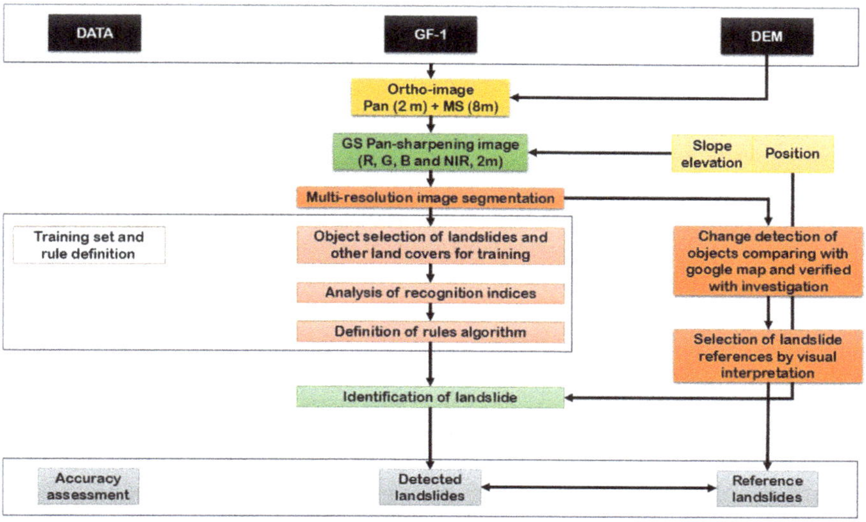

Fig. 1.5 Figure illustrates the Landslide recognition scheme

1.11.2 Thermal Infrared Remote Sensing

Because temperature changes in vegetation and terrain happen on a seasonal and daily basis and are impacted by soil, land-use practices and climate, thermal infrared imagery information used to assess natural hazards is highly ephemeral. In other areas of study, such as the examination of how urbanization affects urban heat, thermal infrared satellite missions appear to be more appealing (Weng et al., 2009). The relative differences in land surface temperature (LST) (Quattrochi et al., 1999) across a slope can be measured using thermal infrared data and may be associated with groundwater circulation and soil moisture. High hydrogeological risk may be indicated by both. Soil humidity observations can be obtained using both airborne and spaceborne TIR sensors (Landsat TM/TM+, AVHRR, MIVIS, HCMM, TIMS, MODIS).

1.11.3 Microwave Sensors

The mainly generally used microwave sensor in landslide investigation is SAR, an active system that can record electromagnetic echo backscattered from the Earth's surface and arrange it in a 2D complex value image map. Such a map's spatial dimensions include the platform's flight direction (Ferretti et al., 2007) and the sensor-target distance, or line-of-sight (LOS). Ground-based, airborne, and space-based platforms all use SAR sensors. Since it is an active sensor, the acquisition of images is not influenced by cloud cover or ambient light.

1.11.4 Airborne and Terrestrial Laser Scanning

Sensors that emit a beam of highly collimated, directional, coherent, and in-phase electromagnetic radiation are referred to as lasers. Airborne laser scanners, also known as LiDAR, are devices that are mounted on aerial platforms and provide extensive coverage. Typically referred to as terrestrial laser scanners, ground-based sensors support local-scale research. ALS HR-DEM-derived hill-shaded visualizations aid in defining the morphological characteristics of landslides (Akbarimehr et al., 2013). The two primary methods for landslide detection are heuristic visual analysis of the topographic surface and semi-automatic recognition of morphometric features.

1.12 Risk Appraisal

The most fundamental and important step in controlling and managing the risks of landslides caused by rainfall is risk assessment (Liu et al., 2022). GIS technology was primarily utilized for the spatial subdivision of the research in order to establish the spatial distribution of the landslides risk caused by rainfall (Menggenang & Samanta, 2017). Multiplication was used to decide the spatial probability and distribution of landslides, the impacted area, or population and the susceptibility. In order to save time and money, remote sensing can gather a large amount of data quickly. This data can then be organized into thematic maps and used to separate the landslide hazard area by analyzing and comparing of different effective factors (Samanta et al., 2012). Risk appraisal has emerged as a crucial instrument in recent years to address the inherent uncertainty of landslide hazards caused by rainfall (Bhandary et al., 2013; Fu et al., 2020). From local-scale site-specific assessment to regional-scale development planning, the rainfall-induced landslides risk assessment technology based on GIS and RS technologies can be used at various stages. This indicates that the study of landslide disasters caused by rainfall has always received attention a lot from academics. The reasonableness of selecting risk assessment factors, research methodologies, and appraisal accuracy will demonstrate a continuous trend of advancement on the basis of GIS and RS technology development.

1.13 Recommendations for Adaptation and Risk-Reduction Strategies

It is widely acknowledged that a combination of adaptation and mitigation tactics operating at various time and location scales, as well as the adoption of a collection of structural and non-structural measures, are necessary for effective risk reduction. Production of physical defenses is implied by structural ("hard") measures. These defenses are designed with the size and type of the expected hazard and a reference return period for the design hazardous event in mind. Most often, a stationary time series of proceedings or triggers is assumed in order to calculate the expected frequency or the return period of the event. The stationary hypothesis may not hold true in the context of a changing climate (Milly et al., 2008), and alternative methods are required for the engineered structures design (Cheng & AghaKouchak, 2014). Another kind of successful non-structural defensive measure that may significantly lower landslide risk, and especially population risk, is the use of early warning systems and landslide monitoring (Stähli et al., 2015). Nevertheless, current systems frequently depend on data and models (such as thresholds for rainfall, displacement, soil moisture, or velocity) that usually do not take into account the anticipated future environmental changes. It is uncertain how to transfer a rainfall threshold set in one climate zone to another that is far away in systems that rely on such thresholds (Guzzetti et al., 2008). Given that global warming will rise the forceful rainfall events frequency and change and, in some cases, expand the areas vulnerable to rapid-moving shallow landslides a mainly deadly type of landslide, we expect that the several people exposed to landslide risk will rise (Guzzetti et al., 2005b).

1.14 Discussion

The objective of historical climate and landslide records is to establish experiential relationships (or the lack of one, Chiarle et al., 2015) between time series of triggers and landslide occurrences. A landslide time series may be viewed as a marked point process, which is a arbitrary component whose values are "point patterns" on a mathematical set, when examined in combination with a equivalent rainfall record (Last & Brandt, 1995). This allows researchers to look for correlations between the rainfall record and landslide events (Gariano et al., 2015). A global hotspot landslide hazard zonation map was proposed by Nadim et al., (2006, 2013), and Petley (2012) illustrated the global distribution of 2620 fatal landslides that were not seismically triggered between 2004 and 2010. It should come as no surprise that precipitation, relief, and population abundance and distribution all have a significant impact on the geographic pattern of both deadly and dangerous landslides. The climate and its fluctuations have an impact on precipitation (directly) and population distribution and abundance (indirectly), which can affect the geographic distribution of landslide hazards and fatal landslides. Geographic LEWSs have shown themselves to be useful instruments for operational landslide forecasting, and through their advisories, they can help prevent fatalities and minimize damage. However, because LEWSs are intricate systems, there are still a number of unresolved problems with their management, implementation, design, and verification. Here, we address what we believe to be the most pertinent unresolved problems with the planning, execution, and administration of LEWSs. While some geotechnical data were measured on-site, others were obtained from the literature. While, some geotechnical data were measured on-site, others were obtained from the literature. It is important to note that rainfall causes landslides

Table 1.3 Summary literature review (Konig et al., 2022)

Study area	Triggered mechanism	Validation	Results
Nova Friburgo, RJ, Brazil	Rainfall	Landslide's inventory	55% landslides in unstable areas
Juiz de Fora, MG, Brazil	Rainfall	Landslide's inventory and susceptibility maps	78.5%landslides in unstable areas
Swabian Alb, Germany	Rainfall, snow melting, earthquakes	Sensitivity analysis	80% landslides in unstable areas
Wudu county, China	Rainfall and earthquakes	Sensitivity analysis	67,5% landslides in unstable areas
Oahu, Hawaii, USA	Rainfall and Volcanism	Landslide's inventory and Susceptibility maps	92% landslides in unstable areas
Sardinia, Italy	Rainfall and earthquakes	ROC analysis	Shalstab had better performance using PEM4PIT method
Taipei county, Taiwan area	Rainfall	Landslide's inventory and variation of pore-water pressure	Variation of pore-water pressure change the soil rupture mechanism
Macon County, North Carolina, USA	Rainfall and hurricane	POD/FAR/CSI	98% landslides in unstable areas
Woomyeon Mountain, Seoul, South Korea	Rainfall	ROC analysis, SI/EI, SC/LP, POD/FAR/CSI	33% landslides in unstable areas

in the majority of the areas under study that are described in the reviewed papers (Table 1.3).

In certain instances, landslides are caused by a combination of volcanic eruptions, earthquakes, and hurricanes or typhoons. Then, known accumulated rainfall values are needed to accurately model the initial soil moisture conditions. The use of RS for mapping susceptibility and analyzing landslide hazards has advanced significantly in recent years. Their goal, which includes mapping landslide inventories and indirect monitoring to predict future failures and, when feasible, lessen their effects, is, in fact, the most crucial one in landslide investigations. However, there is an inherent connection between the three stages of landslide research. Data from landslide detection, primarily landslide inventory maps, which are a stagnant input, is used in landslide hazard assessment. Landslide monitoring data can then be used to validate the hazard model or as a dynamic input. Decision makers can optimize landslide destruction information management, control systems and emergency command by swiftly integrating, processing, and analyzing disaster data with the aid of the most recent information technologies and clever algorithms used in literature review. In order to support landslide destruction management and regional planning, it is recommended that future research concentrate on new intelligent hybrid models that can serve as a guide for policymakers and UN SDGs.

1.15 Limitation of the Study

The capacity of current operational landslide forecasting models to predict landslides is still limited, despite the efforts and clear advancements. Many LEWSs employ forecast models that solely take rainfall into account, ignoring other triggers like irrigation, permafrost melting, and rapid snowmelt. Few models take into account the groundwater and soil moisture levels that regulate the onset of landslides. Given the constraints imposed by the mapping scale and the inherent limitations of data collection and analysis methods, there will always be some subjectivity in the hazard assessment phase of the disaster management cycle. This does not always mean that it is inaccurate. Defined analytical procedures, a quantitative explanation of the factors taken into consideration, and the interpretation of sequential imagery can all significantly increase the assessment's objectivity and reproducibility. However, the analyst's experience is still the most crucial factor, both in terms of the particular conditions of the study area and the different factors involved in hazard surveys. The application of expert systems in hazard assessment has not yet progressed because of the challenge of formalizing expert rules. Beyond the purview of this study, geospatial technology has been applied to address significant problems, like energy concerns, in a variety of locales and diverse contexts (Alausa et al., 2023). A GIS is an application that stores and processes geographic data using computers. It is designed to store, collect, and analyze substance and phenomena, with a focus on four fundamental geographical data-related capabilities: (a) data management, (b) input, (c) data analysis and manipulation, and (d) output (Mustafa et al., 2019). The capacity of GIS to analyze spatial data quickly, accurately, and frequently is one of its applications. GIS is also a tool for creating maps that show possible hazards associated with landslides.

1.16 Conclusion

Finally, we emphasize that a number of natural and human-induced factors are impacted both directly and indirectly by global warming. These factors can then indirectly and directly condition landslide activity and abundance as well as the landslide events frequency. Even in a "stable" climate, the intricate interactions and feedbacks between the natural and man-made drivers are challenging to study and measure. The anticipated changes in landslide activity brought on by climate change may be outweighed by the drivers' magnitude, direction, and effects. In many places, landscapes that have been altered by human activity for a long time will be impacted by the changing climate drivers. All of this increases the uncertainty in assessing how climate change will impact landslides and the constancy of both naturally occurring and artificially created slopes. It is concluded that operational forecasting of landslides caused by weather is currently feasible and possible, and that it can help reduce the risk of landslides, thereby lowering the number of fatalities and financial losses. Nonetheless, operational landslide forecasting is a challenging and unpredictable task, and LEWSs continue to be complex. Both call for advancements in technology and conceptualization. In order to design, implement, manage, and verify geographical. LEWSs, we urge the landslide forecasters community and LEWS managers to come together and suggest common, open standards. We anticipate that these standards will help reduce the risk of landslides and save lives, and we hope that our work will support this effort. In the last, we concede that funding for the LEWSs expansion will probably need to be taken away from another competing risk-reduction and safety initiatives. We expect that, the amount of landslide damage, national, local, and regional priorities, and the resources availability will all influence the expansion and extension of new LEWS in many locations.

References

Ahmed, B., Rahman, M., Islam, R., Sammonds, P., Zhou, C., Uddin, K., & Al-Hussaini, T. (2018). Developing a dynamic Web-GIS based landslide early warning system for the Chittagong Metropolitan Area. Bangladesh. *ISPRS International Journal Geoscience Information, 7*(12), 485. https://doi.org/10.3390/ijgi7120485

Akbarimehr, M., Motagh, M., & Haghshenas-Haghighi, M. (2013). Slope stability assessment of the sarcheshmeh landslide, Northeast Iran, investigated using InSAR and GPS observations. *Remote Sensing, 5*, 3681–3700.

Alausa, O. A., Adaradoun, O. S., Adekunle, G. T., & Priyono, K. D. (2023). Leveraging geospatial technology for enhanced utility management: A case study in electrical distribution power systems. *Forum Geografi, 37*(2), 164–177.

Alfieri, L., Salamon, P., Pappenberger, F., Wetterhall, F., & Thielen, J. (2012). Operational early warning systems for water-related hazards in Europe. *Environmental Science Policies, 21*, 35–49.

Alexander, D. (1993). *Natural disasters*. University College.

Arouri, M., Nguyen, C., & Youssef, A. B. (2015). Natural disasters, household welfare, and resilience: Evidence from rural Vietnam. *World Development, 70*, 59–77.

Bhandary, N. P., Dahal, R. K., Timilsina, M., & Yatabe, R. (2013). Rainfall event-based landslide susceptibility zonation mapping. *Natural Hazards, 69*(1), 365–388.

Barik, M. G., Adam, J. C., Barber, M. E., & Muhunthan, B. (2017). Improved landslide susceptibility prediction for sustainable forest management in an altered climate. *Enginering, Geology, 230*, 104–117.

Baum, R. L., Fischer, S. J., & Vigil, J. C. (2018). *THRESH—Software for tracking rainfall thresholds for landslide and debris-flow occurrence, user manual*. U.S. Geological Survey Techniques and Methods n. 14-A2. Reston, Virginia.

Campbell, R. H. (1975). *Soil slips, debris flows, and rainstorms in the Santa Monica Mountains and vicinity, Southern California*. U.S.G.S. Professional Paper 851. pp. 51.

Cohen, D., & Schwarz, M. (2017). Tree-root control of shallow landslides. *Earth Surface Dynamacis, 5*, 451–477.

Capparelli, G., & Versace, P. (2011). FLaIR and SUSHI: Two mathematical models for early warning of landslides induced by rainfall. *Landslides, 8*, 67–79.

Cieslik, K., Shakya, P., Uprety, M., Dewulf, A., Russell, C., Clark, J., Dhital, M. R., & Amrit Dhakal, A. (2019). Building resilience to chronic landslide hazard through citizen science. *Frontiers in Earth Science, 7*, 278.

Chiarle, M., Coviello, V., Arattano, M., Silvestri, P., & Nigrelli, G. (2015). High elevation rock falls and their climatic control: a case study in the Conca di Cervinia (NW Italian Alps). In: Lollino, G., Manconi, A., Clague, J. J., Shan,W., Chiarle, M. (Eds.), Engineering geology for society and territory. Climate change and engineering geology (Vol. 1, pp. 423–443).

Cheng, L., & AghaKouchak, A. (2014). Nonstationary precipitation intensity-duration-frequency curves for infrastructure design in a changing climate. *Science Report, 4*, 7093.

Chorlton, L.B. (2007). Generalized geology of the world: bedrock domains and major fault in GIS format: a small-scale world geology map with an extended geological attribute database. *Geological Survey of Canada*. https://doi.org/10.4095/223767.

Corominas, J., van Westen, C., Frattini, P., Cascini, L.., Malet, J. P., Fotopoulou, S., & Catani, F., et al. (2014). Recommendations for the quantitative analysis of landslide risk. *Bulletin Engineering Geology Environment, 73*, 209–263.

Coe, J. A., & Godt, J. W. (2012). Review of approaches for assessing the impact of climate change on landslide hazards. In: E. Eberhardt, C. Froese, A. K. Turner, & S. Leroueil (Eds.), *Landslides and Engineered Slopes, Protecting Society Through Improved Understanding: Proceedings 11th International and 2nd North American Symposium on Landslides and Engineered Slopes, Banff, Canada 1. Taylor & Francis Group, London*, 3–8 June (pp. 371–377).

Crozier, M. J. (2010). Deciphering the effect of climate change on landslide activity: A review. *Geomorphology, 124*, 260–267.

D'Orsi, R. N., D'Avila, C., Ortigao, J. A. R., Moraes, L., & Santos, M. D. (1997). Rio-watch: the Rio de Janeiro landslide watch. In: *Proceedings of the 2nd PSL, Pan-Am Symposium on Landslides, 1. Rio de Janeiro, Brazil* (pp. 21–30).

Ferretti, A., Monti-Guarnieri, A., Prati, C., Rocca, F., & Massonet, D. (2007). *InSAR Principles: Guideline for SAR Interferometry Processing and Interpretation*. ESA Publication.

Fowler, H. J., Blenkisop, S., & Tebaldi, C. (2007). Linking climate change modelling to impacts studies: Recent advances in downscaling techniques for hydrological modelling. *International Journal of Climatology, 27*, 1547–1578.

Fu, X., Zhu, Q., Liu, C., Li, N., Zhuang, W., Yang, Z., Lu, H., & Tang, M. (2020). Estimation of landslides and road capacity after August 8, 2017, MS7.0 Jiuzhaigou earthquake using high-resolution remote sensing images. *Advertisement Civil Enginering*, 8828385.

Gariano, S. L., Petrucci, O., & Guzzetti, F. (2015). Changes in the occurrence of rainfall induced landslides in Calabria, southern Italy, in the 20th century. *National Hazards Earth System Science, 15*, 2313–2330.

Gariano, S. L., & Guzzetti, F. (2016). Landslides in a changing climate. *Earth-Science Reviews, 162*, 227–252.

Guzzetti, F., Stark, C. P., & Salvati, P. (2005). Evaluation of flood and landslide risk to the population of Italy. *Environment Managament, 36*(1), 15–36.

Guzzetti, F. (2006). Landslide Hazard and Risk Assessment. Mathematisch-Naturwissenschaftlichen Fakultät der Rheinischen Friedrich-Wilhelms-Universität. Ph.D. Thesis, 389pp. University of Bonn, Bonn, Germany. https://geomorphology.irpi.cnr.it/Members/fausto/PhD-dissertation

Guzzetti, F., Peruccacci, S., Rossi, M., & Stark, C. P. (2008). The rainfall intensity–duration control of shallow landslides and debris flows: An update. *Landslides, 5*(1), 3–17.

Guzzetti, F., Mondini, A. C., Cardinali, M., Fiorucci, F., Santangelo, M., & Chang, K. T. (2012). Landslide inventory maps: New tools for an old problem. *Earth Science Review, 112*, 42–66.

Guzzetti, F., Gariano, S. L., Peruccacci, S., Brunetti, M. T., Marchesini, I., Rossi, M., & Melillo, M. (2020). Geographical landslide early warning systems. *Earth-Science Reviews, 200*, 102973.

Hong, Y., & Adler, R. F. (2007). Towards an early-warning system for global landslides triggered by rainfall and earthquake. *International Journal of Remote Sensing, 28*, 3713–3719.

Hungr, O., Leroueil, S., & Picarelli, L. (2013). The Varnes classification of landslide types, an update. *Landslides, 11*(2), 167–194.

IPCC—Intergovernmental Panel on Climate Change. (2014). Climate Change 2014: Synthesis Report. In *Contribution of Working Groups I, II and III to the Fifth Assessment Report of the Intergovernmental Panel on Climate Change, Geneva, Switzerland* 151 pp.

Kirschbaum, D. B., Adler, R., Hong, Y., Hill, S., & Lerner-Lam, A. (2010). A global landslide catalogue for hazard applications: Method, results, and limitations. *National Hazards, 52*, 561–575.

Kirschbaum, D., Adler, R. F., Hong, Y., Kumar, S., Peters-Lidard, C., & Lerner-Lam, A. (2012). Advances in landslide nowcasting: Evaluation of a global and regional modeling approach. *Environmental Earth Science, 66*, 1683–1696.

Kirschbaum, D., & Stanley, T. (2018). Satellite-based assessment of rainfall-triggered landslide hazard for situational awareness. *Earths Future, 6*, 505–552.

König, T., Hermann, J. H., Kux, A., & Corsi, C. (2022). Advanced models applied for the elaboration of landslide-prone maps, a review. *International Journal of Geosciences, 13*, 174–198.

Last, G., & Brandt, A. (1995). *Marked point processes on the real line. The Dynamic Approach. Probability and its Applications*. Springer.

Liu, H., Du, J., & Yi, Y. (2022). Reconceptualising flood risk assessment by incorporating sediment supply. *Catena, 217*, 106503.

LoPresti, A., Charland, A., Woodard, D., Randerson, J., Diffenbaugh, N. S., & Davis, S. J. (2015). Rate and velocity of climate change caused by cumulative carbon emissions. *Environment Research Letter, 10*, 095001.

Marchesini, I., Ardizzone, F., Alvioli, M., Rossi, M., & Guzzetti, F. (2014). Non-susceptible landslide areas in Italy and in the Mediterranean region. *National Hazards Earth System Science, 14*, 2215–2231.

Marcelino, E. V., Nunes, L. H., & Kobiyama, M. (2006). Banco De Dados De Desastres Naturais: Analise De Dados Globais E Regionais. *Caminhos Da Biogeografia, 7*, 130–149.

Malamud, B. D., Turcotte, D. L., Guzzetti, F., & Reichenbach, P. (2004). Landslide inventories and their statistical properties. *Earth Surface Processes, 29*, 687–711.

Mendes, R. M., Andrade, M. R. Tomasella, M. D., Moraes, J., & Cofield, G. B. S. (2018). Understanding shallow landslides in Campos do Jordao Municipality-Brazil: Disentangling the anthropic effects from Natural Causes in the disaster of 2000. *Natural Hazards and Earth System Sciences, 18*, 15–30.

Mantovani, F., Soeters, R., & van Westen, C. J. (1996). Remote sensing techniques for landslide studies and hazard zonation in Europe. *Geomorphology, 15*, 213–225.

Menggenang, P., & Samanta, S. (2017). Modelling and mapping of landslide hazard using remote sensing and GIS techniques. *Modeling Earth Systems and Environment, 3*(3), 1113–1122.

Milly, P. C. D., Betancourt, J., Falkenmark, M., Hirsch, R. M., Kundzewicz, Z. W., Lettenmaier, D. P., & Stouffer, R. J. (2008). Stationarity is dead: Whither water management? *Science, 319*, 573–574.

Mustafa, J., Sirojuzilam, N., & Sulistiyono, N. (2019). Analisis Tingkat Kerawanan Longsor Dengan Integrasi Analytical Hierarchy Process dan Pemodelan Spasial Sistem Informasi Geografis di Kabupaten Aceh Tenggara. *Serambi Engineering, 4*, 471–481.

Mulyana, A. R., Sutanto, S. J., Hidayat, R., & Ridwan, B. W. (2019). Capability of Indonesian landslide early warning system to detect landslide occurrences few days in advance. *Geophysics Research Abscission, 21*, EGU2019-18102.

Nadim, F., Kjekstad, O., Peduzzi, P., Herold, C., & Jaedicke, C. (2006). Global landslide and avalanche hotspots. *Landslides, 3*(2), 159–173.

Nadim, F., Jaedicke, C., Smebye, H., & Kalsnes, B. (2013). Assessment of global landslide hazard hotspots. In K. Sassa, B. Rouhban, S. Briceno, M. McSaveney, B. He (Eds.), *Landslides: Global risk preparedness* (pp. 59–71). Springer.

Napolitano, E., Marchesini, I., Salvati, P., Donnini, M., Bianchi, C., & Guzzetti, F. (2018). LAND-deFeND – an innovative database structure for landslides and floods and their consequences. *Journal of Environment Management, 207*, 203–218.

Petley, D. (2012). Global patterns of loss of life from landslides. *Geology, 40*(10), 927–930.

Peres, D. J., Cancelliere, A., Greco, R., & Bogaard, T. A. (2018). Influence of uncertain identification of triggering rainfall on the assessment of landslide early warning thresholds. *National Hazards Earth System Science, 18*, 633–646.

Pisano, L., Zumpano, V., Malek, Z., Rosskopf, C. M., & Parise, M. (2017). Variations in the susceptibility to landslides, as a consequence of land cover changes: A look to the past, and another towards the future. *Science Total Environment, 60*, 1147–1159.

Persichillo, M. G., Bordoni, M., & Meisina, C. (2017). The role of land use changes in the distribution of shallow landslides. *Science Total Environment, 574*, 924–937.

Peruccacci, S., Brunetti, M. T., Gariano, S. L., Melillo, M., Rossi, M., & Guzzetti, F. (2017). Rainfall thresholds for possible landslide occurrence in Italy. *Geomorphology, 290*, 39–57.

Quattrochi, D. A., & Luvall, J. C. (1999). Thermal infrared remote sensing for analysis of landscape ecological processes: Methods and applications. *Landslide Ecology, 14*, 577–598.

Sams, C. (2022). Impact of landslides on environment. *Journal of Science and Geosciences, 10*(1), 13–14.

Samanta, S., Pal, D. K., Lohar, D., & Pal, B. (2012). Interpolation of climate variables and temperature modeling. *Theoretical and Applied Climatology, 107*(1–2), 35–45.

Sidle, R. C., & Ochiai, H. (2006). Landslides: processes, prediction, and land use. *Water Resources Monogr Ser, 18*, 312. https://doi.org/10.1029/WM018

Sidle, R. C., Gallina, J., & Gomi, T. (2017). Landscape and urban planning the continuum of chronic to episodic natural hazards: Implications and strategies for community and landscape planning. *Landslide Urban Planning, 167*, 189–197.

Sudmeier-Rieux, K., Jaquet, S., Derron, M. H., Jaboyedoff, M., & Devkota, S. (2012). A case study of coping strategies and landslides in two villages of Central-Eastern Nepal. *Applied Geography, 32*, 680–690. https://doi.org/10.1016/j.apgeog.2011.07.005

Seneviratne, S.I., Nicholls, N., Easterling, D., Goodess, C.M., Kanae, S., Kossin, J., Luo, Y., Marengo, J., McInnes, K., Rahimi, M., Reichstein, M., Sorteberg, A., Vera, C., & Zhang, X. (2012). Changes in climate extremes and their impacts on the natural physical environment. In: C. B. Field, V. Barros, T. F. Stocker, D. Qin, D. J. Dokken, & K. L. Ebi,

Stähli, M., Sättele, M., Huggel, C., McArdell, B. W., Lehmann, P., Van Herwijnen, A., Berne, A., Schleiss, M., Ferrari, A., Kos, A., Or, D., & Springman, S. M. (2015). Monitoring and prediction in early warning systems for rapid mass movements. *National Hazards Earth System Science, 15*, 905–917.

Stokes, A., Douglas, G. B., Fourcaud, T., Giadrossich, F., Gillies, C., Hubble, T., et al. (2014). Ecological mitigation of hillslope instability: Ten key issues facing researchers and practitioners. *Plant and Soil, 377*, 1–23.

UNDRO. (1991). *Mitigating Natural Disasters. Phenomena, Effects and Options.* United Nations Disaster Relief Co-ordinator, United Nations.

UNISDR. (2005). Hyogo framework for action 2005–2015. Building the resilience of nations and communities to disasters. In *World Conference on Disaster Reduction* 18–22 Jan 2005, Kobe, Hyogo, Japan, 55pp.

UNISDR. (2015). Sendai Framework for Disaster Risk Reduction 2015–2030. United Nations Office for Disaster Risk Reduction 37pp.

United Nations. (2016). Report of the Open-ended Intergovernmental Expert Working Group on Indicators and Terminology Relating to Disaster Risk Reduction, 41pp.

Vaciago, G. (2013). The safeland compendium of landslide risk mitigation measures. In C. Margottini, P. Canuti, & K. Sassa (Eds.), *Landslide Science and Practice* (pp. 683–689. Berlin: Springer.

Vanmaercke, M., Ardizzone, F., Rossi, M., & Guzzetti, F. (2017). Exploring the effects of seismicity on landslides and catchment sediment yield: An Italian case study. *Geomorphology, 278*, 171–183.

Van Den Eeckhaut, M., Moeyersons, J., Nyssen, J., Abraha, A., Poesen, J., Haile, M., et al. (2009). Spatial patterns of old, deep-seated landslides: A case-study in the northern Ethiopian highlands. *Geomorphology, 105*, 239–252. https://doi.org/10.1016/j.geomorph.2008.09.027

Vlaeminck, P., Meartens, M., Laeminck, P., Maertens, M., Isabirye, M., Vanderhoydonks, F., et al. (2016). Coping with landslide risk through preventive resettlement. Optimal designing strategies through choice experiments for mount elgon region, Uganda. *Land Use Policy, 51*, 301–311.

Weng, Q. (2009). Thermal infrared remote sensing for urban climate and environmental studies: Methods, applications, and trends. *ISPRS J. Photogramm. Remote Sensing, 64*, 335–344.

Wilson, R. C. (2012). The rise and fall of a debris-flow warning system for the San Francisco Bay region, California. In T. Glade, M. Anderson, M. J. Crozier (Eds.) *Landslide Hazard and Risk* (pp. 493–516). Chichester, West Sussex, England: Wiley.

Zizioli, D., Meisina, C., Valentino, R., & Montrasio, L. (2013). Comparison between different approaches to modeling shallow landslide susceptibility: a case history in Oltrepo Pavese, Northern Italy. *Natural Hazards and Earth System Sciences, 13*, 559–573.

Landslide Sustainability: Effects on Human and Economic

2

S. Aravind Raj and G. Bhaskaran

Abstract

Landslides are one of the most devastating natural disasters, with human and economic consequences, especially in areas with dense populations and vulnerable infrastructure. Landslides have a significant human toll, including fatalities, injuries, displacement, and psychological suffering. These effects are frequently worsened by a lack of effective early warning systems and emergency response procedures, leaving communities exposed to unexpected and catastrophic catastrophes. Economically, landslides impose substantial burdens through both direct and indirect costs. Direct costs include the destruction of homes, roads, and public utilities, necessitating significant expenditures for repair and reconstruction. Indirect costs manifest as loss of productivity, decreased land value, and disruption of economic activities, which can have long-term implications for regional development. Despite ongoing research endeavors, the understanding of landslide impacts remains constrained due to a shortage of systematic studies. Many scholars have attempted to comprehend the multifaceted consequences of landslides; however, the lack of standardized methodologies and comprehensive data hampers the development of effective mitigation strategies. This chapter addresses these challenges by presenting an alternative classification framework that incorporates multiple case studies to elucidate key concepts related to landslide hazards and risks. By adopting a more structured approach, it aims to enhance the understanding of landslide dynamics and inform the development of targeted interventions. Addressing these challenges necessitates comprehensive risk assessments, sustainable land-use planning, and the implementation of effective early warning systems to mitigate the adverse effects of landslides on human lives and economic stability.

S. Aravind Raj (✉) · G. Bhaskaran
Centre for Water Resources Management (CWRM), University of Madras, Chennai, Tamil Nadu, India
e-mail: aravindraj0528@gmail.com

R. M. Yuvaraj et al. (eds.), *Advanced Landslide Susceptibility Modeling*,
https://doi.org/10.1007/978-3-032-14485-0_2

Keywords
Landslide • Human impact • Socioeconomic • Mitigation • Sustainable land use

2.1 Introduction to Landslides

Landslides are sudden movements of rock or soil masses, causes significant hazards and challenges to development in affected areas (Pandey, 2020). They can be triggered by both natural and human-induced factors, with certain geological formations and steep slopes being more susceptible (Pandey, 2020). Landslides are classified based on the nature of the material involved and its movement, with various types identified (Pandey, 2020) (De Blasio, 2011). Landslide hazards are prevalent in many regions, particularly in hilly areas, and require comprehensive investigation, risk assessment, and mitigation strategies (Shroder, 2021). Understanding landslide geometry, causative factors, and monitoring methods is crucial for effective hazard zonation mapping and implementing preventive measures (Pandey, 2020; Yuvaraj & Dolui, 2023). Sustainable mitigation strategies involve hazard assessment mapping, warning systems, and control structures, requiring multidisciplinary approaches and political will (Larsen, 2008). Landslides are integral in the development and evolution of rivers and valleys, impacting their geomorphology, hydrography, and ecology. Landslides destabilize slopes, causing debris and silt to migrate into valleys, leading to their expansion and deepening over time. The buildup of significant amounts of material can modify valley contours and form terraces. Landslides can block river channels, forming natural dams or landslide-dammed lakes that have a considerable impact on river flow and sediment transport. Such barriers can divert rivers, creating new paths and driving the dynamic evolution of valley landscapes. Furthermore, the constant inflow of sand and debris from landslides reshapes river courses and influences their hydrological properties, further altering valley formations (Korup et al., 2010). Weathering act as a driving force in landslides because it compromises the structural integrity of rocks and soils, increasing the chance of slope failure. This process, which includes physical, chemical, and biological factors, reduces solid rock to smaller, less stable bits. Physical weathering, such as freeze–thaw cycles and thermal expansion, causes fractures and weakens rock formations, whereas chemical weathering changes mineral compositions, diminishing strength and cohesiveness. Biological weathering, which involves root penetration and microbial activity, significantly destabilizes slopes. Together, these processes reduce the material's resistance to gravity and other forces like heavy rain, earthquakes, and human activity. Without weathering, many landslide-prone circumstances would not exist because slopes would be more stable and resistant to movement (Crozier, 2010). In the process of cutting material from higher elevations and depositing it in lower places, they aid in the formation of valleys, cliffs, and terraces. Long-term landscape dynamics can be impacted by landslides, which can change slope angles, reveal underlying geological layers, and produce new topographical features. Debris from landslides frequently builds up to create depositional landforms like talus slopes or landslide dams, which can alter drainage patterns and produce makeshift lakes. With landslides serving as a mechanism for both erosion and deposition, these processes throughout time help to dynamically evolve landscapes, modifying landforms in response to environmental and natural changes (Mokarram & Sathyamoorthy, 2018). The study of landslides has advanced significantly, incorporating personal computers for numerical stability analysis, new geotextile products for mitigation, and improved understanding of earth material behavior (Turner & Jayaprakash, 1996). In Sri Lanka, human-induced landslides have increased due to rapid development in hilly areas, prompting the implementation of hazard mapping projects and structural mitigation measures (Sugathapala & Prasanna, 2010). Remote sensing technologies have also been developed to aid in landslide prediction and mitigation efforts (Scaioni, 2013).

2.2 Review of Literature

1. Geosynthetics offer sustainable solutions for landslide stabilization, outperforming traditional methods across various sustainability criteria throughout their lifecycle (Damians et al., 2023). The paper provides an overview of different geosynthetic types, their functions, and applications in landslide mitigation, with a focus on reinforcement. It identifies key sustainability factors to consider when implementing geosynthetic solutions and presents evidence from existing literature demonstrating their superior performance compared to conventional approaches. The authors introduce the value integrated model for sustainable evaluations (MIVES) as a tool for assessing and comparing alternative landslide remediation methods. This comprehensive analysis highlights the potential of geosynthetics in enhancing the sustainability of landslide stabilization projects. The study recommends further research using the MIVES tool to evaluate and compare different landslide mitigation techniques, emphasizing the importance of sustainable approaches in geotechnical engineering (Damians et al., n.d.).
2. Landslides pose significant risks to human life, infrastructure, and the environment in many regions worldwide (Flentje & Chowdhury, 2016). The management of landslides requires a focus on both resilience and sustainability to effectively mitigate their impacts. Resilience involves the ability of communities and systems to withstand and recover from landslide events, while sustainability emphasizes long-term strategies for risk reduction and environmental protection. The authors argue that integrating these concepts into landslide management practices can lead to more comprehensive and effective approaches. This may include implementing early warning systems, improving land-use planning, and developing robust infrastructure designs.

Additionally, the paper likely discusses the importance of community engagement and education in building resilience to landslide hazards. By adopting a resilient and sustainable approach, communities can better prepare for, respond to, and recover from landslide events, ultimately reducing their vulnerability to these natural disasters.

3. Bioengineering techniques offer promising solutions for slope stabilization and landslide mitigation, providing cost-effective and environmentally sustainable approaches to disaster management (Sidle & Ochiai, 2006). These methods aim to prevent loss of life and property by applying modern science and technology in landslide-prone areas. The study emphasizes the importance of implementing appropriate measures and safety precautions at the right time and place to facilitate sustainable development in vulnerable regions. Bioengineering techniques are particularly valuable in young fold mountain chains, where mass movements are an integral part of the geological and geomorphological cycle of landform development. By focusing on affordable and ecologically friendly strategies, these approaches can effectively address the recurring natural phenomenon of landslides while minimizing environmental impact. The research draws on extensive field observations and literature review to develop practical and sustainable landslide management solutions (Sidle & Ochiai, 2006).
4. Rainfall-triggered landslides pose significant risks to human life and property in densely populated mountainous regions (Larsen, 2008). As global population growth continues in steep hillslope areas, the potential for landslide-related losses is expected to increase. Mitigating these hazards involves a multifaceted approach, including hazard assessment mapping, warning systems, control structures, and policy development (Larsen, 2008). For sustainable hazard mitigation, it is crucial to align natural resource management with local economic and social interests. Successful strategies require a combination of multidisciplinary scientific and engineering approaches, as well as political commitment at various levels, from local communities to national governments (Larsen, 2008). The implementation of these strategies is essential for reducing the human and economic costs associated with rainfall-triggered landslides and ensuring the safety of populations living in vulnerable areas.
5. Remote sensing has emerged as a crucial tool in landslide investigations, offering significant potential for predicting and mitigating these hazardous geological events. The impact of landslides on society, both in terms of human casualties and economic losses, has driven extensive research efforts to develop sustainable solutions (Scaioni, 2013). This research encompasses various aspects, including geological and geo-statistical analysis, geotechnical modeling, and the design of effective mitigation and protection structures. Additionally, the development of advanced sensors has played a key role in enhancing landslide detection and monitoring capabilities. The integration of remote sensing techniques into landslide research has facilitated a transition from theoretical studies to practical applications, enabling more effective risk assessment and management strategies. This shift toward applied

research highlights the growing importance of remote sensing in addressing the challenges posed by landslide hazards and improving overall public safety.

6. Landslides, a type of geological disturbance, have gained increased attention due to their significant socioeconomic impacts, particularly in mountainous regions (Thakur & Sharma, 2022). These events can be triggered by various factors, including changes in pore water pressure, loss of vegetation, erosion, heavy rainfall, earthquakes, and human activities such as blasting and earthwork (Thakur & Sharma, 2022). Climate change has also contributed to the increased occurrence of landslides. The issue is of particular concern in the Himalayan region of India, where local populations are heavily dependent on native resources and are vulnerable to landslide events (Thakur & Sharma, 2022). As a result, landslides have become a major focus for government organizations, authorities, geotechnical engineers, and urban planners, who are working to ensure both public and structural safety in these high-risk areas (Thakur & Sharma, 2022).
7. This paper explores the integration of sustainability and resilience concepts in slope stabilization engineering. The authors argue that while numerous models and metrics exist for assessing sustainability in engineered systems, the concept of resilience is often overlooked or inadequately addressed (Das et al., 2016). The study emphasizes the importance of considering both sustainability and resilience in slope stabilization projects to ensure long-term effectiveness and adaptability to changing conditions. The researchers propose a framework that combines sustainability assessment tools with resilience analysis methods, aiming to provide a more comprehensive approach to slope stabilization design and implementation. By incorporating resilience factors, such as the ability to withstand and recover from extreme events, the authors suggest that engineers can develop more robust and adaptable solutions for slope stabilization projects, ultimately enhancing their overall sustainability and long-term performance (Das et al., 2016).
8. Anderson and Holcombe (2006) present a community-based approach to reducing landslide risk in developing countries, where vulnerable populations often inhabit unplanned settlements in high-risk areas. The study focuses on a pilot project in St Lucia, West Indies, which demonstrates the effectiveness of low-cost, sustainable measures for landslide risk reduction. The authors describe three key elements of their approach: a mapping methodology for assessing landslide risk, identification of primary slope instability factors, and implementation of affordable, sustainable interventions. The project involved establishing a government management team that worked in collaboration with local communities. The success of this pilot project suggests that such low-cost approaches are feasible and can significantly contribute to minimizing landslide risk in vulnerable communities, offering a promising solution for similar situations in other developing countries (Anderson & Holcombe, 2006).

9. The concept of multifunctional landscapes has emerged as a response to increasing demands on land and natural resources. Helming and Wiggering (2013) emphasize the need for new approaches to land development that can address multiple functions simultaneously. They argue that traditional single-purpose land-use strategies are no longer sufficient to meet the complex challenges of sustainable landscape management. The authors highlight the importance of developing tools to assess the impacts of land management practices on landscape sustainability. These tools are crucial for understanding the trade-offs and synergies between different landscape functions and for informing decision-making processes. While the abstract does not provide specific examples of such tools or their applications, it underscores the significance of this research area for promoting sustainable land use and resource management in the face of evolving societal needs and environmental pressures.
10. This study proposes an integrated Analytic Hierarchy Process (AHP) and VIKOR methodology for sustainable design of anti-slide piles in slope stability projects (Tuskan & Başarı, 2023). The approach considers environmental, economic, safety, and technical factors, evaluating five main criteria: slope stability, sustainability, anti-slide pile capacity, cost, and ease of construction. A case study validates the methodology, revealing slope stability as the highest weighted criterion (0.404), followed by anti-slide pile capacity (0.283), sustainability (0.129), cost (0.146), and ease of construction (0.038). The research demonstrates that incorporating sustainability criteria leads to anti-slide pile alternatives within specific ranges ($\xi = 0.1$–0.3 and $s/D = 2.0$–3.0), differing from scenarios focused solely on economic and technical aspects. The study concludes that a pile diameter of 1.00 m is the most sustainable option, balancing slope safety, pile capacity, cost, and construction ease requirements (Tuskan & Başarı, 2023).
11. Landslides have become increasingly frequent and severe, causing significant damage to life and property and posing challenges to human development (Yadav et al., 2025). This increase is attributed to a combination of geological, morphometric, climatic, and anthropogenic factors that contribute to slope instability. While many landslides are natural events, human intervention has often exacerbated their occurrence and impact. Uncontrolled development and encroachment into unsafe areas have transformed natural processes into disasters. The paper emphasizes the need for sustainable development through effective disaster management, particularly in the context of landslides. It suggests that understanding the concept and philosophy of landslide management is crucial for mitigating risks and reducing the negative consequences of these events. The study's findings are based on extensive field observations and a comprehensive review of literature from secondary sources (Yadav et al., 2025).
12. This study examines the economic impacts of landslides using the Global Fatal Landslide Database, covering up to 150 countries from 2004 to 2017 (Robson et al., 2022). The research identifies geographic, economic, and

institutional factors influencing landslide fatalities, suggesting that countries with challenging geography can mitigate risks through increased government spending, economic openness, and political stability (Robson et al., 2022). Notably, the study reveals a surprising short-term positive effect of landslides on economic growth, stimulating activity by 0.5–1.2 percentage points (Robson et al., 2022). The authors emphasize the importance of understanding landslide impacts, given their increasing frequency due to climate change, and highlight the previous lack of reliable data on this natural disaster (Robson et al., 2022). This research contributes to filling the knowledge gap on landslide consequences and provides insights for policymakers to develop effective mitigation strategies.

13. The 1999 landslide in Western Washington had significant socioeconomic impacts on affected property owners, as revealed by a study conducted two years after the event (Burke et al., 2002). The research found that victims experienced substantial personal financial losses and received minimal financial assistance for recovery. Notably, 93% of property owners received no relief from insurance policies, while only 7% obtained temporary rental assistance. The landslide resulted in various monetary and personal losses, which were associated with emotional distress among the affected individuals. However, some participants also reported positive outcomes, including new life perspectives. The study highlights the need for mechanisms to mitigate the economic consequences of landslides and suggests areas for future research. These findings underscore the long-term socioeconomic ramifications of landslides on families and communities (Burke et al., 2002).
14. Landslides and floods can have significant socioeconomic impacts even without causing injuries or fatalities, particularly by disrupting road networks and isolating communities (Winter et al., 2019). These impacts can be categorized as direct economic, direct consequential economic, and indirect consequential economic impacts. The study examines several debris flow events and a flood event in Scotland to quantify these economic impacts. The research introduces the concept of a "vulnerability shadow," which extends far beyond the immediate area affected by the event and is determined by the transport network rather than the event's physical footprint. This shadow represents the widespread impact of such events on access to services, markets, employment, healthcare, education, and social activities. By analyzing these events, the paper demonstrates the extensive economic consequences of landslides and floods on road networks and surrounding communities (Winter et al., 2019).
15. Landslides pose significant socioeconomic challenges that are often underestimated, necessitating better understanding by policymakers and government agencies (Schuster, 1996). The economic impact of landslides is substantial, with annual losses estimated at $4 billion in Japan, the most severely affected nation, and between $1 billion to $2 billion in countries like the USA, Italy, and India (Schuster, 1996). In terms of human cost, global data from 1971–1974 indicates that landslides claim approximately 600 lives annually worldwide, with 25 to 50 deaths occurring yearly in the USA alone (Schuster,

1996). Despite these negative impacts, the chapter also acknowledges some positive effects of landslides. The study emphasizes the need for increased awareness of landslide risks and their socioeconomic implications, suggesting that future landslide activity and its consequences should be a focus of concern for relevant authorities (Schuster, 1996).

16. This study examines the economic impacts of landslides in the Samigaluh sub-district of Kulonprogo, Indonesia, using a hedonic pricing approach (Saptutyningsih & Dewanti, 2023). The research, based on data from 300 residential households in landslide-prone areas, reveals that structural characteristic such as building age, number of rooms, and size correlate with property prices. Notably, there is a positive relationship between property prices and distance from landslide points, indicating that properties further from landslide-prone areas are more valuable. The study also found that distance to rivers impacts property prices, and surprisingly, more frequent landslides are associated with higher property values. These findings can help property owners understand factors contributing to property devaluation due to landslides. The authors suggest developing landslide insurance programs to address this significant issue for real property owners (Saptutyningsih & Dewanti, 2023).
17. Landslides in the Western Hemisphere are causing increasing economic losses and casualties despite advancements in prediction and mitigation techniques (Schuster & Highland, 2001). This trend is attributed to growing population pressures leading to development in landslide-prone areas. The authors highlight significant socioeconomic impacts across the Americas. Landslides also have substantial environmental effects, including altering the Earth's topography both above and below sea level, affecting rivers, forests, and grasslands, and disrupting habitats for various species on land and in aquatic ecosystems (Schuster & Highland, 2001). These environmental disturbances contribute to the overall degradation of the Earth's surface through gravitational mass wasting and erosion processes. The study emphasizes the need for improved land-use planning and management strategies to address the growing risks associated with landslides in vulnerable regions.
18. Landslides pose a significant threat to communities in the Kigezi highlands of South Western Uganda, with increasing occurrence and severe socioeconomic impacts (Nseka et al., 2021). A study conducted between 2018 and 2020 identified 85 visible landslide scars in the Rukiga uplands, revealing an increase in spatial–temporal distribution over the past 40 years (Nseka et al., 2021). These landslides have resulted in reduced land quality, loss of lives, and destruction of infrastructure, farmlands, and crops. The research emphasizes the need for sustainable prevention measures, including total landscape reforestation with deep-rooted trees and implementation of preparedness and mitigation policies. Additionally, proper soil and water conservation techniques are recommended to enhance soil strength against landslide hazards. The study's findings underscore the importance of developing reliable strategies to address the growing

threat of landslides and protect community livelihoods in the Kigezi highlands (Nseka et al., 2021).

19. This study proposes a methodology to estimate economic losses in rural lands due to landslides, focusing on two case studies in Italy and Romania (Zumpano et al., 2018). The research considers both current and future landslide susceptibility scenarios for 2050, assuming similar socioeconomic conditions. In the Italian case study, the estimated loss of land value due to landslide susceptibility was 10.45% in 2007 and 9.90% in 2050. The Romanian case study showed higher losses, with 29.60% in 2010 and 29.81% in 2050 (Zumpano et al., 2018). The authors emphasize that landslides significantly impact rural economies, often underestimated in less populated areas where agriculture and pastoral activities are primary sources of income. The proposed methodology aims to provide a valuable approach for assessing landslide-induced economic losses, which can be utilized in spatial planning and decision-making processes (Zumpano et al., 2018).
20. His study provides a comprehensive economic assessment of landslide damage in a 2910 km^2 area west of Brussels, Belgium. Despite only 2.3% of the land being used for dwellings, roads, and railways, these land-use types account for the majority of economic damage from landslides (Vranken et al., 2013). The annual direct damage cost is estimated at €688,148, with €550,740 attributed to houses, while indirect damage amounts to €3,020,049 annually, including €2,007,375 for real estate (Vranken et al., 2013). Interestingly, the study reveals that the increase in amenity value of forests and grasslands due to landslides outweighs the production value loss (Vranken et al., 2013). This research offers valuable quantitative data for future risk assessments and provides crucial information for government officials regarding the costs associated with different land-use types in landslide-prone areas (Vranken et al., 2013).
21. Landslides, defined as the downslope movement of soil and rock, can result from natural or human-induced causes and manifest in various forms such as falls, slides, topples, lateral spreads, and flows (Highland & Geertsema, 2019). While often occurring as secondary effects of severe weather, volcanic activity, or earthquakes, recent research has begun to focus on the health impacts of landslides beyond the typical geological and socioeconomic analyses. These health effects include mortality, compromised water quality, agricultural disruptions affecting both crops and livestock, disturbance of hazardous waste sites, and significant mental stress on affected populations (Highland & Geertsema, 2019). The long-term consequences of landslides are becoming increasingly apparent, with communities facing prolonged periods of recovery and reconstruction. This emerging emphasis on health outcomes in landslide research highlights the need for a more comprehensive understanding of these events' full impact on human populations.
22. Landslides pose a significant threat to human life, causing thousands of deaths annually. Pollock and Wartman (2020) developed a data-driven tool to estimate individual mortality risk based on landslide intensity, addressing the

limitations of previous subjective methods. They found that human behavior is the primary factor influencing mortality for landslide depths between 1–6 m. The study revealed a strong correlation between landslide vulnerability and regional economic development, but unlike other natural hazards, landslide losses are not significantly stratified by gender or age. Simple actions, such as moving to higher floors or prepared refuge spaces, can increase survival odds by up to 12 times. The researchers emphasize the importance of community-scale hazard awareness programs and citizen first responder training to maximize survival rates. This research provides valuable insights for improving landslide risk assessment and developing effective risk reduction strategies.

23. This study investigates the causes of a major landslide disaster in the Almora District of Uttarakhand, India, in 2010. Contrary to local beliefs attributing the disaster to human impacts or geological weaknesses, the research found that these factors were not significantly correlated with landslide volume (Haigh & Rawat, 2011). Instead, the primary factors influencing landslide size were the height of roadcuts, channel incision undermining roadcuts, and the steepness of upslope hillsides. While human activities did not directly affect landslide volumes, a comparison with a forested road showed that human impacts in Almora have prevented landslide healing over time. The study concludes that the main factors contributing to the disaster were the inherent instability of the hillsides and road engineering practices, rather than recent urban development or deforestation (Haigh & Rawat, 2011).
24. The study by (Chung et al., 2018) presents an analytical procedure for assessing the impact of large-scale landslides on nearby human settlements in Taiwan. Following the devastating effects of Typhoon Morakot in 2009, which caused numerous deaths and landslides, the Central Geological Survey identified over 1000 potential large-scale landslides, with 103 potentially affecting human settlements. The researchers integrated various technologies, including remote sensing image interpretation, hydrogeological investigation, and numerical analysis, to evaluate potential failure scenarios and landslide scale. Using the Xinzhuang landslide as a case study, they employed GeoStudio and RAMMS analysis modes to assess failure mechanisms, influence zones, and post-failure mass migration and accumulation. The study resulted in the creation of a landslide hazard and evacuation map, demonstrating the procedure's ability to quantitatively estimate potential threats to human settlements and its applicability for disaster prevention and evacuation planning in other villages.
25. This study investigated the distribution and causes of landslides in Loano Sub-District, Indonesia, from 2004 to 2008 (Sartohadi, 2008). The research mapped 291 landslide locations and classified the area into different hazard levels using field surveys, laboratory analysis, and GIS techniques. Notably, the entire study area was found to be at risk of landslides, with no completely safe zones identified (Sartohadi, 2008). The research revealed that human activities, rather than natural conditions, were the primary drivers of landslides in the region. These anthropogenic factors included slope cutting

for road construction, housing, agriculture, and mining activities (Sartohadi, 2008). Furthermore, the study found a strong correlation between landslide occurrences and road networks, with over 90% of landslides occurring in close proximity to roads (Sartohadi, 2008). These findings highlight the significant impact of human intervention on landslide risk in the area.

26. This study presents a comprehensive analysis of global fatal non-seismic landslides from 2004 to 2016, revealing that 55,997 people were killed in 4862 distinct events (Froude & Petley, 2018). The spatial distribution of landslides is uneven, with Asia being the most affected region. While there is significant interannual variation in landslide occurrence, the authors note that longer-term datasets are needed to establish clear links between climate modes and landslide patterns. Importantly, the research highlights an increasing trend in human-induced landslides, particularly those related to construction, illegal mining, and hill cutting activities. This finding suggests that human disturbance may play a more significant role in future landslide occurrences than climate factors. The study underscores the importance of high-quality landslide databases for risk assessment and emphasizes the need for further research to better understand the complex interplay between human activities and landslide hazards (Froude & Petley, 2018).
27. Landslides are significant geological processes that shape landscapes and can become natural disasters when combined with human activities (Sidle & Ochiai, 2006). These mass movements, both large episodic events and smaller chronic occurrences, leave lasting impacts on mountains, hills, valleys, and waterways. The book "Landslides: Processes, Prediction, and Land Use" explores the mechanisms of landslide initiation, their human and economic consequences, and available hazard assessment and prediction methods. It also examines how land-use practices, such as timber harvesting, road construction, and urban development, influence the spatial and temporal distribution of landslides. Additionally, the authors investigate the effects of land use and climate change on landslide occurrence. This comprehensive work provides valuable insights into understanding and managing landslide risks in the context of human activities and environmental changes (Sidle & Ochiai, 2006).
28. Human-induced landslides (HIL) are landslide events triggered or exacerbated by anthropogenic activities, such as modifying topography, altering water circulation, changing land use, and infrastructure aging (Jaboyedoff et al., 2018). The authors identify seven types of human-induced changes that can increase landslide hazards, noting that these categories are not mutually exclusive. Many HIL case studies are complex, and environmental changes affecting slope stability are often overlooked, potentially leading to underestimation of landslide risks. To address this issue, (Jaboyedoff et al., 2018) propose seven guiding questions about future environmental changes and their relationship to slope stability over time. These questions form a conceptual model, though the authors acknowledge that a robust scientific foundation for this model is currently lacking. The paper emphasizes the importance of considering anthropogenic factors in landslide hazard assessment and management.

29. Landslides pose significant risks to public safety and infrastructure, resulting from both natural factors like intense rainfall and steep slopes, as well as human activities such as deforestation and poor construction practices (See-Sew Gue, 2023). Effective mitigation strategies are crucial to minimize the impacts of landslides, particularly during periods of heavy rainfall. The paper proposes a comprehensive approach to landslide management, emphasizing education, research, planning, and collaboration among various stakeholders (See-Sew Gue, 2023). A key recommendation is the establishment of a centralized slope agency to serve as a One Stop Centre, providing expertise, guidance, and control for slope-related projects. This agency would be responsible for risk assessment, mitigation measures, and research and development in slope engineering. The success of landslide mitigation efforts relies heavily on collaboration between government agencies, academic institutions, industry professionals, and local communities (See-Sew Gue, 2023).
30. Human-induced landslides encompass a diverse range of topics, as explored in the paper by (Fell., 2018). The study covers various aspects, including landslide susceptibility, hazard and risk zoning for both natural and constructed slopes, and the management of hazards associated with large landslides in dam reservoirs. Additionally, the paper addresses rock fragmentation in rock falls and the design of rock slopes. Fell (2018) also discusses the monitoring of landslides through remote imaging and in-situ instruments, as well as geotechnical characterization and stabilization of individual landslides. The wide-ranging nature of these topics highlights the complexity of human-induced landslides and the challenges in presenting a cohesive analysis. The paper incorporates case studies to illustrate relevant themes and includes insights from the author's personal experience to supplement the discussion on various topics (Fell, 2018).

2.3 Aim and Objective

This chapter aims to critically examine the Landslide sustainability and its effects on human and economic impacts of landslides. The primary objective of this book chapter is to carry out an extensive study of previous landslides in order understand their diverse effects on human lives and economy. The chapter examines documented cases that elucidate on the direct and indirect consequences of landslides, such as fatalities, injuries, displacement, and psychological trauma, as well as economic repercussions such as infrastructure damage, loss of livelihoods, and long-term financial burdens for affected communities. Furthermore, the chapter seeks to assess the efficiency of current mitigation techniques and make concrete recommendations to improve resilience and promote sustainable development in areas susceptible to landslides.

2.4 Methodology

This chapter reviews the literature on landslides' human and economic repercussions. Numerous studies have analyzed how natural disasters not only cause displacement, injuries, displacement, and psychological trauma, but also severe economic and human disruptions such as infrastructure damage, property loss, loss of life and long-term economic instability. Researchers have used a variety of models and approaches to evaluate and mitigate these impacts. Scientometric analyses have been employed to elucidate global trends in landslide risk assessment and economic losses, underscoring the necessity for integrated approaches that account for both physical and socioeconomic factors (Marín-Rodríguez et al, 2024). The physical and psychosocial effects of landslides, highlighting the importance for specific strategies for solving mental health challenges among survivors (Yadav et al., 2025). Additionally, this chapter has highlighted the necessity of understanding economic vulnerability and human susceptibility to landslides, including factors such as behavior, knowledge, and readiness, in order to reduce mortality and morbidity. This chapter attempts to give a detailed knowledge of the multifaceted human and economic impact of landslides and the approaches used to mitigate them via integrating numerous research initiatives.

2.5 Result and Discussion

2.5.1 Human Impacts on Landslide

Human activities significantly impact landslide occurrence and risk. Anthropogenic factors like urban development, deforestation, and agricultural terrace collapse can contribute to landslides, though natural factors like slope angle and road-cut height remain important (Haigh & Rawat, 2011). Human-induced landslides are primarily caused by modifications to topography, alterations in water circulation, changes in land use, and the aging of infrastructure. Activities such as deforestation, excavation, construction, and inadequate drainage systems can destabilize slopes, increasing the likelihood of landslide occurrences. Over time, aging infrastructure without proper maintenance further exacerbates these risks, highlighting the need for sustainable land management and proactive mitigation strategies. The historical examples and statistics of landslide impacts on human life, drawing from multiple studies. Between 2004 and 2010, 2620 fatal landslides caused 32,322 deaths worldwide, primarily in Asia (Petley, 2012). In Latin America and the Caribbean, 611 landslides resulted in 11,631 fatalities from 2004 to 2013 (Sepúlveda & Petley, 2015). Calabria, Italy: Over the past 70 years, Calabria has seen a significant reduction in landslide fatalities, from 196 deaths in the 1950s to just 20 in the 2010s. This reduction is attributed to improved housing and changes in population habits, although the number of people involved in incidents has increased due to better information dissemination (Petrucci, 2024).

British Columbia, Canada: A study estimated that several hundred to a few thousand people live with a landslide risk exceeding 100 micromorts annually. The region experiences a historical fatality rate of 0.4 to 1.4 deaths per year, highlighting the ongoing risk in landslide-prone areas (Strouth, 2024). Global Perspective: Between 1900 and 2019, landslides worldwide resulted in 70,728 deaths and affected 14.5 million people. This underscores the global scale of landslide impacts and the need for comprehensive risk management strategies. A study in Indonesia found that landslide-displaced individuals experienced three stages: evacuation, temporary housing, and relocation, each with distinct psychological challenges (Burrows et al., 2021). These psychological impacts were further exacerbated by limited of institutional capacity and insufficient aid underscoring the need for comprehensive mental health support and effective disaster response strategies (Baseler & Hennig, 2023; Fernando, 2020). Landslides can lead to the release of harmful substances into the environment, contaminating soil and groundwater (Li, 2021) (Göransson et al., 2012). In regions near industrial activities, such as uranium mining, this risk is heightened. For instance, in Tummalapalle, Andhra Pradesh, India, uranium mining operations have been associated with the leaching of radioactive materials into the groundwater, adversely affecting both agriculture and public health (Rana et al., 2022). Landslides can also generate large amounts of suspended sediment, increasing turbidity in rivers. This can alter water quality, making it unsafe for drinking and recreational use. High turbidity can also disrupt aquatic ecosystems, further complicating water management (Göransson et al., 2012; Thapa et al., 2024). The sediment and debris from landslides can remain in the environment for extended periods, contributing to poor air quality. This is particularly problematic in areas with limited ventilation or where landslides occur in urban or densely populated regions (Nikolic, 2014). Landslide survivors often experience PTSD, with prevalence rates as high as 27% in some studies. Factors such as the severity of the disaster, loss of family members, and displacement increase the likelihood of developing PTSD (Kennedy et al., 2015; Cahyani et al., 2023) and psychological trauma can lead to anxiety, depression, and other mental health disorders. These conditions are often exacerbated by the loss of livelihoods, homes, and social support networks. Even when communities are relocated together, the strain on social relationships can persist (Baseler & Hennig, 2023; Burrows et al., 2021). On August 2, 2014, a disastrous landslide occurred on the right bank of Nepal's Sunkoshi River. The gigantic rockslide, believed to be 5 million cubic meters, blocked the river, creating a dam and a vast lake upstream. This disaster claimed the lives of approximately 155 people, destroyed several homes, and severely damaged the Araniko Highway, a critical commerce route connecting Nepal and China. The landslide also flooded a hydroelectric facility, emphasizing the fragility of infrastructure in landslide-prone areas (Priyom Roy et al., 2014). The Krishna Bhir landslide is a recurring disaster in Nepal, causing loss of lives and property damage. Factors contributing to this landslide include steep slopes, loose soil, weathering, poor drainage, intensive rainfall, and human activities such as deforestation and inadequate infrastructure development.

Despite mitigation efforts involving civil engineering structures and bioengineering techniques, the area continues to experience instability, underscoring the need for ongoing maintenance and monitoring (Maskey, 1999). In February 2021, a glacier collapsed in Uttarakhand's Chamoli district, triggering an avalanche and resulting in landslide that destroyed two hydropower installations (Martha et al., 2021). The incident highlighted the risks connected with glacial instability in the Indian Himalayas, especially in the light of climate change and increased human activities in environmentally vulnerable areas (Thayyen et al., 2022).

2.5.2 Economic Impacts on Landslide

The damage incurred from landslides can disrupt local economies, increase repair costs, and necessitate investment in mitigation strategies. Assessing network vulnerability is crucial, especially in mountainous regions with limited redundancy (Schlögl & Matulla, 2018). Researchers have developed methods to identify critical road segments susceptible to landslides and evaluate their potential impacts. These approaches integrate factors such as terrain, geology, and historical data to create landslide susceptibility maps (Zhang et al., 2020; Schlögl & Matulla, 2018). For instance, a study on Oahu Island identified that approximately 34% of the area faced high or very high landslide likelihood, emphasizing critical highway segments that require monitoring (Zhang et al., 2020). The socioeconomic losses are exacerbated in regions where urban development encroaches on unstable slopes (Miele et al., 2021). The consequences of landslides extend to businesses, as disrupted transportation networks can hinder supply chains and access to markets, leading to economic downturns (Lentini et al., 2019). Studies have quantified both direct and indirect damages in various regions. In Belgium, annual costs for direct damage to houses were estimated at €550,740, with total indirect damages reaching €3,020,049 (Vranken et al., 2013). Land loss due to landslides, and food security is critical for rural economies. Landslides not only threaten agricultural productivity but also exacerbate food insecurity, particularly in vulnerable regions (Riahi & Zare, 2021). In Nepal, landslides have resulted in livelihood insecurity, affecting food availability and employment opportunities for rural households (Shrestha et al., 2023). Insurance often provides little relief, with 93% of property owners in a Washington state landslide receiving no insurance compensation (Burke et al., 2002). Annual costs for landslide repairs on US state highways averaged $68.5 million in contracts and $37.4 million in maintenance from 1986 to 1990 (Walkinshaw, 1992). While government and insurance costs are substantial, they can also stimulate economic recovery through reconstruction investments, which may offset some initial losses (Xie et al., 2014). One of the most immediate effects of landslides is the damage to infrastructure, particularly roads and transportation networks. For instance, in the case of the Felbertauern landslide in Austria, the destruction of a major road led to significant traffic disruptions, which in turn affected the regional economy, including tourism (Pfurtscheller & Genovese, 2019). Similarly, in Nepal, landslides along the Kathmandu Kyirong

Highway caused frequent road blockages, making it difficult for tourists to reach popular destinations (Dhakal et al., 2020). Landslides often result in the closure of tourist attractions, either due to direct damage or safety concerns. For example, in Bali, the areas around Lake Beratan and Pura Ulun Danu Beratan were found to be prone to landslides, leading to the closure of these attractions during high-risk periods (Sunarta et al., 2018). In the Indian Himalayas, the 2013 landslide events caused widespread destruction of tourist infrastructure, including temples and trails, which were critical to the local tourism industry (Lorenz & Dittmer, 2021). On the other hand, landslides and the associated risks can deter tourists from visiting affected regions. In the aftermath of the 2015 Gorkha Earthquake in Nepal, tourist arrivals in the affected areas dropped significantly, with some hotels experiencing a 50–100% reduction in bookings (Dhakal et al., 2020). Similarly, in the Philippines, natural disasters, including landslides, have been shown to reduce tourist arrivals and travel expenditure (Biardeau & Sahli, 2024). In Nepal, the income loss for hotels in the affected towns was found to be 50–100% in the first year following the landslide events (Dhakal et al., 2020). Landslides can also disrupt employment opportunities, particularly in regions where tourism is a major source of income. In the Indian Himalayas, the destruction of tourist infrastructure led to a loss of livelihoods for many locals who depended on tourism-related activities (Lorenz & Dittmer, 2021). The Kathmandu Kyirong Highway in Nepal experienced frequent landslides between 2015 and 2018, leading to significant disruptions in tourism. The number of road-damaging landslides was high, with an average density of 1.46/km and 3.63/km in the main and new tracks, respectively. The dominant landslide types were debris slides and rock falls, which were concentrated in areas with specific geological and climatic characteristics. The impact on tourism was severe, with a 50–100% reduction in tourist arrivals in the first year following the landslides (Dhakal et al., 2020). The Felbertauern landslide in Austria caused the destruction of approximately 100 m of road, including an avalanche gallery, leading to a complete blockage of the Felbertauern Road for several weeks. The economic impact on tourism was significant, with losses estimated at €9–10 million. The study highlighted the importance of cost–benefit analysis in determining the feasibility of alternative routes and the need for rapid decision-making to minimize economic losses (Pfurtscheller & Genovese, 2019). The Indian Himalayas, particularly the state of Uttarakhand, have experienced severe landslide events that have impacted the tourism industry. The 2013 landslide events, triggered by heavy rainfall, caused widespread destruction of tourist infrastructure, including temples and trails. The study revealed that tourism-related vulnerabilities were exacerbated by the disaster, leading to long-term impacts on the local economy and livelihoods (Lorenz & Dittmer, 2021; Chauhan and Dixit, 2023).

2.6 Sustainable Mitigations and Recommendations

Proper land-use planning and zoning regulations are crucial for identifying and avoiding high-risk areas, as well as implementing slope stabilization measures to prevent landslides (Patel & Bakshi, 2024). A global review suggests the creation of a National Slope Master Plan to streamline policy and institutional frameworks for landslide mitigation, ensuring political commitment and resource allocation (See-Sew et al., 2009). Effective mitigation requires a comprehensive approach involving education, research, planning, and collaboration between various stakeholders (Gue, 2023). Successful implementation requires political will and community engagement at local and national levels (Larsen, 2008). To ensure long-term sustainability, mitigation strategies should consider socioeconomic factors and human settlement planning needs. Landslide hazard mapping has proven reliable for planning purposes, with recent landslides occurring in areas identified as medium to high hazard. An interdisciplinary approach is essential, incorporating advanced analysis methods, understanding of soil mechanics, airborne electromagnetics, ground geoelectrical measurements, hydrological and geological mapping and risk assessment (Chowdhury & Flentje, 2014; Supper et al., 2008). Satellite-based radar interferometry has emerged as a reliable method for landslide detection and prediction, offering high accuracy and temporal coverage (Casagli et al., 2020). Modern systems utilize wireless sensor networks and Internet of Things (IoT) technology to overcome limitations of traditional cable-based methods (Bhosale et al., 2017). These systems employ various geosensors, including inclinometers, crackmeters, and piezometers, to measure parameters like ground acceleration, soil displacement, and water levels (Chaulya et al., 2024; Fosalau et al., 2016). Autonomous software agents embedded in wireless nodes continuously collect and analyze data, enabling real-time monitoring and early warning capabilities (Georgieva et al., 2012). Advanced systems can measure soil displacements with sub-millimeter resolution and estimate subsurface layer movement (Fosalau et al., 2016). Data is transmitted wirelessly to remote servers for further analysis and storage, with some systems capable of generating automated alerts via SMS, email, and audio-visual means (Chaulya et al., 2024). While EWS have shown success in reducing fatalities and economic losses, challenges remain in accurately identifying potential landslides and improving warning effectiveness (Tang et al., 2012). Ongoing technological advancements, such as MEMS-based wireless sensors and satellite InSAR, promise to enhance the accuracy and reliability of landslide EWS (Yang et al., 2017; Casagli et al., 2020). These technologies significantly enhance the ability to predict and respond to potential landslide events. Involving local residents in the monitoring process can enhance the effectiveness of EWS. For example, communities can be trained to observe and report early signs of landslides, such as changes in soil moisture or unusual sounds (Casagli et al., 2020). To enhance our understanding of the relationship between climate variables and landslide activity, it is imperative to establish comprehensive ground observation networks (Pei et al., 2023). These networks should integrate various monitoring techniques, including visual inspections, remote sensing technologies, and geological investigations, to observe and analyze factors such as crack development, ground displacement, and hydrological conditions in landslide-prone areas. By systematically collecting and analyzing data from these ground-based observations, researchers and policymakers can develop more accurate models to predict landslide occurrences and implement effective mitigation strategies (Wu et al., 2024).

2.7 Conclusion

In conclusion, this chapter has examined the multifaceted human and economic impacts of landslides, highlighting their profound effects on communities and economies. Human consequences encompass fatalities, injuries, displacement, and psychological trauma, often exacerbated by inadequate early warning systems and emergency response mechanisms. Economically, landslides impose substantial direct costs through the destruction of infrastructure and indirect costs such as loss of productivity and disruption of economic activities. To address these challenges, the implementation of early warning systems (EWS) has emerged as a critical component in disaster risk reduction. Effective EWS provide timely alerts, enabling communities to take proactive measures to minimize harm. The success of these systems heavily relies on the active participation of local communities. Engaging residents in the design, operation, and maintenance of EWS fosters a sense of ownership and ensures that the systems are tailored to the specific needs and contexts of the communities they serve. Such participatory approaches enhance the effectiveness of EWS by incorporating local knowledge and ensuring that warnings are culturally and contextually appropriate. Technological advancements have further bolstered the capabilities of EWS. The integration of remote sensing, Geographic Information Systems (GIS), and Internet of Things (IoT) devices allows for real-time monitoring and data collection, improving the accuracy of landslide predictions. These technologies facilitate the rapid dissemination of warnings and enable authorities to implement timely evacuation plans. By combining technological innovations with community engagement, EWS can significantly reduce the adverse impacts of landslides.

In summary, mitigating the human and economic toll of landslides necessitates a multifaceted approach that includes comprehensive risk assessments, sustainable land-use planning, effective early warning systems, and active community participation. By embracing these strategies, societies can enhance their resilience to landslides and safeguard both lives and livelihoods.

References

Anderson, M. G., & Holcombe, L. (2006, March). Sustainable landslide risk reduction in poorer countries. In *Proceedings of the Institution of Civil Engineers-Engineering Sustainability* (Vol. 159, No. 1, pp. 23–30). Thomas Telford Ltd.

Baseler, T., & Hennig, J. (2023). JEL Classifications: Q54, J61, H84, O15, R23 keywords: displacement, natural disasters, climate refugees, forced migration. Development Research.

Bhosale, A., Nimbore, P., Shitole, S., & Govindwar, O. (2017). Landslides monitoring system using IoT. *Imperial Journal of Interdisciplinary Research (IJIR), 3*(4).

Biardeau, L. T., & Sahli, M. (2024). Investigating the non-linear impacts of seven types of natural disasters on inbound tourism: Insights from the EM-DAT database. *Tourism Economics, 30*(4), 900–923.

Burke, T. J., Sattler, D. N., & Terich, T. (2002). The socioeconomic effects of a landslide in Western Washington. *Global Environmental Change Part B: Environmental Hazards, 4*(4), 129–136.

Burrows, K., Desai, M. U., Pelupessy, D. C., & Bell, M. L. (2021). Mental wellbeing following landslides and residential displacement in Indonesia. *SSM-Mental Health, 1*, 100016.

Cahyani, H. E., Warsini, S., & Raymomdalexas, C. M. (2023). Post-traumatic stress disorder (PTSD) among landslide survivors: A descriptive study. *Psychiatry Nursing Journal (Jurnal Keperawatan Jiwa), 5*(1).

Casagli, N., Tofani, V., Sassa, K., Bobrowsky, P. T., & Takara, K. (Eds.). (2020). *Understanding and reducing landslide disaster risk: Volume 3 monitoring and early warning*. Springer Nature.

Chauhan, V., & Dixit, J. (2023). Geomorphic anomalies in Uttarakhand, India: A GIS-based approach for active tectonics. *Journal of Earth System Science, 133*(1), 2.

Chaulya, S. K., Mishra, P. K., Kumar, N., Kumar, V., & Rawani, V. K. (2024). Landslide monitoring and prediction system using geosensors and wireless sensor network. *Discover Geoscience, 2*(1), 6.

Chowdhury, R., & Flentje, P. (2014). *Mitigation of landslide impacts, strategies and challenges for the 21st century*.

Chung, M. C., Chen, C. H., Lee, C. F., Huang, W. K., & Tan, C. H. (2018). Failure impact assessment for large-scale landslides located near human settlement: Case study in southern Taiwan. *Sustainability, 10*(5), 1491.

Crozier, M. J. (2010). Landslide geomorphology: An argument for recognition, with examples from New Zealand. *Geomorphology, 120*(1–2), 3–15.

Damians, I. P., Miyata, Y., Rimoldi, P., Touze, N., & Kraus, J. (2023). Sustainability of geosynthetics-based landslide stabilization solutions. In *Progress in landslide research and technology, Volume 1 Issue 1, 2022* (pp. 197–205). Springer International Publishing.

Das, J. T., Puppala, A. J., Bheemasetti, T. V., Walshire, L. A., & Corcoran, M. K. (2016, December). Sustainability and resilience analyses in slope stabilisation. In *Proceedings of the Institution of Civil Engineers-Engineering Sustainability* (Vol. 171, No. 1, pp. 25–36). Thomas Telford Ltd.

De Blasio, F. V. (2011). *Introduction to the physics of landslides: Lecture notes on the dynamics of mass wasting*. Springer Science & Business Media.

Dhakal, S., Cui, P., Rijal, C. P., Su, L. J., Zou, Q., Mavrouli, O., & Wu, C. H. (2020). Landslide characteristics and its impact on tourism for two roadside towns along the Kathmandu Kyirong Highway. *Journal of Mountain Science, 17*(8), 1840–1859.

Fell, R. (2018). *Human induced landslides. Landslides and engineered slopes. Experience, theory and practice* (pp. 171–199).

Fernando, R. L. S. (2020). Artificial intelligence and disaster management in Sri Lanka: problems and prospects. In *AI and robotics in disaster studies* (pp. 149–166).

Flentje, P., & Chowdhury, R. (2016, December). Resilience and sustainability in the management of landslides. In *Proceedings of the Institution of Civil Engineers-Engineering Sustainability* (Vol. 171, No. 1, pp. 3–14). Thomas Telford Ltd.

Fosalau, C., Zet, C., & Petrisor, D. (2016, October). Implementation of a landslide monitoring system as a wireless sensor network. In *2016 IEEE 7th Annual Ubiquitous Computing, Electronics & Mobile Communication Conference (UEMCON)* (pp. 1–6). IEEE.

Froude, M. J., & Petley, D. N. (2018). Global fatal landslide occurrence from 2004 to 2016. *Natural Hazards and Earth System Sciences, 18*(8), 2161–2181.

Georgieva, K., Smarsly, K., König, M., & Law, K. H. (2012). An autonomous landslide monitoring system based on wireless sensor networks. In *Computing in Civil Engineering* (pp. 145–152).

Gue, S. S. (2023). Landslides: How, why and the way forward. *IEM Journal, 84*(2).

Göransson, G., Larson, M., Bendz, D., & Åkesson, M. (2012). Mass transport of contaminated soil released into surface water by landslides (Göta River, SW Sweden). *Hydrology and Earth System Sciences, 16*(7), 1879–1893.

Haigh, M., & Rawat, J. S. (2011). Landslide causes: Human impacts on a Himalayan landslide swarm. *Belgeo. Revue Belge De Géographie, 3–4*, 201–220.

Helming, K., & Wiggering, H. (Eds.). (2013). *Sustainable development of multifunctional landscapes*. Springer Science & Business Media.

Highland, L., & Geertsema, M. (2019). Human health effects of landslides. *Encyclopedia of Environmental Health*, 554–557.

Jaboyedoff, M., Michoud, C., Derron, M. H., Voumard, J., Leibundgut, G., Sudmeier-Rieux, K., Nadim, F., & Leroi, E. (2018). Human-induced landslides: toward the analysis of anthropogenic changes of the slope environment. In *Landslides and engineered slopes. Experience, theory and practice* (pp. 217–232). CRC Press.

Kennedy, I. T., Petley, D. N., Williams, R., & Murray, V. (2015). A systematic review of the health impacts of mass earth movements (landslides). *PLoS currents, 7* (ecurrents-dis).

Korup, O., Densmore, A. L., & Schlunegger, F. (2010). The role of landslides in mountain range evolution. *Geomorphology, 120*(1–2), 77–90.

Larsen, M. C. (2008). Rainfall-triggered landslides, anthropogenic hazards, and mitigation strategies. *Advances in Geosciences, 14*, 147–153.

Lentini, V., Distefano, G., & Castelli, F. (2019). Consequence analyses induced by landslides along transport infrastructures in the Enna area (South Italy). *Bulletin of Engineering Geology and the Environment, 78*(6), 4123–4138.

Li, F., Torgoev, I., Zaredinov, D., Li, M., Talipov, B., Belousova, A., Kunze, C., & Schneider, P. (2021). Landslide-induced mass transport of radionuclides along transboundary Mailuu-Suu river networks in Central Asia. *Remote Sensing, 13*(4), 698.

Lorenz, D. F., & Dittmer, C. (2021). Disasters in the 'abode of gods'—Vulnerabilities and tourism in the Indian Himalaya. *International Journal of Disaster Risk Reduction, 55*, 102054.

Marín-Rodríguez, N. J., Vega, J., Zanabria, O. B., González-Ruiz, J. D., & Botero, S. (2024). Towards an understanding of landslide risk assessment and its economic losses: A scientometric analysis. *Landslides, 21*(8), 1865–1881.

Martha, T. R., Roy, P., Jain, N., Khanna, K., Mrinalni, K., Kumar, K. V., & Rao, P. V. N. (2021). Geospatial landslide inventory of India—An insight into occurrence and exposure on a national scale. *Landslides, 18*(6), 2125–2141.

Maskey, S. (1999). A case study of the Krishna Bhir slope failure disaster: past and present scenario at a glance. *International Journal of Rock Engineering and Mechanics, 2*(1–11).

Miele, P., Di Napoli, M., Guerriero, L., Ramondini, M., Sellers, C., Annibali Corona, M., & Di Martire, D. (2021). Landslide awareness system (laws) to increase the resilience and safety of transport infrastructure: The case study of pan-American highway (Cuenca–Ecuador). *Remote Sensing, 13*(8), 1564.

Mokarram, M., & Sathyamoorthy, D. (2018). A review of landform classification methods. *Spatial Information Research, 26*(6), 647–660.

Nikolic, T. (2014). Direct and indirect impact of landslide on environment. *Engineering geology for society and territory-volume 5: Urban geology, sustainable planning and landscape exploitation* (pp. 1237–1241). Springer International Publishing.

Nseka, D., Mugagga, F., Opedes, H., Ayesiga, P., Wasswa, H., Mugume, I., Nimusiima, A., & Nalwanga, F. (2021). The damage caused by landslides in socio-economic spheres within the

Kigezi highlands of South Western Uganda. *Environmental & Socio-Economic Studies, 9*(1), 23–34.
Pandey, H. K. (2020). Introduction to landslides. In *Techniques for disaster risk management and mitigation* (pp. 181–195).
Patel, A. B., & Bakshi, V. (2024). Determining land induced factors for landslide susceptibility in Indian cities. In *Landslide: Susceptibility, risk assessment and sustainability: Application of geostatistical and geospatial modeling* (pp. 243–266). Springer Nature Switzerland.
Pei, Y., Qiu, H., Yang, D., Liu, Z., Ma, S., Li, J., Cao, M., & Wufuer, W. (2023). Increasing landslide activity in the Taxkorgan River Basin (eastern Pamirs Plateau, China) driven by climate change. *CATENA, 223*, 106911.
Petley, D. (2012). Global patterns of loss of life from landslides. *Geology, 40*(10), 927–930.
Petrucci, O. (2024). Changes in human vulnerability to flood and landslide: Evidences from historical data. *Sustainability, 16*(16), 6715.
Pfurtscheller, C., & Genovese, E. (2019). The Felbertauern landslide of 2013 in Austria: Impact on transport networks, regional economy and policy decisions. *Case Studies on Transport Policy, 7*(3), 643–654.
Pollock, W., & Wartman, J. (2020). Human vulnerability to landslides. *GeoHealth, 4*(10), e2020GH000287.
Priyom Roy, P. R., Martha, T. R., & Kumar, K. V. (2014). Assessment of the Sunkoshi (Nepal) landslide using multitemporal satellite images. *Current Science.*
Rana, B. K., Jha, S. K., Molla, S., Dhumale, M. R., & Kulkarni, M. S. (2022). Assessment of environmental radioactivity and groundwater quality around Tummalapalle uranium mining site, Andhra Pradesh. *India. Radiation Protection and Environment, 45*(3 & 4), 170–182.
Riahi, V., & Nasire Zare, S. (2021). Evaluating the vulnerability of agricultural land use to the landslide risk in rural areas (Case study: Tarom county). *Journal of Research and Rural Planning, 10*(2), 17–32.
Robson, M., Radulescu, R., & Petley, D. (2022). The economic consequences of landslides: Evidence from a global database. *Available at SSRN 4282912.*
Saptutyningsih, E., & Dewanti, D. S. (2023, May). Measuring the economic impact of landslide hazard: Hedonic price approach. In *IOP Conference Series: Earth and Environmental Science* (Vol. 1180, No. 1, p. 012048). IOP Publishing.
Sartohadi, J. (2008, December). The landslide distribution in loano sub-district, Purworejo district, Central Java province, Indonesia. *Forum Geografi, 22*(2), 129–144.
Scaioni, M. (2013). Remote sensing for landslide investigations: From research into practice. *Remote Sensing, 5*(11), 5488–5492.
Schlögl, M., & Matulla, C. (2018). Potential future exposure of European land transport infrastructure to rainfall-induced landslides throughout the 21st century. *Natural Hazards and Earth System Sciences, 18*(4), 1121–1132.
Schuster, R. L. (1996). Socioeconomic significance of landslides. In *Landslides: Investigation and mitigation* (Vol. 247, pp. 12–35).
Schuster, R. L., & Highland, L. M. (2001). *Socioeconomic and environmental impacts of landslides in the western hemisphere* (No. 2001–276).
See-Sew, G., Dwikorita, K., & Shiao-Yun, W. (2009). *Policy and institutional framework for landslide mitigation and risk reduction. In Landslides—Disaster risk reduction* (pp. 531–554). Springer Berlin Heidelberg.
Sepúlveda, S. A., & Petley, D. N. (2015). Regional trends and controlling factors of fatal landslides in Latin America and the Caribbean. *Natural Hazards and Earth System Sciences, 15*(8), 1821–1833.
Shrestha, S., Pradhan, P. M., & Shrestha, H. K. (2023). Assessment of rainfall-induced shallow landslides in Kavre district. *Nepal. SCITECH Nepal, 17*(1), 84–90.
Shroder, J. F. (2021). *Landslide hazards, risks, and disasters.* Elsevier.
Sidle, R., & Ochiai, H. (2006). Processes, prediction, and land use. In *Water resources monograph* (Vol. 525, p. 870). American Geophysical Union, Washington

Strouth, A. B. (2024). Quantitative life-loss risk evaluation for landslides. *Doctoral Dissertation, University of British Columbia.*
Sugathapala, K., & Prasanna, J. (2010). *Essential human settlement planning considerations for sustainable landslide mitigation: With special reference to Padiyapelella landslide area.* National Building Research Organisation.
Sunarta, I. N., Susila, K. D., & Kariasa, I. N. (2018, February). Landslide hazard analysis and damage assessment for tourism destination at Candikuning Village, Tabanan Regency, Bali, Indonesia. In *IOP Conference Series: Earth and Environmental Science* (Vol. 123, No. 1, p. 012006). IOP Publishing.
Supper, R., Römer, A., Jochum, B., Bieber, G., & Jaritz, W. (2008). A complex geo-scientific strategy for landslide hazard mitigation—From airborne mapping to ground monitoring. *Advances in Geosciences, 14*, 195–200.
Tang, Y. M., Zhang, M. S., Xue, Q., & Bi, J. B. (2012). Landslide monitoring and early-warning: An overview. *Geological Review, 58*(3), 533–541.
Thakur, N., & Sharma, D. S. (2022). Study of sustainable techniques for effective risk management and control: A review. *International Journal for Research in Applied Science and Engineering Technology, 10*(6), 1688–1696.
Thapa, P. S., Daimaru, H., Ichion, E., & Yanai, S. (2024). Impacts of sediment transported downstream from the 2015 deep-seated landslide in Mt. Hakusan, Japan. *Earth Surface Processes and Landforms, 49*(4), 1273–1288.
Thayyen, R. J., Mishra, P. K., Jain, S. K., Wani, J. M., Singh, H., Singh, M. K., & Yadav, B. (2022). Hanging glacier avalanche (Raunthigad–Rishiganga) and debris flow disaster on 7 February 2021, Uttarakhand, India: A preliminary assessment. *Natural Hazards, 114*(2), 1939–1966.
Turner, A. K., & Jayaprakash, G. (1996). Chapter 1 Ā introduction. *Landslides: Investigation and Mitigation. Transportation Research Board Special Report, Issue*, (247), 3Ā11.
Tuskan, Y., & Basari, E. (2023). Evaluation of sustainable slope stability with anti-slide piles using an integrated AHP-VIKOR methodology. *Sustainability, 15*(15), 12075.
Vranken, L., Van Turnhout, P., Van Den Eeckhaut, M., Vandekerckhove, L., & Poesen, J. (2013). Economic valuation of landslide damage in hilly regions: A case study from Flanders, Belgium. *Science of the Total Environment, 447*, 323–336.
Walkinshaw, J. (1992). Landslide correction costs on US state highway systems. *Transportation Research Record*, 36–36.
Winter, M. G., Peeling, D., Palmer, D., & Peeling, J. (2019). Economic impacts of landslides and floods on a road network. *Auc Geographica, 54*(2), 207–220.
Wu, Z., Yang, H., Cai, Y., Yu, B., Liang, C., Duan, Z., & Liang, Q. (2024). Intelligent monitoring applications of landslide disaster knowledge graphs based on chatglm2. *Remote Sensing, 16*(21), 4056.
Xie, L., Sun, C., & Ren, X. (2014). Analysis of reservoir landslide deformation mechanism based on the three-dimensional fluid-structure coupling. *Electronic Journal of Geotechnical Engineering, 19.*
Yadav, A. K., Sharma, S. K., Kumar, P., Singh, T., Mamtani, H., Singh, N. K., Goyal, A., Manikappa, S. K., & Shukla, M. (2025). Shifting grounds: Exploring the physical and psychosocial impact of landslides. *IJPR, 7*(1), 28–32.
Yang, Z. J., Qiao, J. P., Uchimura, T., Wang, L., Huang, D., Lei, X. Q., & Shi, L. L. (2017). Explore on hydro-mechanical threshold for early warning of rainfall induced shallow landslides. In *Advancing culture of living with landslides: Volume 4 diversity of landslide forms* (pp. 193–199). Springer International Publishing.
Zhang, Q., Yu, H., Li, Z., Zhang, G., & Ma, D. T. (2020). Assessing potential likelihood and impacts of landslides on transportation network vulnerability. *Transportation Research Part D: Transport and Environment, 82*, 102304.
Zumpano, V., Pisano, L., Malek, Ž, Micu, M., Aucelli, P. P., Rosskopf, C. M., Balteanu, D., & Parise, M. (2018). Economic losses for rural land value due to landslides. *Frontiers in Earth Science, 6*, 97.

Yuvaraj, R. M., & Dolui, B. (2023). Geographical assessment of landslide susceptibility using statistical approach. *Quaternary Science Advances, 11*, 100097.

Overview of Landslides in Tamil Nadu and Its Effects

3

K. S. Banu and B. Vijayakumari

Abstract

This is a detailed review of the landslides in the hilly areas of the state of Tamil Nadu and especially Nilgiris areas, the cause, distribution, effects and the current methods of mitigation. Geomorphological conditions (slope gradient, geology, soil composition), hard monsoon rains and human activities like deforestation and development of infrastructural facilities are the major causes of landslides. GIS-based susceptibility analysis based on techniques such as Analytical Hierarchy Process (AHP), logistic regression, neural networks, and fuzzy logic has classified almost 50% of Nilgiris as medium to very high risk. There were documented landslide incidences in 1902, 1978–79, 1993, 2009, and afterward in Cyclone Gaja in 2018, which led to infrastructure damage, loss of lives and degradation of the environment. But mainly missing are obsolete hazard maps, limited field validation, and no real-time monitoring and early-warning systems. We suggest revising zoning maps with high-resolution data confirmation, introducing real-time sensors, enhancing slope stability by geotechnical examinations and training the community. These measures are necessary to mitigate vulnerability to landslides and encourage safe development in Tamil Nadu business sensitive areas.

Keywords

Landslides • Tamil Nadu • Nilgiris • GIS susceptibility mapping • Monsoon rainfall • Early-warning systems

K. S. Banu (✉) · B. Vijayakumari
Department of Geography, Queen Mary's College (A), Chennai, Tamil Nadu, India
e-mail: k.s.banu.aqua68@gmail.com

R. M. Yuvaraj et al. (eds.), *Advanced Landslide Susceptibility Modeling*,
https://doi.org/10.1007/978-3-032-14485-0_3

3.1 Introduction

The landslides, which can be described as the down movement of soil, rocks and debris through the action of gravity, are features that are widespread and damaging natural hazards in the highlands and hilly areas of Tamil Nadu. Although the Nilgiris district of the Western Ghats is the area most at risk—due to its steep and changing topography, fractured geology, and heavy monsoon-related precipitation (Kumar et al., 2017)—recent occurrences have demonstrated that risks are far beyond these classic areas (Elangovan & Shanthi, 2021). An example is the Thiruvannamalai district, which is not generally considered a landslide-prone area, where a tragic landslide happened in late 2024 when an outcrop of boulders fell after Arunachala hill got strong rains due to Cyclone Fengal. This created two consecutive landslides between December 1–2, 2024, leveling houses in VOC Nagar and killing seven relatives including five children, which underscores the fact that even stable landscapes are prone to landslides during extreme weather conditions (NDTV News Desk, 2024; New Indian Express, 2024a, 2024b, 2024c; The Federal, 2024).

The other districts have also experienced rising slope instability during the monsoon and cyclone seasons in addition to Nilgiris and Thiruvannamalai district. Late in 2023 and early 2024, landslides in Coimbatore (especially in and around Kunjappa-Panai) and slope failures in Madurai, Theni, Dindigul, and Tiruppur led to school closures and the temporary discontinuation of services on the Nilgiris Mountain Railway (Business CNBC-TV18, 2023; Standard, 2023). Vulnerability is also present in the Eastern Ghats as in October 2024, rockslides occurred in Jawadhu Hills in Tirupattur district which disrupted vehicular traffic to Pudur Nadu (The New Indian Express, 2024a, 2024b, 2024c) shown in Table 3.1.

Table 3.1 Major landslide events in Tamil Nadu (2000–2024)

Year	Location	Cause	Impact	Source
2009	Nilgiris district	Northeast monsoon	43 Deaths, major road damage	Elangovan and Shanthi (2021)
2018	Nilgiris and delta	Cyclone gaja	45+ Deaths, rail and road blockades	India Today (2024)
2023	Coimbatore	Heavy monsoon rains	Landslide near ghat road; school closures	CNBC-TV18 (2023)
2024	Tiruvannamalai	Cyclone fengal	7 deaths, 3 houses crushed in VOC nagar	India Today (2024)
2024	Tirupattur	Intense rainfall	Road blocked, vehicle damage	New Indian Express (2024a, 2024b, 2024c)
2025	Nilgiris district	Pre-monsoon instability	275 families evacuated	Times of India (2025a, 2025b)

These events are a result of the combination of environmental and anthropogenic events. First, there are extreme weather conditions that soils are quickly saturated through monsoonal or cyclonic rains, which decrease the shear strength of the soils and destabilize the slopes (Elangovan & Shanthi, 2021; NDTV News Desk, 2024). Second, the geological structure of the region, which is a combination of schists, gneisses, charnockites, or remnants of lateritic soils or weathered profiles, adds to instability in the circumstances of long-term moisture input (Sharma & Sanjeevi, 2015). Third, this slope is worsened by human activities like deforestation to plantations, quarrying, cutting of slopes to build roads and buildings, and poor drainage (Guru et al., 2017). Noticeably, even hunting grounds have been defied by rockslides such as the lethal Sivaganga quarry collapse in 2025 which despite being caused by blasting highlights the rock mass unsteadiness within the state (Times of India, 2025a, 2025b).

These events have far reaching socio-economic implications. In Nilgiris, road and railway damage, destruction of property, displacement of shelter and loss of agriculture is a common monsoon event- as was the case in May of 2025 with 275 families shifting to relief camps (Times of India, 2025a, 2025b). In December 2024, Thiruvannamalai tragedy led to the loss of lives, destruction of homes, and hindered the rescue operations because of the continuous rains and unstable slopes (NDTV News Desk, 2024; The Indian Express, 2024). The transportation connections were cut off in such districts as Tirupattur and Coimbatore, affecting the daily lives of residents and emergency services (CNBC-TV18, 2023; New Indiam Express, 2024a, 2024b, 2024c). Taken together, these events indicate the way landslide risks in Tamil Nadu have moved beyond common geographic limits and are no longer limited to the traditional Western Ghats hotspots but now extend to Eastern Ghats and plateau margins. Knowledge of this distribution can be used to revise the risk framework and develop interventions that are specific to the distribution, such as slope reinforcement, carefully planned land-use development, real-time observation, and localized early warning, even in spaces that have been thought of as low risk.

3.2 Rationale of the Study

Although landslide susceptibility in Nilgiris and other Western Ghats locations has been widely studied, a number of gaps are still present. First, the problem with numerous current zonation maps and models is that they have a poor spatial resolution or have obsolete datasets, which do not apply to dynamic landscapes (Elangovan & Shanthi, 2021). Moreover, it is noted that most studies are based on extensive usage of GIS and remote-sensing without extensive ground truthing; geotechnical data and slope failures are not properly checked in the field (Sharma et al., 2015). Second, the research on landslides is mostly centered around Western Ghats areas of the past, whereas more susceptible areas of Eastern Ghats like Thiruvannamalai, Kolli Hills, and Jawadhu Hills are overlooked. Third, the increase in severe weather such as the deadly landslide in Tiruvannamalai as a result of the Cyclone Fengal in December 2024 also shows how even the areas that were deemed as low risk are becoming more dangerous (New Indian Express, 2024a, 2024b, 2024c; India Today, 2024). Nevertheless, real-time rainfall sensors and slope monitors, which can form the components of an early-warning system, are still poorly integrated at the regional or village scale (The News Minute, 2024). Lastly, there is no multi-disciplinary integration, including hydrology, geology, community resilience and policy frameworks. The majority of research is confined to technical GIS or lab-based research and fails to transfer research on the policy level to concrete risk reduction measures or community education initiatives (Guru et al., 2017). This paper seeks to address these gaps by integrating GIS mapping with recent case events (e.g., Tiruvannamalai), assessing field-based validation studies (e.g., IIT-Madras sampling) and providing suggested integrated early-warning and community-engagement models. This argument highlights the pressing necessity of a more comprehensive, validated, and proactive method of landslide risk assessment in Tamil Nadu.

3.3 Materials and Methods

A three pronged methodology was applied in the study; (1) literature-based GIS susceptibility analysis, (2) recent case-event synthesis, and (3) geotechnical field validation.

Susceptibility Review Based on GIS: We went through the seminal studies on the basis of familiar methodologies like weighted overlay, logistic regression, fuzzy logic, and ensemble machine learning applied to Nilgiris and other hilly areas (Beigh et al., 2024; Elangovan & Shanthi, 2021). Layers were slope, elevation, lithology, lineament and drainage density, land use, soil type and rainfall intensity. Comparison was done in terms of model accuracy measures (ROC-AUC), spatial resolution, and validation methods.

Case-Event Documentation: We summarized that the landslide events happened in December 2024 (Tiruvannamalai), Coimbatore, Madurai and Tirupattur, 2023–24 landslide events, and sites on the Eastern Ghats (New Indian Express, 2024a, 2024b, 2024c; India Today, 2024). In the case of Tiruvannamalai, we referred to several sources reporting reasons, rescuing activities, geological samples, and precipitation levels (New Indian Express, 2024a, 2024b, 2024c; The News Minute, 2024). In order to emphasize the expansion of Eastern Ghats zones, the IndiaCore susceptibility map (IIT-Delhi) was referred (Mongabay, 2024).

Field and Geotechnical Validation: We examined on-site sampling reports done by the IIT-Madras after Fengal and geotechnical studies at Kolli Hills (Sampath, 2024). The undisturbed samples, soil mechanics testing (modified Bishop method) and calculation of Newmark displacement were assessed concerning slope failure limit.

Analytic description of the key processes that affect landslides events in Tamil Nadu, especially in steep and mountainous areas such as the Western and Eastern Ghats. Of these, slope gradient is the most significant factor as it contributes about 37.7% to the landslides susceptibility in the GIS-based models. The reason is that the steepness of the slopes enhances the strength of gravity that makes the mass of soil and rocks more susceptible to falling down particularly following rainfall saturation. The second factor that is most influential is geological structure (28.1%). The high occurrence of fractured rocks like charnockite, schist and gneiss in areas like Nilgiris, Kolli Hills, and Tiruvannamalai forms structural weakness in the terrain. These are easily tapped in case of heavy rains or in case of seismic activities leading to slope failures. Where several lineaments overlap, the closeness to lineaments also contributes to greater instability, which explains almost 11.9% of susceptibility. These fractures occur naturally and promote infiltration of water, decrease slope cohesion and act as planes of failure when water falls heavily. Drainage density, which is the level of surface water channel coverages, adds 7.1%. When the drainage density is high, surface runoff and erosion are there and these processes can undermine slopes and destabilize soil structures. In the meantime, monsoonal bursts and cyclonic storms are also a major trigger, as they reduce soil shear strength and cause landslides. It has a lower weight of 6.2 but when combined with another terrain factor it is a catalyst. Finally, land use, especially deforestation and agricultural slope alteration to build new housing, construction, or quarrying, also play a variable role depending on the local circumstances. Human activities such as the deposition of slope stability, distortion of natural drainage patterns, aggravate the landslide risk.

In general, Table 3.2 shows that landslide vulnerability in Tamil Nadu is due to a complex interplay of natural topographical and geological factors and human actions. This knowledge is essential in developing area-specific mitigation policies and setting priorities in areas to be monitored and mitigated.

Data synthesis involved comparison of GIS model performance and occurrences of events and quality of validation. The gaps in the implementation of early-warning detection were discovered with respect to literature and auditing of the location. The new strategy allows assessing the changing hazard profile, particularly in the low-risk Eastern Ghats landscapes in the past, and understanding the preparedness to mitigate based on community-based information.

3.4 Results and Discussion

3.4.1 GIS Susceptibility Modeling Performance

The Nilgiris studies with weighted overlay, logistic regression, and fuzzy models showed high ROC-AUC scores (0.82–0.91), which showcase a good predictive ability in historical landslide events (Chandrasekaran & Kumar, 2019; Elangovan & Shanthi, 2021). In the IIT-Delhi map, ensemble models include very highly susceptible regions that encompass 4.75% of India, including emerging areas in

Table 3.2 Key factors contributing to landslides in Tamil Nadu

Factor	Description	Weight (%) in GIS models	Source
Slope gradient	Steep terrain increases gravitational pull	37.7%	Elangovan and Shanthi (2021)
Geology	Charnockite, gneiss, schist—highly fractured and weathered	28.1%	Kumar et al. (2017)
Proximity to lineaments	Geological faults and fractures promote instability	11.9%	Guru et al. (2017)
Drainage density	High density leads to soil erosion and undercutting	7.1%	Elangovan and Shanthi (2021)
Rainfall intensity	Sudden cloudbursts and cyclones trigger slope saturation	8.6%	Yuvaraj and Dolui (2021)
Land use/ deforestation	Plantation, construction, and quarrying disturb slope stability	Varies regionally	Guru et al. (2017)

the Eastern Ghats (Sharma et al., 2024). These results confirm the generalizability of the advanced methods in all terrains, however, site-specific calibration is required.

Table 3.3 provides a summary of the most important landslide susceptibility modeling studies carried out in Tamil Nadu with the emphasis made on the range of approaches and their relative accuracy. The literature is mainly centered in the Nilgiris district and the surroundings where landslides are traditionally of great concern. Different statistical and calculation methods, including weighted overlay analysis, logistic regression, and remote sensing combined with GIS have been used to define and forecast vulnerable areas.

The application of the Analytical Hierarchy Process (AHP) weighted overlay developed by Elangovan and Shanthi (2021) showed a high predictive validity of the weighted overlay with a ROC-AUC value of 0.87, which was verified by a field test, and so it could be used to map the areas of landslides. Equally, Guru et al. (2017) used the logistic regression in Coonoor and yielded an accuracy of 0.82 using historic data on landslide inventory, which validated the effectiveness of statistical models to describe the relationship between conditioning factors and the occurrence of landslides.

The study of Beigh et al. (2024) compared remote sensing data with GIS, and with an accuracy of 0.85, they stated that incorporation of satellite imagery enhances the sensitivity and spatial ability of the susceptibility mapping. It is worth noting that the 2024 national susceptibility map-based in IIT-Delhi is based on

Table 3.3 GIS-based landslide susceptibility models reviewed

Author(s)	Study area	Method used	Accuracy (ROC-AUC)	Validation type
Yuvaraj and Dolui (2023)	Nilgiris district	Frequency ratio (AHP)	0.86	Field verification
Guru et al. (2017)	Coonoor	Logistic regression	0.82	Historic inventory mapping
Chandrasekaran and Kumar (2019)	Coonoor	Remote sensing + GIS	0.85	Ground truth points
IIT-Delhi	All-India (incl. Eastern Ghats)	Ensemble AI models	~0.91	National cross-validation

more sophisticated models of ensemble AI, which is the most accurate (around 0.91) and covers more areas than the traditional hotspots, which is an improvement in both the level of sophistication of the models and the coverage.

The table reveals that most models have high accuracy, but selection of method will be dependent on data availability, nature of study area, and purpose of use. Field validation is also essential to validate model predictions. This move toward ensemble and machine learning methods is an indicator of a significant change to more resilient, data-driven hazard estimates that can change with the environmental conditions, such as new landslide-prone regions in Tamil Nadu.

3.4.2 Expansion of Vulnerable Zones

More recent mapping has located some hidden high-risk areas in Eastern Ghats areas like Kolli, Jawadhu and Tiruvannamalai. According to the IIT-Delhi susceptibility map, 95% of accuracy is confirmed in these regions, which brings out risk ignorance (Mongabay, 2024). Extreme rainfall and the slope instability of sandstone charnockites and subsequent lateritic soils has predisposed Eastern Ghats to landslides.

3.4.3 Case of Tiruvannamalai Landslide (Dec 2024)

The heavy rain brought about by Cyclone Fengal caused a giant boulder (between 40 and 80 tons) to break loose in the Annamalaiyar Hill and wiped out homes in VOC Nagar, killing seven people. NDRF rescue operations and IIT-Madras assessments after the event gave post-event geotechnical soil analysis. Delayed rescue because of saturated soil and unstable debris highlighted the importance of pre-event hazard monitoring. This occurrence is an example of how excessive cyclonic rainfall may initiate high magnitude slides in areas that have no previous history of landslides.

3.4.4 Other Events Outside Nilgiris

Landslides caused by monsoons in Coimbatore (Kunjappa-Panai) and in the Eastern Ghats (Jawadhu Hills, Tirupattur) in 2023–24 blocked transport and led to school/railway closures. Geotechnical investigations in the Kolli Hills showed that cut slopes in ghat roads had low safety factors which justified vulnerability on a basis of lithological and structural considerations (Kumar et al., 2021).

3.4.5 Drivers and Conditioning Factors

In each of the areas Western and Eastern Ghats, such important conditions as steep slopes, fractured geology (charnockites, schists, sandstone), soil saturation, hydrological pressure, and anthropogenic changes (deforestation, road cuts, quarrying) are central (Kumar & Annadurai, 2013). Mass failures are caused by extreme precipitation events which significantly reduce shear strength (Chellamuthu & Ganapathy, 2024).

3.4.6 Model Validation and Field Calibration

GIS models have been proven to be accurate, but there is limited field verification with geotechnical measurements. Recent efforts include the IIT-Madras project of Tiruvannamalai and Kolli Hills soil samples. Such attempts allowed slope stability tests with the help of Modified Bishop and Newmark techniques—the key to determining real weak zones.

3.4.7 Gaps in Early Warning and Community Response

Real-time warning systems (rainfall, slope sensors) are not significantly implemented even though technically advanced. News Minute was an account of delayed rescue in Tiruvannamalai, which indicated social administrative problems. The Eastern Ghats communities have low hazard awareness and evacuation and thus respond to the hazard rather than act on it.

3.4.8 Socio-economic and Infrastructure Impacts

Transport connection through ghat roads, mountain railways and plantation lands is occasionally interrupted. The landslide in Tiruvannamalai resulted in the death of seven people flattening houses and costing ₹500,000 in ex-gratia payments (Kumar et al., 2021). In Coimbatore and Nilgiris, school closings, economic loss and displacement were also reported.

Table 3.4 Consequences of recent landslides (2023–2025)

Location	Disruption type	Sector affected	Government response
Tiruvannamalai	Three houses destroyed, seven deaths	Housing, emergency response	₹5 lakh ex-gratia, NDRF rescue, IIT-M analysis
Coimbatore	Ghat road blocked, school closures	Transport, education	Temporary closures, public alerts
Kolli Hills	Road instability, landslip risk	Tourism, mobility	Geological survey, slope reengineering plans
Nilgiris	275 families relocated (2025)	Community safety	Relief camps, temporary shelters

3.4.9 Synthesis and Interpretation

The worsening of vulnerability further east of Nilgiris, through major events such as Tiruvannamalai and fledgling Eastern Ghats at risk, requires a re-calibration of disaster planning. Higher-resolution models need to be accompanied by field validation, advance warning, and social involvement. By combining geotechnics, GIS, and policy frameworks, a robust foundation will be developed against the effects of climate on landslides.

Table 3.4 demonstrates the various and broad-based impacts of the recent landslides in the state of Tamil Nadu in 2023 and 2025. One of the most destructive landslides was the Tiruvannamalai landslide of December 2024, which resulted due to Cyclone Fengal. It led to the falling down of huge boulders of the Annamalai Hill and total destruction of houses in VOC Nagar killing seven people. The event necessitated the involvement of the National Disaster Response Force (NDRF) to assist with the emergency response and forensic geotechnical investigations conducted by IIT-Madras and compensation of 5 lakh rupees per family. This accident, on a low-risk zone, highlighted the importance of active hazard surveillance of non-traditional zones.

On 2023, heavy rains in Coimbatore led to landslides along the ghat roads, leading to the closure of schools and causing severe transport and education service disruptions. The susceptibility of infrastructure is evident in these interruptions, particularly in areas where the road alignments are steep and where drainage is inadequate.

In the same manner, Kolli Hills, which is a section of the Eastern Ghats were exposed to frequent risks of landslips because of slope failures as well as uncontrolled construction. The instability of roads impacted the local tourism and movement, and geological surveys and slope reengineering plans were initiated. These incidents explain why livelihoods and regional economies are also under threat by frequent landslides alongside lives.

With a historical landslide-prone Nilgiris, the 2025 pre-monsoon instability resulted in 275 families being evacuated and temporary shelters and relief camps

Table 3.5 Recommended actions for landslide risk reduction

Domain	Recommended action	Expected benefit
Geotechnical	Field soil tests (Bishop/Newmark methods)	Accurate slope stability assessment
Monitoring	Install real-time sensors (rainfall, inclinometers)	Early warning before failure
Community engagement	Awareness campaigns, evacuation drills	Localized, fast response
Planning and regulations	Enforce slope-cutting rules, ban quarrying on weak slopes	Prevent human-triggered failures
Policy integration	Use susceptibility maps in urban and rural development planning	Sustainable land use, reduced exposure

being set up. This is a more organized government reaction than previous incidents, but it also points to the magnitude of displacement and the frequency of such disasters in the area.

All these cases collectively indicate landslides in Tamil Nadu not only affect human lives but also critical infrastructure, education, tourism and general socio-economic stability of the communities affected. They emphasize the criticality of multi-hazard preparedness, better land-use control, and prompt mitigation of known and newly identified landslide areas.

Table 3.5 highlights a multi-sectoral approach to address landslide risks in Tamil Nadu, particularly as the area of hazard is increased moving beyond Western Ghats. In the geotechnical field, more precise evaluation of susceptible slopes will be possible, using detailed field-based soil and slope stability analyses of the Bishop and Newmark techniques. This aids in replacing remote-sensing dependency with data-based local vulnerability tests. With regards to surveillance systems, real-time sensors like rain gauges, piezometers and inclinometers should be installed to give early alerts as per rainfall levels and slope movement, especially in highly prone sites like Tiruvannamalai and Kolli hills.

The community-engagement aspect highlights the fact that disaster preparedness must be inclusive. Local training, awareness campaigns and evacuation exercises could improve the response time of the community and reduce the number of casualties, particularly in rural and hilly settlements where the population has minimal access to formal alert mechanisms. Under city and rural planning models, the anthropogenic causes of landslides can be avoided by enacting regulations on slope-cutting, prohibition of quarrying in unstable areas, and installation of appropriate drainage systems. The measures are especially applicable in areas where haphazard construction has caused instability of land, as in both Nilgiris and newly developing Eastern Ghats sites.

Lastly, by combining scientific susceptibility maps with government planning tools (such as Master Plans and village-level development plans) risk-sensitive infrastructure projects, agricultural expansions, and tourism developments will be implemented. Such policy integration makes it possible to have risk reduction also being part of sustainable development and not a response to a disaster. Therefore, the recommendations target not only disaster response, but long-term risk prevention and resilience building across the fluid landslides regions of Tamil Nadu.

3.5 Limitations

- Predominantly literature-based, with limited primary fieldwork conducted by us.
- Case-event data depends on media reports and may lack standardized geotechnical detail.
- GIS model comparisons were indirect; inconsistency in metrics hinders direct evaluation.
- Socio-economic impacts are qualitatively assessed; no household-loss surveys were administered.
- Early-warning feasibility was discussed conceptually without pilot implementation.

3.6 Recommendations

1. **Geo-Validation Extension**: Conduct systematic sampling and slope stability testing (e.g., Bishop/Newmark) in Eastern Ghats zones to complement geospatial models (Kolli, Tiruvannamalai).
2. **Real-Time Monitoring**: Deploy automated sensors (rainfall, piezometers, inclinometers) at identified hotspots for threshold-based early warning.
3. **Community Training**: Establish local disaster volunteer teams; run awareness workshops and evacuation drills.
4. **Regulatory Enforcement**: Implement slope-cutting guidelines, drainage control, reforestation, and stability checks for construction and quarrying projects.
5. **Integrated Hazard Planning**: Mandate linking of zonation maps with local governance, agrarian, and infrastructure planning decisions.

3.7 Conclusion

Landslides in Tamil Nadu have not only spread to areas considered as hotspots in Western Ghats, but also in Eastern Ghats and plateau fringe districts. New GIS modeling capabilities offer high-resolution hazard maps but unless they are validated in the field and combined with early-warning, communities are not at a disadvantage. This vulnerability is highlighted in the Tiruvannamalai tragedy. To minimize future disaster risk, a comprehensive multi-disciplinary, proactive approach is required, incorporating geotechnical validation, monitoring infrastructure, community preparedness and regulatory oversight. This integrated model will deliver scientific rigor as well as practical resilience of developing safer development pathways in morphologically complex terrains of Tamil Nadu.

Acknowledgements The author(s) acknowledge the use of AI-based tools for grammar and language correction during the preparation of this manuscript.

References

Beigh, I. H., Bukhari, S. K., & Patil, A. S. (2024). GIS-based landslide susceptibility assessment using a statistical approach: a case study of Srinagar-Bandipora highway, Kashmir Himalaya India. In *Landslides in the Himalayan region: Risk assessment and mitigation strategy for sustainable management* (pp. 237–257). Springer Nature Singapore.

Business Standard. (2023, November 9). *Heavy rain triggers landslides in Tamil Nadu, schools shut in 5 districts*. Business Standard.

Chandrasekaran, R., & Kumar, P. S. (2019). The assessment and prediction of landslide susceptibility zonation (LSZ) by relative effect method (Re) and GIS in Coonoor block, Nilgiri district, Tamil Nadu, India. *International Journal of Engineering Research & Technology*, *7*(11).

Chellamuthu, S. N., & Ganapathy, G. P. (2024). Quantifying the impact of changing rainfall patterns on landslide frequency and intensity in the Nilgiris district of Western Ghats, India. *Progress in Disaster Science, 23*, 100351.

CNBC-TV18. (2023, November 9). *Heavy rains lash part of Tamil Nadu, landslide reported in several areas*.

Elangovan, K., & Shanthi, S. (2021, December). Evaluation of quantitative landslide susceptibility zonation (LSZ) method for Nilgiri district. In *ICCAP 2021: Proceedings of the First International Conference on Combinatorial and Optimization, ICCAP 2021, December 7–8 2021, Chennai, India* (p. 99). European Alliance for Innovation.

Guru, B., Veerappan, R., Sangma, F., & Bera, S. (2017). Comparison of probabilistic and expert-based models for landslide susceptibility mapping using GIS: A case study from Coonoor, Nilgiris, Tamil Nadu. *Spatial Information Research, 25*(6), 757–768. https://doi.org/10.1007/s41324-017-0143-y

India Today. (2024, December 3). Panic grips Tamil Nadu's Tiruvannamalai as 7 die in landslides from Cyclone Fengal. *India Today*. ijert.orgthenewsminute.com+6indiatoday.in+6independent.co.uk+6

Kumar, M. K., & Annadurai, R. (2013). Mapping of landslide susceptibility using geospatial technique-a case study in Kothagiri region, Western Ghats, Tamil Nadu, India. *International Journal of Engineering Research & Technology (IJERT), 2*(12), 1–12.

Kumar, M., Krishnaveni, V., & Muthukumar, S. (2021). Geotechnical investigation and numerical analysis of slope failure: A case study of landslide vulnerability zone in Kolli hills, Tamil Nadu. *Journal of the Geological Society of India, 97*(5), 513–519.

Kumar, N., Balasundaram, N., & Meenambal, T. (2017). A Study On Factors Influencing Landslides In Nilgiris, Tamilnadu, India. *International Journal of Civil Engineering and Technology, 8*(8), 1011–1018.

Mongabay. (2024, June 26). Eastern Ghats debuts in new landslide susceptibility map. *India Climate Connections*. https://india.mongabay.com/

NDTV News Desk. (2024, December 2). Family of 7 killed in landslide in Tamil Nadu's Tiruvannamalai. *NDTV*.

New Indian Express. (2024, December 2). Bodies of seven people killed in Tiruvannamalai landslide recovered. *The New Indian Express*. newindianexpress.com+1timesofindia.indiatimes.com+1

New Indian Express. (2024, December 2). Tiruvannamalai landslide: Five bodies recovered, two still trapped under debris; operations intensify.

New Indian Express. (2024, October 23). Landslide disrupts traffic near Jawadhu Hills in Tirupattur.

Sampath, R. (2024, December 3). Rajalakshmi Sampath reporting soil analysis by IIT-Madras experts. *New Indian Express.* https://newindianexpress.com/

Sharma, G., & Sanjeevi, S. (2015). Landslide hazard zonation using remote sensing, ground penetrating radar surveys and geographical information system in Katteri. *International Journal of Current Engineering and Technology, 5*(2), 1160–1169.

Sharma, N., Saharia, M., & Ramana, G. V. (2024). High resolution landslide susceptibility mapping using ensemble machine learning and geospatial big data. *CATENA, 235*, 107653.

The Federal. (2024, December 2). TN | Another landslide hits Tiruvannamalai near spot where 7 of family trapped.

The Indian Express. (2024). 'Couldn't get her out quickly': 7 lives snuffed out in Tamil Nadu's Tiruvannamalai as boulder crushes house.

The News Minute. (2024, December 4). Tiruvannamalai landslide: residents allege delayed response. *The News Minute.* https://thenewsminute.com/

Times of India. (2024, December 3). Bodies retrieved after 20 hours in Tiruvannamalai landslide. *Times of India.* https://timesofindia.indiatimes.com/

Times of India. (2025a, May 23). Accident fallout: Govt directive to inspect quarries.

Times of India. (2025b, May). 275 families put up in relief camps in the Nilgiris.

Yuvaraj, R. M., & Dolui, B. (2021). Statistical and machine intelligence based model for landslide susceptibility mapping of Nilgiri district in India. *Environmental Challenges, 5*, 100211.

Yuvaraj, R. M., & Dolui, B. (2023). Geographical assessment of landslide susceptibility using statistical approach. *Quaternary Science Advances, 11*, 100097.

Part II

Contemporary Issues of Geographical and Geological Aspects of Landslide, Various Susceptibility Modeling Techniques and Accuracy Assessment

Terrain-Based Modeling of Landslide Susceptibility

4

R. M. Yuvaraj

Abstract

Landslides occur frequently in Gudalur Taluk, Nilgiri District, Tamil Nadu with its steep terrains and high rainfall and weak geological features. The proposed research seeks to come up with the detailed landslide susceptibility map based on Shannon Entropy model combined with both Remote Sensing and GIS techniques. Soil type, soil depth, slope, lithology, rainfall, land use/land cover, NDVI, and lineament characteristics were some of the main causative factors analyzed. The values of entropy were obtained to show relative relevance of each factors and soil depth was found to be the most relevant followed by land use/land cover and soil type. A map of landslide vulnerability was developed and classified into low, moderate, and high-risk areas. The ROC curve was used to validate, with the prediction accuracy in terms of Area Under the Curve (AUC) values of 0.738 and 0.762 being obtained on the training and test datasets, respectively. The findings indicate that areas in the middle, south and southeast parts of the study area are prone to landslides with northern areas being less prone. The integrated approach will help offer a reliable tool to manage the disaster risk and help in land-use planning and mitigation measures in the areas with landslides.

Keywords

Landslide susceptibility • Shannon entropy model • Remote sensing • GIS • Gudalur Taluk

R. M. Yuvaraj (✉)
Department of Geography (IDE), University of Madras, Chennai, India
e-mail: yuvaraj1023@gmail.com

R. M. Yuvaraj et al. (eds.), *Advanced Landslide Susceptibility Modeling*,
https://doi.org/10.1007/978-3-032-14485-0_4

4.1 Introduction

Landslides are severe and common geomorphic hazard, which have a high potential risk to infrastructure, agriculture, and communities in mountainous terrain. The Nilgiri District of Tamil Nadu in India is particularly susceptible, with its sharp slopes, heavy rainfall, and complicated geological structure. The situation presents the importance of landslide susceptibility mapping in addressing and controlling the effects of such natural hazards.

Landslides occur in the Gudalur taluk of The Nilgiri District, its rugged terrain and heavy rainfall during certain seasons cause movement downwards of soil, rocks, and debris under the effect of gravity. This is worsened by high rainfall, steep slopes, deforestation, and certain soil structures, which contribute to the occurrence and intensity of landslides. The occurrence of landslides causes massive destruction of important infrastructures including roads and bridges and endangers life and livelihood of people here, which raises the role of landslide risk evaluation and its efficient mitigation measures.

Factors such as natural and man-made factors have increased the occurrence and severity of landslides in the Gudalur taluk in the last few decades. Rapid growth and population increase have increased the urgency of more accurate and detailed landslide susceptibility mapping that will aid in the identification of hazard prone regions and assist policy makers in devising effective ways of managing risks.

Geographic Information System (GIS) and Remote Sensing technologies have significantly improved the mapping of landslide susceptibility with strong spatial analysis power being able to give researchers a better understanding of complex terrains. Satellite images and aerial photography collected by the Remote Sensing methodology are a source of helpful information that can be processed to identify land cover, vegetation, soil moisture, and other important variables to evaluate the risk of landslides. GIS allows this information to be combined with others providing researchers with a more accurate visualization and interpretation of landslide-related information.

Integration of Remote Sensing and GIS has been useful in producing detailed landslide susceptibility maps which take into consideration terrain features, land use as well as hydrological features. The integration can provide precise, timely information that could enhance forecasts and estimations of lands likely to experience landslides. Furthermore, GIS can be used to process big data sets and thus it is possible to analyze many variables over a large area. This is necessary when investigating areas such as the Gudalur taluk where the topography is diverse and therefore landslides occur frequently.

Shannon Entropy model is an information theory based tool that can help in evaluating landslide susceptibility. This model was derived by Claude Shannon in 1948 and measures the uncertainty and randomness in a dataset and researchers can measure the effect of different factors on landslides (Shannon, 1948). The Shannon Entropy model is useful in landslide susceptibility analysis to enable researchers to comprehend the relative significance of causative factors slope, aspect, soil type, land cover, and distance to faults by determining their information contribution or entropy values. These values determine which factors are of critical interest in the landslide incidence and hence they can be used to better delineate landslide prone areas (Yilmaz, 2009).

The application of the Shannon Entropy model in the Gudalur taluk can aid an objective method of measuring landslide risk, by taking an objective weight on each causative factor. The values of entropy of such aspects as the amount of precipitation, slope gradient, and the types of land covers can provide information on the environmental and geological factors contributing to the vulnerability of lands to landslides in particular regions. The method is particularly useful where landslide history is limited, since it is based on the distribution and the properties of known causative factors, and not on historic landslides itself (Park et al., 2013).

The combination of the Shannon Entropy model, Remote Sensing and GIS data can be used to produce a detailed susceptibility map. By associate weights of each of the causative factors with the entropy values, the prioritization of high-risk areas is possible. Not only does this approach improve allocation of resources in landslide management, but also offers a strong base upon which informed decisions can be made in disaster preparedness and response plans.

This paper seeks to create a landslide susceptibility map of the Gudalur Taluk, The Nilgiri District using the Shannon Entropy model with both Remote Sensing and GIS data. The objectives are to (1) determine the main causes of landslides in the district, (2) estimate the effects of each of these factors to landslides, and (3) generate an effective susceptibility map that can facilitate effective land-use planning and risk avoidance. By achieving these, the study will give useful information to the policymakers and helps to minimize the effects of landslides on the area.

4.2 Study Area

Gudalur Taluk, situated in the western part of the Nilgiri district in Tamil Nadu, lies between 11.45° N and 11.70° N latitude and 76.25° E to 76.65° E longitude, covering an area of approximately 1243 square kilometers. It is an integral part of the ecologically significant Western Ghats and is uniquely positioned at the tri-junction of Tamil Nadu, Kerala, and Karnataka. Gudalur shares its borders with Chamarajanagar district of Karnataka to the north, Wayanad district of Kerala to the west, Udhagamandalam (Ooty) Taluk to the east, and Pandalur Taluk to the south. The region is known for its rich biodiversity, dense forests, and widespread tea and coffee plantations. Figure 4.1 represents the geographical location of the study area.

However, Gudalur is very susceptible to landslides because it is very steep, its ecosystem is fragile and it also receives heavy rain during the southwest monsoon (June–September). They are between 900 and 2500 m with steep slopes and highly carved gullies that enhance the risk of slope failure. The rocks beneath the region are predominantly charnockites and gneisses and readily eroded by the rain. Other forms of human activities that destabilize slope are deforestation, road-building, and unplanned settlements. They are also mainly lateritic and red loamy

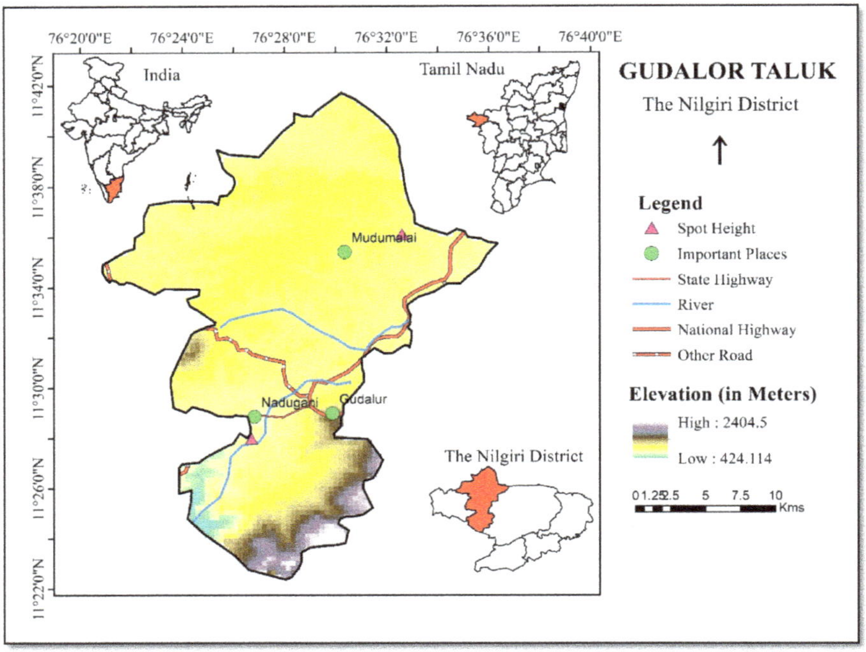

Fig. 4.1 Location of the study area

soils that retain water though are unstable when saturated, particularly in poorly drained areas. These have contributed to recurrent landslides which have damaged roads, farm lands and houses especially in the village of Pandalur, Cherambadi, and Devala. A good landslide management system and a sustainable land use are essential to mitigating the risk of disaster in the area.

4.3 Data and Methodology

The measure of abnormality, changeability, unstable behaviour and uncertainty in every system is performed by Entropy (Pourghasemi et al., 2012). In general, the concept of entropy identities the amount of abnormality of casus and result or in determination with regard to different subjects. An entropy index is an approximation of the disparity between the average shares of single groups within the entire system. Entropy index is considered as the measure of the average difference in the proportion of unit groups, relative to the total system.

In various areas of science (including studies of storms, droughts, debris flows, and landslides), entropy has been applied with various purposive definitions including instability, amount of information, and degree of unpredictability (Pourghasemi et al., 2012). Shannon also gave his personal definition of entropy with respect to the Boltzmann H-theorem (Shannon, 1948), in which the quantity of information of each event is a random variable whose expected value (average) is referred to as ShE or the index of entropy. When one or two events are more likely to occur than the others, then observation of that event is less informative, and the ShE will be small. Entropy merely reflects the likelihood of observing a particular event, the information it represents is therefore information about the distribution underlying these events, not the meaning of the events. The latter permits the applicability of ShE in other subjects. The mathematical description of the above states are as follows (Shannon, 1948):

$$Y_{ij} = \frac{\mathrm{FR}}{\sum_{j=1}^{r_j} \mathrm{FR}}.$$

$$Tj = -\sum_{i=1}^{r_j} Y_{ij} \log_2 Y_{ij} j = 1, 2, \ldots .n$$

$$T_{j\,\max} = \log_2 r_j$$

r_j is the number of classes

$$O_j = \frac{T_{j\,\max} - T_j}{T_{j\,\max}} O = (0, 1), j = 1, 2, \ldots ..n$$

$$S_j = O_j Y_{ij}$$

Y_{ij} is the probability density, T_j and $T_{j\max}$ = represent entrop values, O_j = Information coefficient, and S_j represents the resultant weight value for each parameters.

$$\mathrm{LSM_{SE}} = \sum_{i=1}^{n} \frac{a}{b_i} XDXS_j$$

$\mathrm{LSM_{SE}}$ represents the landslide susceptibility map based on Shannon Entropy, i is the number of map of particular conditioning factors, a is the number of classes within the conditioning factors with the greatest number of classes, b_i is the number of classes within the conditioning factors, D is the value of the class after secondary classification, and S_j is the weight of the each conditioning factors.

The Shannon entropy usually indicates to quantification of the expected amount of information enclosed by a message. At present, it has been widely used to determine the weighted index in natural hazards, and in integrated assessment of natural-environmental processes such as debris flows, droughts, and sandstorms. In the case of landslides, it measures the dissimilarity or diversity in the environment, indicating the potential of each factor in causing landslides. In other words, the entropy of landslides refers to the extent the various factors influence landslide. Greater is the entropy index, greater is the influence of the factor in causing landslide. Finally, it also should be mentioned that various landslide-related criteria are not the same regarding their attributes and dimension. Therefore, it is not possible to conduct a direct comparison between those mentioned criteria which are applied in a LSM process.

4.4 Result and Discussion

One of the major causes of landslides is soil. Sandy soils located in slopes that are not stable and exposed to high rainfalls are quite prone to landslides and can cause serious land damage (Yuvaraj and Dolui, 2021; Patanakango, 2001). In a study conducted by Wieczorek et al. (1996), investigating the role of soil types in landslides, it was determined that landslides are prevalent in regions with unconsolidated till and clay soils. As presented in Fig. 4.2, the study area soil types will be categorized into four types; sandy clay loam, sandy clay, rock land, and clay loam.

The soil depth also is of significant importance in assessing land stability and landslide susceptibility. Darker soils are more likely to absorb moisture hence, less surface run-off. On the contrary, shallow soils are less stable and more vulnerable to landslides. In this research, the depth of soil has been categorized into four smaller categories.

Another significant control on slope stability is slope (Anbalagan, 1992; Pachauri et al., 1998; Saha et al., 2002; Yalcin, 2008), steeper slopes providing greater impetus to mass movements. Slope is generally considered as an important factor which influences landslide potential. Slope values in the study area vary between 0 and 50 degrees with steep inclines prevailing in most parts of the area. The slope map has been classified into five classes: 05, 515, 1530, 3050, and above.

Rainfall, in particular, the process of interception is an important factor of landslides (Martelloni et al., 2011; Yuvaraj & Dolui, 2023). It is one of the most important as well as one of the simplest to measure correctly with rain gauges. Owing to this fact, early warning systems based on rainfall can be both low-cost and simple. These systems are based on measurements of rainfall and set threshold values to determine the occurrence of possible landslides (Aleotti, 2004; Keefer et al., 1987). When the precise increase in rainfall level that causes landslides in a particular area is known, it can then be applied to an area to provide warnings when the predicted rainfall is likely to be more than the predicted increase in rainfall. In this eventuality, threats may be sent to regional officials, the local community, and emergency response staff. Hence, accurate and timely predictions of rainfalls are critical in the effective management of landslides. According to Lee and Talib (2005) and Pradhan and Lee (2009), Normalized Difference Vegetation Index (NDVI) is a landslide predicting parameter. The vegetation cover is also low, which contributes to the appearance of landslides in mountainous areas (Dahigamuwa et al., 2016). Heavy vegetation cover can slow down the soils erosion rate on slopes (Gomez & Kavzoglu, 2005; Weerasinghe et al., 2011). Soil erosion is

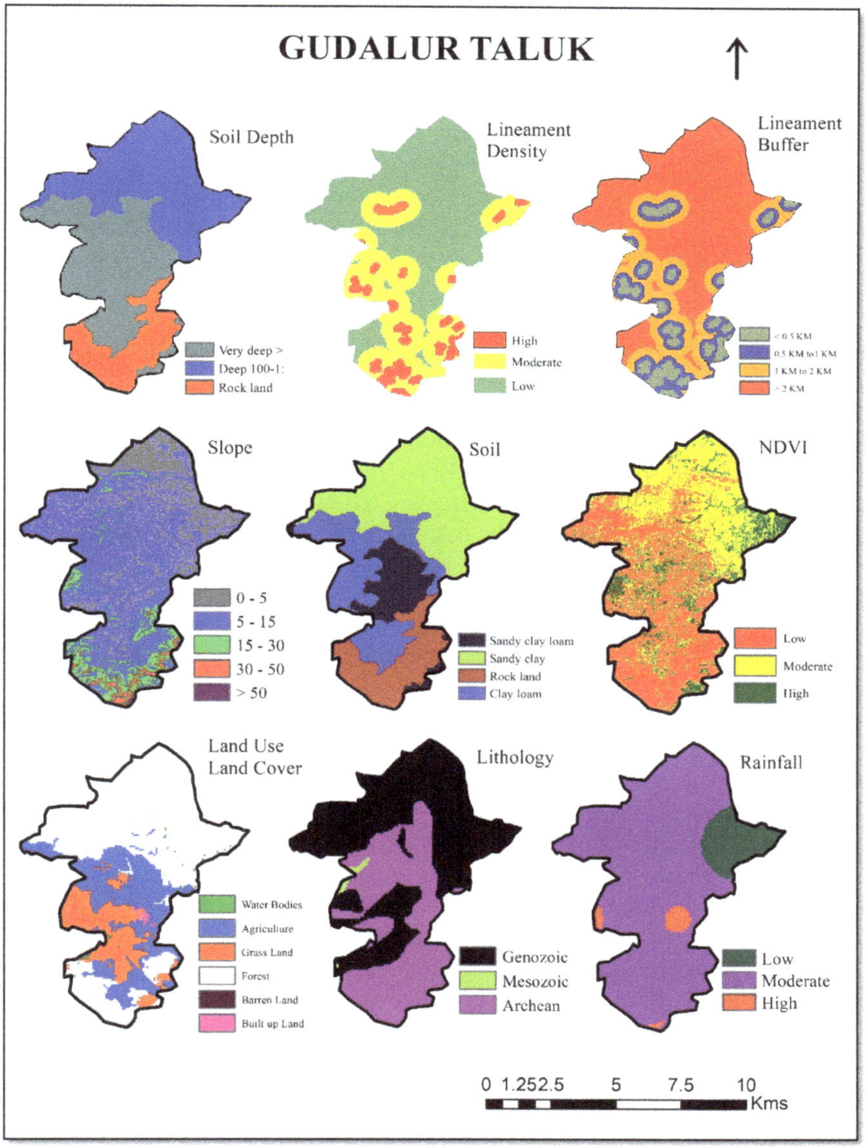

Fig. 4.2 Controlling variable of landslide

low in regions with significant vegetation, terrain degradation is slow, and slope damage is not so serious. The greater the NDVI value, the more vegetation density.

Lithology is also an important factor in landslide prone area study (Dai et al., 2001). It is a common knowledge that the physical properties of rock and soil types, including their strength and permeability, are greatly determined by their characteristics, which also determines the potential of landslides. The interaction between lithology and slope gradient is very important in the occurrence of type and intensity of landslides. This paper divides lithological formations into three subtypes, namely Cenozoic, Mesozoic, and Archean.

Land use is a major factor that affects occurrence of landslides in the study area. It usually affects stability of slope materials and regulates moisture content in slope. Especially vegetation strengthens the slopes via root systems which is an inherent support mechanism. In addition, land cover also aids in the absorption of soil moisture and, as a consequence, lowers the risk of landslides. The land use/ land cover map is categorized in this study into six different categories.

Significant factors affecting stability of rock formations are also lineaments that are linear terrain features like faults, fractures, ridges, and major geological discontinuities (Sarkar et al., 1995). The role of lineaments in landslide susceptibility research is generally accepted (Anbalagan & Singh, 1996; Nagarajan et al., 2000; Saha et al., 2002). Where the land is covered with soil, as in such areas as the Gudalur taluk, major drainage patterns frequently follow large lineaments, and the straight valley formations frequently indicate their presence. Spatial magnitudes of the controlling variables provided in Fig. 4.2.

The closer the distance between lineament, the greater the facture of rock masses and the extent of weathering thus higher the likelihood of landslides. Lineaments are faults and fracture zones that are seen on satellite images.

The outcome of entropy of Shannon using relationship between occurrence of landslide and landslide factors are presented in Table 4.1. The table shows different parameters and the weights attached to them, and these weights are reflective of the relative significance that they have in a spatial analysis, most likely associated with groundwater potential zoning or site suitability assessment. The most weighted parameters include soil depth (0.20006), land use land cover (0.19805), and soil type (0.165024), meaning that they are deemed to be the most important when it comes to determining the suitability or potential of a particular area. The reason is that deeper soils are able to hold more water, land use influences infiltration run-off patterns, and the soil properties impact on water permeability and storage. Others that are highly weighted are lithology (0.153839) and rainfall (0.138742), the two factors that are of critical importance in groundwater recharge and availability. Conversely, the weight given to parameters like NDVI (0.01477), lineament density (0.04387), and more importantly the lineament buffer (0.006872) is lower implying that they play a relatively small role in the analysis.

Table 4.2 assists in the analysis and comparison of all the contributory factors in order to establish the one that has the most influence. A larger value indicates that the depth of the soil is the most crucial value among all the others that are taken into consideration and that is represented in red in the graph presented in

Table 4.1 Calculation of Shannon's entropy for different parameters

Variables	Class	FR	Y_{ij}	T_j	$T_{j\max}$	O_j	S_j
Land use land cover	Water body	0	0	0.34874	1.2851	0.44817	0.19805
	Agriculture	1.87	0.22847				
	Grass land	0.42	0.05168				
	Forest	0.13	0.0158				
	Barren land	0	0				
	Built up land	5.77	0.70405				
Soil depth	Very deep > 150	1.29	0.81008	0.21111	2.0959	0.44247	0.20006
	Deep 100–150	0	0				
	Rock land	0.3	0.18992				
NDVI	Low	0.75	0.23105	0.46183	2.0959	0.96795	0.01477
	Moderate	1.05	0.32218				
	High	1.46	0.44677				
Soil	Sandy clay loam	1.44	0.70551	0.32621	1.66096	0.54183	0.16503
	Rock land	0.48	0.23727				
	Clay loam	0.12	0.05722				
	Sandy clay	0	0				
Slope	0–5	0.57	0.17615	0.54968	1.43068	0.78641	0.07878
	May-15	1.15	0.35305				
	15–30	1.22	0.3752				
	30–50	0.31	0.0956				
	> 50	0	0				
Rainfall	Low	1.69	0.73408	0.29408	2.0959	0.61637	0.13874
	Moderate	0.54	0.23379				
	High	0.07	0.03212				
Lithology	Genozoic	0	0	0.27365	2.0959	0.57355	0.15384
	Mesozoic	0.67	0.32436				
	Archean	1.39	0.67564				
Lineament	Low	1.14	0.50995	0.42245	2.0959	0.88542	0.04387
Density	Moderate	0.81	0.36362				
	High	0.28	0.12643				
Lineament buffer	0–500	0.92	0.21843	0.59625	1.66096	0.99035	0.00687
	500–1000	1.17	0.27903				
	1000–2000	1.26	0.30039				
	> 2000	0.85	0.20215				

Table 4.2 Weights of conditioning factors obtained from Shannon's entropy model

Parameters	Weight
Land use	0.198050
Lineament buffer	0.006872
Lineament density	0.043870
Lithology	0.153839
NDVI	0.014770
Rainfall	0.138742
Slope	0.078782
Soil	0.165024
Soil depth	0.200060

Fig. 4.3. Land use/land cover is the second factor with the most impact and the third is soil type, lithology and lastly is the rainfall with slope being the least factor in the study area.

The last landslide prone map was obtained by summing up the weights allocated to each factor contributing in the process as indicated in Fig. 4.4. This map is divided into three levels of susceptibility: low, moderate and high. Red means areas with low risk of landslides, dark-yellow means moderate-risk areas, and green indicates high-risk areas. The map shows that the majority of the northern, central, and part of the western areas have low susceptibility. In the northeast and southeast of the study area, moderate risk of landslides is witnessed. The high-susceptibility zones of landslide occurrence are known to be located at the borders of the central, southern and southeast regions of the Gudalur Taluk. Figure 4.5 shows the percentage proportion of area of each susceptibility categories according to the Entropy model landslide delineation by Shannon.

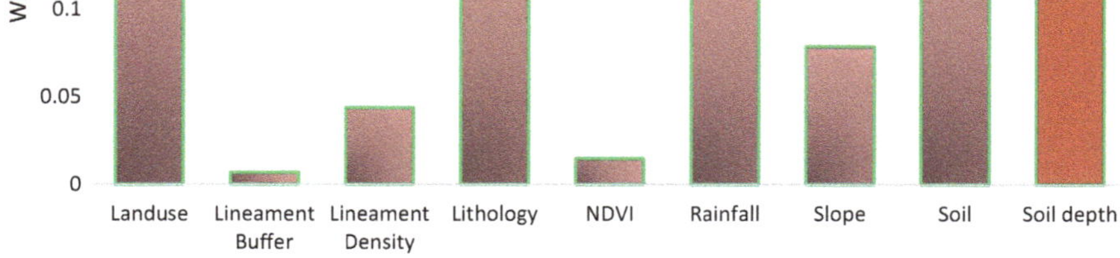

Fig. 4.3 Percentage shares of parameters

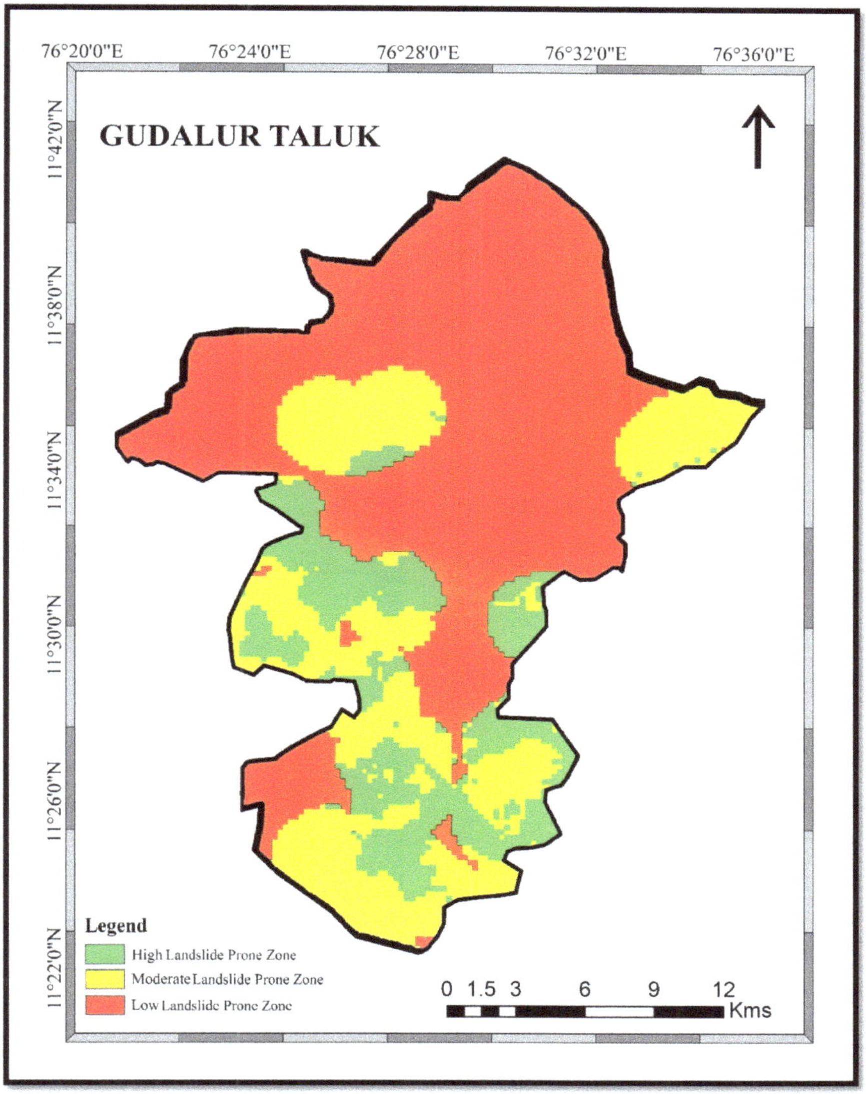

Fig. 4.4 Landslide susceptible map using Shannon's entropy

4.5 Validation

Validation step is believed to be one of the most crucial steps in the process of development of susceptibility maps as it evaluates their capability to forecast future events as it is an important attribute in any natural hazard study and management. The predictive capability of any given landslide susceptibility map (LSM) and its

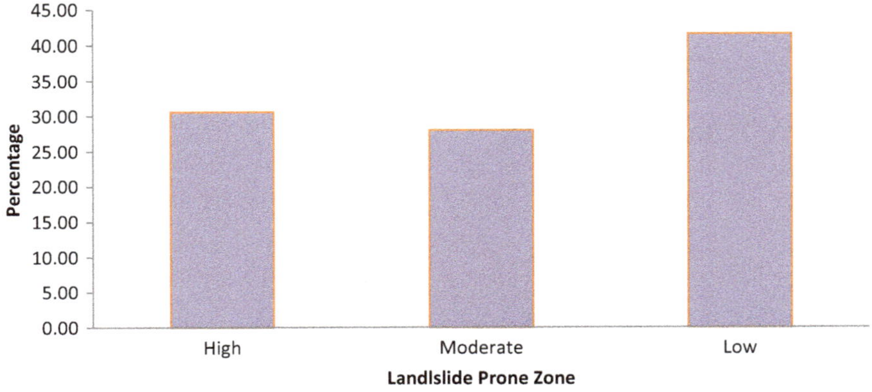

Fig. 4.5 Percentage of susceptibility classes

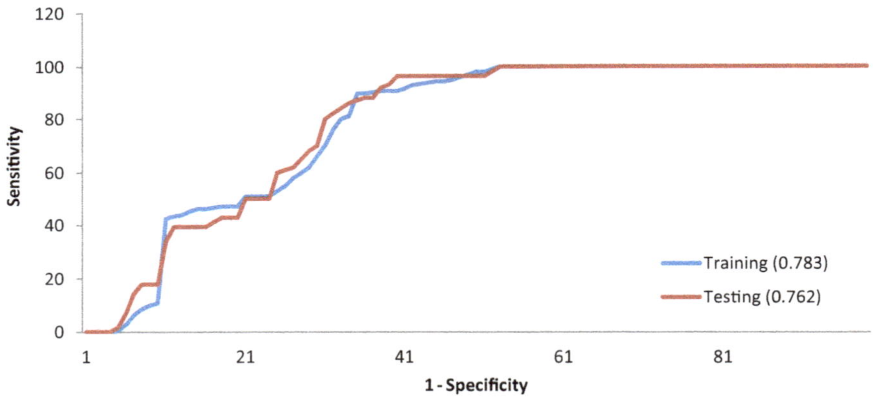

Fig. 4.6 Accuracy assessment for Shannon entropy

results is normally evaluated with independent landslide occurrence data which were not used in the model-building exercise (i.e., a test sample of the landslide inventory). Therefore, training and test data are employed. The precision of the developed LSM in this study was also measured through the Relative Operating Characteristic (ROC) curve as well as the percentage of known land slope events in various susceptibility categories, basing on the test data. The results of Area Under the Curve (AUC) are depicted in Fig. 4.6 in reference to the model used in the study area. In the training data, the AUC value is 0.738, whereas in the validation data AUC was a little bit higher at 0.762. All in all these findings state that the Shannon Entropy model has a good amount of accuracy in landslide susceptibility mapping.

4.6 Conclusion

The association between landslides incidents and topological aspects was examined with different parameters. The topographical factors identified included the most influential one of soil depth, then soil, lithology and slope. The hydrological factor, rainfall received a substantial weight, which means that it was prominent as a cause of landslides. Other contributory factors are land use land cover, lineament density, NDVI and lineament buffer. Shannon Entropy model discusses that soil depth is the most vital parameter determining landslides susceptibility, then LULC, soil, lithology, rainfall, and slope is not of much concern in this area. The spatial analysis indicates that most of the eastern part, central, and some areas in the western region are under high landslide susceptibility zone. The northeastern and southeastern areas are moderately prone, and the north border going towards the north and eastern areas were low landslides-prone.

Acknowledgement The author acknowledge the use of AI-based tools for grammar and language correction during the preparation of this manuscript.

References

Aleotti, P. (2004). A warning system for rainfall-induced shallow failures. *Engineering Geology, 73*(3–4), 247–265.

Anbalagan, R. (1992). Landslide hazard evaluation and zonation mapping in mountainous terrain. *Engineering Geology, 32*(4), 269–277.

Anbalagan, R., & Singh, B. (1996). Landslide hazard and risk assessment mapping of mountainous terrains—A case study from Kumaun Himalaya India. *Engineering Geology, 43*(4), 237–246.

Dahigamuwa, T., Yu, Q., & Gunaratne, M. (2016). Feasibility study of land cover classification based on normalized difference vegetation index for landslide risk assessment. *Geosciences, 6*(4), 45.

Dai, F. C., Lee, C. F., Li, J. X. Z. W., & Xu, Z. W. (2001). Assessment of landslide susceptibility on the natural terrain of Lantau Island Hong Kong. *Environmental Geology, 40*(3), 381–391.

Gomez, H., & Kavzoglu, T. (2005). Assessment of shallow landslide susceptibility using artificial neural networks in Jabonosa River Basin Venezuela. *Engineering Geology, 78*(1–2), 11–27.

Keefer, D. K., Wilson, R. C., Mark, R. K., Brabb, E. E., Brown, W. M., Ellen, S. D., & Zatkin, R. S. (1987). Real-time landslide warning during heavy rainfall. *Science, 238*(4829), 921–925.

Lee, S., & Talib, J. A. (2005). Probabilistic landslide susceptibility and factor effect analysis. *Environmental Geology, 47*(7), 982–990.

Martelloni, G., Massaro, E., & Bagnoli, F. (2011). Particle based method for shallow landslides: modeling sliding surface lubrication by rainfall. *CMSIM International Journal Nonlinear Science, 1*, 147–158.

Nagarajan, R., Roy, A., Kumar, R. V., Mukherjee, A., & Khire, M. V. (2000). Landslide hazard susceptibility mapping based on terrain and climatic factors for tropical monsoon regions. *Bulletin of Engineering Geology and the Environment, 58*(4), 275–287.

Pachauri, A. K., Gupta, P. V., & Chander, R. (1998). Landslide zoning in a part of the Garhwal Himalayas. *Environmental Geology, 36*(3), 325–334.

Park, S., Choi, C., Kim, B., & Kim, J. (2013). Landslide susceptibility mapping using frequency ratio, analytic hierarchy process, logistic regression, and artificial neural network methods at the Inje area Korea. *Environmental Earth Sciences, 68*(5), 1443–1464.

Patanakanogo, B. (2001). *Landslide hazard potential area in 3 dimension by remote sensing and GIS technique*. Land Development Department.

Pourghasemi, H. R., Mohammady, M., & Pradhan, B. (2012). Landslide susceptibility mapping using index of entropy and conditional probability models in GIS: Safarood Basin Iran. *CATENA, 97*, 71–84.

Pradhan, B., & Lee, S. (2009). Landslide risk analysis using artificial neural network model focussing on different training sites. *International Journal of Physical Sciences, 4*(1), 1–15.

Saha, A. K., Gupta, R. P., & Arora, M. K. (2002). GIS-based landslide hazard zonation in the Bhagirathi (Ganga) valley Himalayas. *International Journal of Remote Sensing, 23*(2), 357–369.

Sarkar, S., Kanungo, D. P., & Mehrotra, G. S. (1995). Landslide hazard zonation: a case study in Garhwal Himalaya, India. *Mountain Research and Development, 1*, 301–309.

Shannon, C. E. (1948). A mathematical theory of communication. *Bell System Technical Journal, 27*(379–423), 623–656.

Weerasinghe, K. M., Gunaratne, M., Ratnaweera, P., & Arambepola, N. M. S. I. (2011). Upgrading of the subjective landslide hazard evaluation scheme in Sri Lanka. *Civil Engineering and Environmental Systems, 28*(2), 99–121.

Wieczorek, G. F. (1996). Landslides: Investigation and mitigation. Chapter 4-Landslide triggering mechanisms. *Transportation Research Board Special Report*, (247).

Yalcin, A. (2008). GIS-based landslide susceptibility mapping using analytical hierarchy process and bivariate statistics in Ardesen (Turkey): Comparisons of results and confirmations. *CATENA, 72*(1), 1–12.

Yilmaz, I. (2009). Landslide susceptibility mapping using frequency ratio, logistic regression, artificial neural networks, and their comparison: A case study from Kat landslides (Tokat-Turkey). *Computers & Geosciences, 35*(6), 1125–1138.

Yuvaraj, R. M., & Dolui, B. (2021). Statistical and machine intelligence based model for landslide susceptibility mapping of Nilgiri District in India. *Environmental Challenges, 5*, Article 100211.

Yuvaraj, R. M., & Dolui, B. (2023). Geographical assessment of landslide susceptibility using statistical approach. *Quaternary Science Advances, 11*, Article 100097.

Landslide Susceptibility Modeling Using Integrated Geospatial and AHP Approaches: A Case Study in Idukki District, India

5

Kongeswaran Thangaraj, Prabakaran Kulandaisamy, Sivakumar Karthikeyan, Muthuramalingam Rajendran, Muruganantham Arumugam, Perumal Velmayil, Venkatramanan Senapathi, and Bangaru Priyanga Sundaram

Abstract

This study delves into the assessment of landslide susceptibility in a monsoon-prone region within the Western Ghats of Kerala, India. The research employs an integrated approach utilizing the Analytical Hierarchy Process (AHP), remote sensing, and GIS to generate a comprehensive susceptibility map. Nine crucial factors, including geomorphology, slope, proximity to roads, drainage, soil composition, land use patterns, and precipitation, are meticulously weighted through AHP. Based on AHP analyses, the study area (4612 km^2) is classified into five risk categories: Very Low Risk (678.90 km^2, 15.60%), Low Risk (684.61 km^2, 15.73%), Moderate Risk (1055.68 km^2, 24.26%), High Risk (1142.69 km^2, 26.26%), and Very High Risk (790.02 km^2, 18.15%). The last three categories together encompass 68.67% of the area, identifying the most hazardous regions which require immediate mitigation measures for slope protection. Validation through correlation with historical landslides establishes the reliability of the map for future land use planning. It notably identifies various types of landslides, including rock falls and debris flows, which pose substantial risks such as road blockages, disrupted supply lines, and reservoir sedimentation. Persistent rainfall and unsustainable land use are significant vulnerability factors. This research underscores the ongoing necessity for the evaluation

K. Thangaraj · P. Kulandaisamy · S. Karthikeyan (✉) · M. Rajendran · M. Arumugam · P. Velmayil · B. P. Sundaram
Department of Geology, Faculty of Science, Alagappa University, Karaikudi 630003, Tamil Nadu, India
e-mail: siva.karthi90@yahoo.com

V. Senapathi
PG and Research Department of Geology, National College (Autonomous), Tiruchirappalli 620001, Tamil Nadu, India

R. M. Yuvaraj et al. (eds.), *Advanced Landslide Susceptibility Modeling*,
https://doi.org/10.1007/978-3-032-14485-0_5

and mitigation of landslide hazards in this susceptible region, emphasizing the critical role of informed land use planning to mitigate potential risks.

Keywords

Landslide • Susceptibility • AHP • GIS • Idukki • Kerala • Disaster

5.1 Introduction

Life and infrastructure are at risk from landslides, these devastating natural hazards. The Analytical Hierarchy Process (AHP) and Geographic Information Systems (GIS) provide a powerful synergy for achieving these goals. AHP provides a structured framework for incorporating expert knowledge and weighting the relative importance of various factors contributing to landslide susceptibility. This expertise guides the integration of diverse spatial data within GIS, creating a robust platform for analyzing, modeling, and mapping landslide-prone areas. This combined approach leverages AHP's decision-making strength with GIS's spatial analysis capabilities, leading to more accurate and insightful assessments of landslide risk. By empowering us to identify vulnerable areas and prepare preventative measures, this synergy allows us to mitigate the impact caused by landslides and foster safer communities (Demir, 2019; Lei et al., 2020; Ju et al., 2015). Landslides are a common occurrence in the Idukki district of Kerala, India, especially during the monsoon season. This study focuses on identifying the most vulnerable zones in this region. Idukki, the second largest district in Kerala, is characterized by its mountainous terrain and extensive forests, resulting in a low population density (KSDMA, 2018). Landslides in this region are often triggered by man-made factors, including activities such as deforestation, terracing, and blockage of watercourses, and inappropriate agricultural practices in hilly terrain. Various factors, including geomorphology, geology, climate, land use and rainfall, contribute to landslide risk. Road construction can also destabilize slopes, increasing the likelihood of landslides (Khan et al., 2022; Deoja & Purohit, 2018).

This study uses a GIS-based spatial multi-criteria analysis to create a landslide susceptibility map. Factors such as land cover, lithology and proximity to roads, faults, streams, drainage density, elevation, aspect, and slope are considered. The Analytical Hierarchy Process (AHP) is used to weight these variables according to their impact on landslides (Bajracharya & Tani, 2022; Das et al., 2022; Dey & Mandal, 2021; Jaiswal et al., 2014; Kuriakose & Yesodhara, 2015). The susceptibility map is divided into five different categories: very low, low, moderate, high and very high. Landslide hazard maps play a central role in both landslide disaster preparedness and mitigation. They provide valuable information on the probability of landslides in specific regions (Mersha & Meten, 2020; Ye et al., 2019; Tapas et al., 2012; Hagen et al., 2013). This research considers elements such as slope gradient, aspect, curvature, lithology, geological composition, and drainage characteristics as factors contributing to landslide occurrence. External factors such as earthquakes, rainfall and erosion also play a role (Vakhshoori & Zare, 2016). The concept of vulnerability is complex because it refers to the potential damage to the elements that are at risk during a landslide event. Vulnerability can vary depending on factors such as exposure, population resilience and susceptibility to landslides. Idukki is extremely prone to landslides due to its location in the Western Ghats. As people settle in hilly areas, it is important to understand landslide hazards and apply appropriate land use planning and engineering practices (Rifat & Liu, 2020; Proag, 2014; Shashidhar et al., 2021). While some causes of landslides cannot be ruled out, geological surveys, effective land use regulations and engineering practices can reduce the risk of landslides. In summary, landslides are a significant geological risk in the Western Ghats of India, particularly in Idukki. Understanding the causes and risks of landslides is essential for effective disaster management and planning in this vulnerable area (Chawla et al., 2018). In recent decades, GIS-based approaches have been used to map landslide susceptibility. These include but are not limited to Landslide Hazard Zonation (Biju Abraham & Shaji, 2013; Ajin et al., 2022), Causative factors (Wadhawan & Balmukund, 2020), Landslide Susceptibility Zonation (Thomas et al., 2021; Jones et al., 2021), Pilot Assessment (Amal George et al., 2023), Artificial Neural Network (Jacinth Jennifer & Saravanan, 2022). Based on the best of knowledge of the authors, no studies using one of the above approaches in landslide susceptibility mapping have been conducted in the Idukki district, Kerala. Thus, for this study, a GIS-based Analytical Hierarchy Process (AHP) model was employed in landslide susceptibility mapping in this area.

The research presented in this paper offers valuable insights into landslide susceptibility in the Idukki district, Kerala, India. However, certain research gaps warrant further exploration. The study primarily provides a snapshot of susceptibility at specific time points, leaving a gap in understanding temporal dynamics and seasonal variations. Additionally, there is room for finer-scale analyses focusing on localized variations within the district. The integration of dynamic factors such as changing weather patterns and human activities, along with the inclusion of social and economic aspects, would contribute to a more comprehensive and nuanced understanding of landslide vulnerability in the region. Addressing these gaps would enhance the practical applicability of the findings for effective land use planning and hazard mitigation.

The selection criteria for the study were as follows:

- Investigate and evaluate landslide susceptibility in the monsoon-prone Western Ghats region, specifically Idukki district, Kerala, India using integrated AHP, remote sensing, and GIS techniques.
- Consider nine key factors (geomorphology, slope, proximity to roads, drainage, soil, land use, precipitation) and assign weights using AHP for a comprehensive susceptibility map.

- Generate a detailed susceptibility map with five risk zones, validated through correlation with historical landslides, and identify diverse landslide types (rock falls, debris flows) and associated risks.
- Explore vulnerability factors such as continuous rainfall and unsustainable land use, emphasizing the ongoing need for hazard evaluation and providing insights for informed land use planning.

5.2 Study Area

Idukki, the Western Ghats, comprises the second largest district in Kerala, known for its rugged and mountainous landscape. It stretches between latitudes 9°16'24"N to 10°21'37"N and longitudes 76°37'27"E to 77°24'50"E. The region is a high plateau with rugged terrain, numerous river valleys and deep gorges (IDA, 2023). The northern part of Idukki forms a sub-plateau that rises above the surrounding district (Fig. 5.1). The western area of the district is mostly covered by forests and hills and forms the border with the eastern regions of Ernakulam and Kottayam districts. Prominent places like Neriyamangalam, Vannapuram, Thommankuthu, Moolamattom, Vagamon, and Kuttikanam are located in this area. Within the district, the catchment area of the Periyar River includes areas such as Vandiperiyar, Ayyapankoil, Rajakkad, Idukki, Cheruthoni, and Adimaly, which form a high plateau crisscrossed by river valleys, nestled in the towering peaks of the western and eastern parts of the district. The landmark of Idukki is the Idukki Dam, which is considered the largest arch dam in Asia. Located in the municipality of Idukki, the dam is situated where the Periyar River flows through a gorge formed between two massive rocks known as "Kuravan" and "Kurathi" (KT, 2023).

5.2.1 Geology

The Peninsular Gneissic Complex, located in of the Charnockite Group within the Charnockite Group of Rocks. This group includes various rock types such as pyroxene granulite, magnetite quartzite and charnockite, with charnockite being the most widespread. Pyroxene granulite and magnetite quartzite can be observed as linear bands within the gneissic terrain, oriented parallel to the stratification trend. On the other hand, the charnockite forms a continuous rock unit extending over the hilly regions to the south and northwest (Goudarzi et al., 2019). It exhibits a range of compositions from massive to gneissic varieties and may be medium to acidic in composition. The central, northeastern, and southeastern parts of the district are dominated by the migmatitic complex, which consists of rocks such as biotite gneiss and hornblende-biotite gneiss (composite gneiss). The composition of the south-central region, which includes Pynavu and Thodupuzha, is predominantly biotite gneiss, while the northeastern and southeastern areas consist of hornblende-biotite gneiss (Fig. 5.2a). Within the biotite gneiss, remnants of

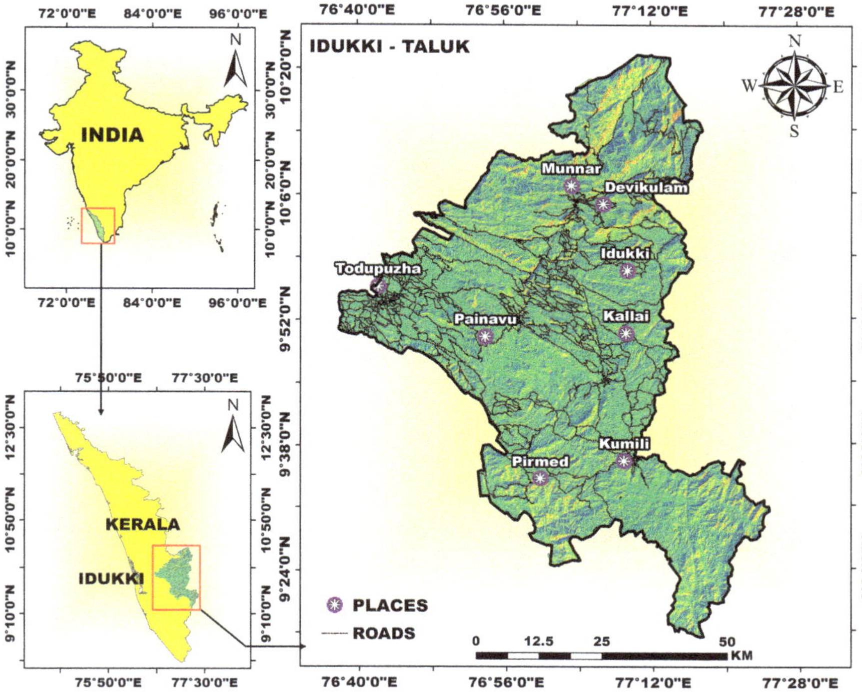

Fig. 5.1 Location map of the study area

earlier rocks of the khondalite and charnockite group are often found in the form of lenticular and linear formations that adhere to the schistosity.

5.2.2 Geomorphology

The midland region lies on average between 40 and 60 m above sea level and is characterized by a rugged topography with small hills separated by deep valleys. The area as a whole slopes toward the west. The foothill region, a narrow strip of land that acts as a transition between the midland region and the plateau, lies between 80 and 500 m above sea level and has very steep slopes, often exceeding 30% and occasionally up to 80%. The width of the foothills region varies between 2 and 8 km. The plateau region is the most important physiographic unit in the district. It is characterized by large land masses with moderate slopes of less than 30% and elevations of less than 1500 m above sea level (Kale & Wategaonkar, 2016). Most of the district lies in this region, which is crisscrossed by numerous deeply incised streams. Based on the maturity of the landscape, the orientation of the peaks and the remaining cover, the presence of two former leveling areas was determined. The older area lies at an altitude of more than 1200 m, while the

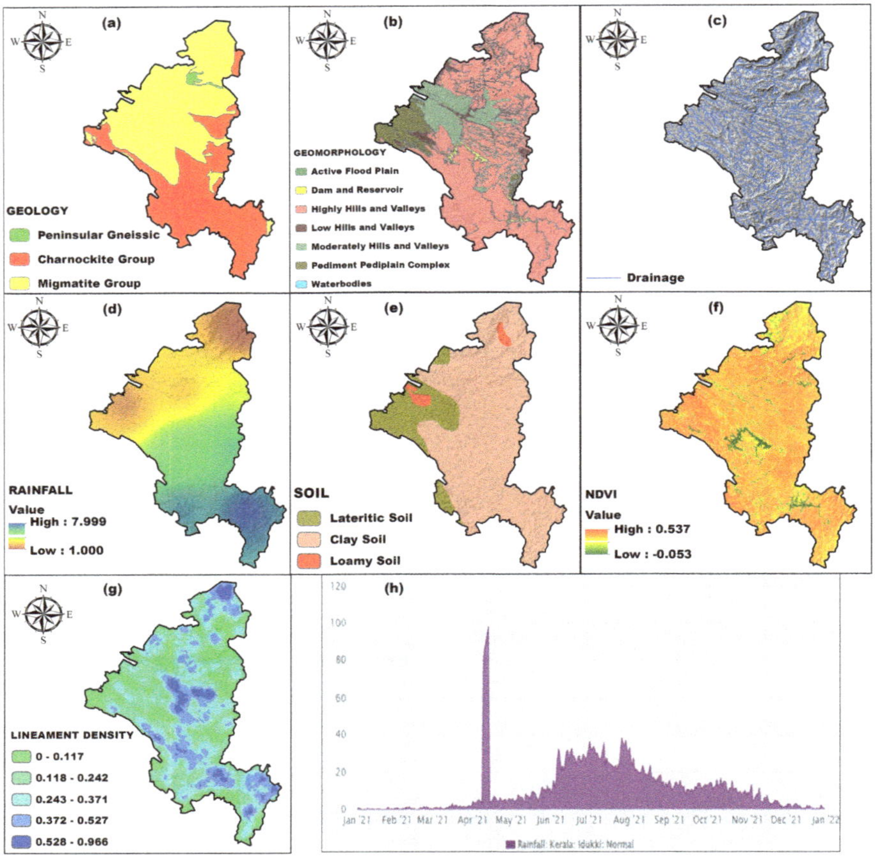

Fig. 5.2 Study area **a** geology **b** geomorphology **c** drainage **d** rainfall map **e** soil **f** NDVI **g** lineament density **h** rainfall data. *Source* https://WWW.CEICDATA.COM

younger area lies between 900 and 1200 m. Along the western edge of the plateau, shallow valleys have developed as a result of dissection by rivers. The high ranges, located in the northeastern part of the district, are characterized by high mountains with elevations of more than 1500 m above sea level. Anaimudi, the highest peak in South India, is 2693 m above sea level and is located in the northcentral part of the district. Other prominent peaks are Kewarikkat Mala (2523 m) and Chunduvarai Mala (2525 m) (Fig. 5.2b). This region is characterized by steep terrain and several impressive waterfalls (Nair & Achuthan, 2018; Preetha & Nayak, 2020).

5.2.3 Drainage and Hydrogeology

The district of Kerala is drained by the river Periyar, which is the largest river in the region and has its source on the south-eastern border. It is accompanied by several major tributaries, including the Mullayar, the Panniyar, the Puyankuttiyar, the Perinjamkuttiyar, and the Muthirapuzhayar, all of which are aligned along major fault zones and have good structural control. The Muvattupuzha River, on the other hand, has two major tributaries, Kaliyar and Thodupuzhayar, which originate in the Thodupuzha and Nagarampara reserve forests of the district (Fig. 5.2c). The tributaries of the Amaravathi, namely the Pambar and the Tenyar, flow toward the east and drain the northeastern part of the district. Finally, a small area in the southern part of the district is drained by the Azhutha River, a tributary of the Pamba River (Sankaran Nair & Achuthan, 2011).

5.2.4 Rainfall

The district receives an average annual rainfall of about 3677 mm, though the amount varies in different areas. For example, Marayoor, Kanthaloor, Chinnar, and the northeastern regions of Anamudi receive less than 1000 mm, while Peerumedu, Neriyamangalam and other areas receive about 5000 mm. The rainfall increases gradually from east to west as the eastern part of the district lies in the rain shadow of the Western Ghats. The main source of rainfall is the southwest monsoon, which occurs from June to September and accounts for about 60% of the total annual rainfall. The northeast monsoon, which occurs from October to December, contributes 24% of the annual rainfall, while the remaining amount falls between January and May. Although the district has not experienced severe or extreme droughts in the last ten years, the absence of summer showers and rainfall from the northeast monsoon from 2006 to 2021 has affected crop growth and the availability of drinking water in the rain shadow region of the eastern part (Fig. 5.2d and h). In Idukki, the month of April is the wettest. During periods of heavy rainfall, there is a significant risk of landslides due to the triggering factors (Abraham et al., 2019).

5.2.5 Soil Types

Approximately 50% of the district is covered by clayey soils, which are the result of rock weathering under the forest cover. These soils are characterized by a dark, reddish-brown to black color and contain a high proportion of organic matter. They tend to be acidic and have high nitrogen content, but lack bases due to heavy leaching. The texture of these soils is typically loamy to silty-loamy. In areas where forests have been cleared, leaching and the deposition of humus in the lower layers are common. Lateritic soils are mainly found in the Elamdesam and Thodupuzha blocks of the district. These soils are formed from laterites and are well drained.

However, they have a low content of plant nutrients and organic matter, which leads to low soil fertility. The available nitrogen and phosphorus content in these soils is generally low. Brown hydromorphic soils are found in valley sections in hilly terrain. They are formed by the transport and sedimentation of material from the adjacent hills. These soils are brownish-black in color and vary in surface texture from sandy loam to clay. Alluvial soils are mainly found in narrow strips along the river banks in the district, being more common along the Thodupuzha River (Fig. 5.2e). These soils are fertile and have a surface texture ranging from sandy loam to clay, making them well suited for agriculture due to their fertility (Gopinathan & Thomas, 2015).

5.2.6 NDVI

The normalized difference vegetation index (NDVI) is essential remote sensing tool that assesses the density and condition of vegetation cover in a given land area. It measures the contrast between the reflection of visible and near-infrared light by plants (Doan et al., 2023). NDVI values are usually between − 1 and 1, with each value providing important insights into the state of vegetation in a particular region. For the Idukki district, the reported NDVI value is high at 0.537976 and low at − 0.0535451 (Fig. 5.2f).

5.2.7 Transportation

The district's road network covers an estimated 8000 km and is an important means of transportation for residents and visitors. One of the major national highways passing through Idukki is the Kollam-Theni route. In addition, there are other important routes like Adimaly-Kumily and Adoor-Vandiperiyar. The district is also traversed by important state highways including Munnar-Kumily, Muvattupuzha-Theni, and Alappuzha-Madurai. These state highways play an important role in connecting Idukki with neighboring districts and states. Notably, NH 185 is an important highway in Idukki, which runs from Adimali to Kumily via Kattapana and ends at Kumily Attapallam village in Kerala. With a length of 98 km, it is the largest national highway that passes through the district. Another notable national highway is SH 43, which begins in Muvattupuzha and ends in Theni Town in Tamil Nadu. With a length of 105 km, this highway facilitates traffic between Idukki and the neighboring regions. Moreover, the Kottayam-Kumily Road serves as an important expressway connecting the districts of Kottayam and Idukki in Kerala, aiding the transportation of goods and people between these two districts. In addition to these major highways, Idukki has an extensive network of district and village roads connecting various villages, towns and taluks within the district. These roads are essential for local transportation and provide access to remote areas (KTD, 2023).

5.2.8 Lineament and Lineament Density

Lineaments are linear features in a landscape that indicate the presence of underlying geological structures such as faults, fractures, or fissures. These features are often used when analyzing remote sensing data to identify structures or fractures. The lineament density of Idukki can be determined using the lineament density tool in ArcGIS, which categorizes the measurement into five categories based on the density of lineaments present in the area, ranging from very low to very high (Fig. 5.2g) (Solanki et al., 2019).

5.2.9 Meteorological Parameters

The plateau region is subject to seasonal temperature fluctuations, with the higher temperatures occurring from March to May and the lower temperatures in January and February. The average monthly maximum temperature is between 25.1 and 31.5 °C, while the minimum temperature is between 18.6 and 14.0 °C. The relative humidity varies throughout the day, with higher values in the morning and lower values in the evening. In the morning, the relative humidity is between 85.0 and 98.45%, while in the evening it varies between 40.8 and 87.3%. Evaporation rates are influenced by the time of year, with higher rates generally observed in the summer months from March to May. On average, evaporation is between 1.11 and 5.0 mm per day. During the southwest monsoon season, evaporation is between 1.11 and 2.13 mm per day. Sunshine duration is also subject to seasonal fluctuations, with the highest in February. Conversely, the months of June to August generally have the least sunshine due to the cloudy skies. The months of December to May generally have a high number of hours of sunshine, with the highest number recorded in February. The wind speed varies throughout the year, with higher speeds occurring in certain months. Wind speeds range between 3.90 and 7.95 km per hour and tend to be higher from March to July and November to January.

5.2.10 Population

The district of Idukki covers a total area of 4612 km^2, of which 35 km^2 are designated as urban and 4321 km^2 as rural. Of the district's 1,230,962 inhabitants, 52,045 people live in urban areas, while 1,056,929 people live in rural areas. The number of households in the urban areas is 12,604, while the rural areas are home to 267,208 households. In terms of literacy, there are 44,799 people living in urban areas that can read and write, while there are 878,211 people living in rural areas that can read and write.

5.3 Materials and Methods

The Analytical Hierarchy Process (AHP) is a widely used method for landslide susceptibility assessment. It involves a structured process of pairwise comparisons to prioritize and rank the factors that contribute to landslide susceptibility. Here is a detailed methodology for using the AHP in landslide susceptibility assessment (Fig. 5.3):

Step 1: Problem Definition and Data Collection
Define the study area and objectives for landslide susceptibility assessment. Collect spatial data and information, including geologic, geomorphologic, climatic, and land use data relevant to landslide occurrence.

Step 2: Selection of Factors
Identify the factors that contribute to landslide susceptibility based on available data and expert knowledge. These may include slope gradient, geology, land use, precipitation, soil type, vegetation cover, and previous landslide events.

Step 3: Weight Assignment
Create a pairwise comparison matrix for each factor to determine its relative importance to landslide susceptibility. Experts or stakeholders weight each pair of factors using a numerical scale based on their judgment and knowledge. These weightings represent the relative importance of one factor compared to another. The pairwise comparisons should be consistent, and the consistency ratio should be checked to ensure the reliability of the judgments.

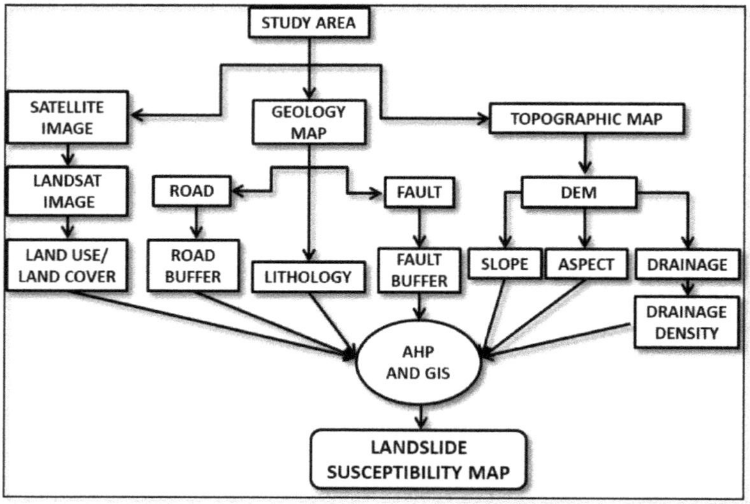

Fig. 5.3 Framework for landslide susceptibility model

Step 4: Normalization
Normalize the weights by dividing each weight by the sum of the weights for that particular factor. This will ensure that the weights are on a consistent scale, with values between 0 and 1.

Step 5: Data Preparation
Convert all relevant data layers to a common format and resolution. This usually involves reclassifying the data into categories and assigning numerical values.

Step 6: Overlay and weighted sum
Overlay the normalized data layers using GIS software. Multiply each normalized data layer by its corresponding normalized weight. Add the weighted layers to create a map of vulnerability. This map represents the overall susceptibility to landslides in the study area.

Step 7: Validation
Validate the susceptibility map against historical landslide data or field investigations to assess its accuracy. Statistical measures such as the area under the Receiver Operating Characteristic Curve (AUC-ROC) can be used to evaluate model performance.

Step 8: Sensitivity Analysis
A sensitivity analysis is performed to test the robustness of the model by varying the input weights and assessing the impact on the susceptibility map.

Step 9: Interpretation and Reporting
Interpret the susceptibility map to identify areas with different susceptibility to landslides.

Prepare a report that includes the methodology, results, validation, and recommendations for land use planning and risk mitigation.

Step 10: Continuous Monitoring
Landslide susceptibility is dynamic, so continuous monitoring of environmental conditions and updates to the model may be required to account for changes in susceptibility over time.

5.4 Results and Discussion

In the study, Landsat 8 satellite images were used to capture the visual data of the study area for the creation of thematic maps. In addition, a digital elevation model (DEM) was created from USGS Earth Explorer data. The study considered nine different factors that are recognized as potential landslide triggers. The weighting of each of these factors was determined using the AHP criteria, which includes parameters such as slope, elevation, lithology, aspect, land use/land cover, road buffer, stream buffer, disturbance buffer, and drainage density. Landslides, also known as landslides, occur when gravity combined with other shear stresses exceed the shear strength of the materials that make up the slope. Various processes, such as excessive steepening of the slope base, increased water inflow, a rising water table or the accumulation of debris on the surface of the slope, can contribute to the build-up of shear stresses. To avoid deaths and economic losses from landslides, it is often recommended to avoid building cities near steep terrain. However, this is not always practical or feasible, especially in areas with rapid population growth or historic towns. Therefore, the analysis and management of regional landslide risk has become an important task for urban planners and authorities. Various influencing factors were used for the study, including lithology, slope and aspect, land use/land cover, distance to roads, distance to streams and distance to faults, elevation and drainage density. These factors were integrated and analyzed using ArcGIS 10.8 to create a comprehensive map of geofactors. Additionally, the AHP was applied to further enhance the analysis.

Landslide Triggering Factors

5.4.1 Slope

The term “slope” can be defined as the angular measure, inclination, steepness, or gradient of a straight line. It is commonly used to describe the degree of steepness found on the surface of the ground. Slope can be quantified by measuring the gradient, which refers to the increase in height in a given unit of measurement, divided by the distance, which represents the horizontal distance measured in the same units as the slope. In Idukki district, the terrain is characterized by a steep slope, with slope angles ranging from 5.75° to 77.29°. These angles are further divided into six ranges: 0°–7.93°, 7.93°–14.39°, 14.39°–21.15°, 21.15°–28.20°, 28.20°–37.89°, and 37.89°–74.90° (Table 5.1) (Fig. 5.4a).

Table 5.1 Slope angle

Si. no	Slope angle	Area in km^2	%
1	0°–7.93°	859.54	19.75
2	7.93°–14.39°	1117.26	25.67
3	14.39°–21.15°	1053.68	24.21
4	21.15°–28.20°	760.83	17.48
5	28.20°–37.89°	448.62	10.31
6	37.89°–74.90°	111.99	2.57
	Total	4351.92	100

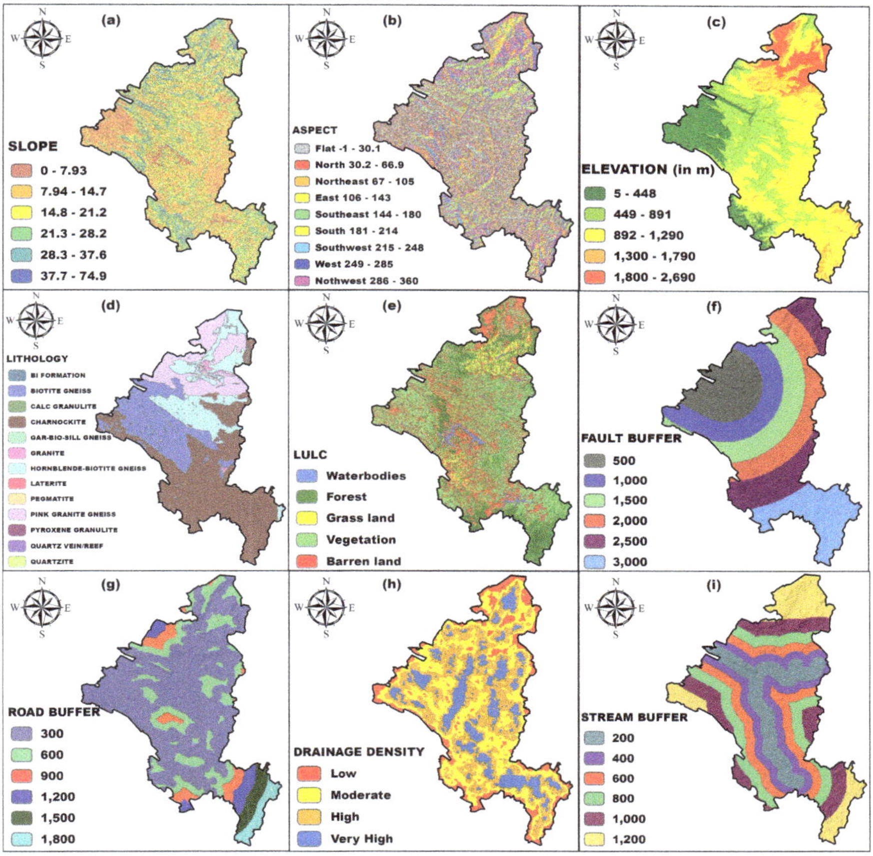

Fig. 5.4 Study area **a** slope **b** slope aspect **c** elevation **d** lithology **e** land use/land cover **f** fault buffer **g** road buffer **h** drainage density and **i** stream buffer

5.4.2 Slope Aspect

The ten categories of aspects are as follows: "Flat" (− 1), "North" (0°–22.5°), "Northeast" (22.5°–67.5°), "East" (67.5°–112.5°), "Southeast" (112.5°–157.5°), "South" (157.5°–202.5°), "Southwest" (202.5°–247.5°), "West" (247.5°–292.5°), "Northwest" (292.5°–337.5°), and "North" (337.5°–360°) (Table 5.2) (Fig. 5.4b).

5.4.3 Elevation

Elevation, the geometric height of points on the earth's surface, is of great importance in geographical descriptions. The highest elevation in the region reaches about 2694 meters. The Anamudi in the Western Ghats is the highest peak in both

Table 5.2 Slope aspect

Si. no	Aspect	Area in km^2	%
1	Flat (– 1)	456.9	10.5
2	North (0–22.5)	469.78	10.79
3	Northeast (22.5–67.5)	387.69	8.91
4	East (67.5–112.5)	387.49	8.9
5	Southeast (112.5–157.5)	422.46	9.71
6	South (157.5–202.5)	494.8	11.37
7	Southwest (202.5–247.5)	518.16	11.91
8	West (247.5–292.5)	406.69	9.35
9	Northwest (292.5–337.5)	387.42	8.9
10	North (337.5–360)	420.54	9.66
		4351.93	100

the Western Ghats and South India with an elevation of 2694 meters (8842 feet) and an impressive topographical elevation of 2479 meters (8133 feet) (Fig. 5.4c). Conversely, the lowest elevation in the region is 5 m.

5.4.4 Lithology

Geologically, the district can be divided into three large belts running north–south. The northern region is characterized by the Peninsular Gneissic Complex with granite gneiss as the representative rock. These rocks have a pronounced schistosity and have undergone regional folding. In the southern part of the district, the predominant and widespread rock type is charnockite, which belongs to the charnockite group. Other rock types within this group, such as pyroxene granulite and magnetite quartzite, occur as linear bands that generally parallel the folding trend within the gneissic terrain. The area between the northern Peninsular Gneissic Complex and the southern Charnockite Group is known as the Migmatitic Complex (Fig. 5.4d). This complex consists of rocks such as biotite gneiss and hornblende-biotite gneiss, collectively referred to as composite gneiss. Biotite gneiss is prevalent in the central, northeastern and southeastern parts of the district, while hornblende-biotite gneiss is more common in the northeastern and southeastern regions. Biotite gneiss often contains remnants of older rocks of the khondalite and charnockite group, which appear as lenticular and linear bodies oriented with the schistosity trend.

Table 5.3 Land use/land cover

Si. no	LULC	Area in km^2	Area in %
1	Water bodies	6.31	0.15
2	Forest	454.11	10.43
3	Grassland	857.77	19.71
4	Vegetation	1513.71	34.78
5	Barren land	1520.02	34.93
	Total	4351.93	100

5.4.5 Land Use/Land Cover

The distribution of land cover within the study area is characterized by different categories. Water bodies make up a relatively small proportion of the total area, at around 0.15%. The largest land cover category is barren land, which covers about 34.93% of the region. In contrast, forests cover a considerable portion, accounting for 10.43% of the landscape, while grasslands cover about 19.71% of the area. Vegetation, which includes various forms of plant cover, is the most extensive land cover category, covering 34.78% of the total area (Table 5.3) (Fig. 5.4e). This diversity of land cover types plays an important role in shaping the ecological and environmental characteristics of the study area.

5.4.6 Fault Buffer

The fault line, an important factor for the occurrence of landslides, was divided into six different buffer zones. These buffer zones vary in width and range from 500 to 3000 m, with intervals of 500 m. The 500 m buffer zone covers an area of 725.321 km^2, while the 1000 m buffer zone covers 725.559 km^2. The 1500-meter buffer zone covers an area of 725.321 km^2 and the 2000 m buffer zone covers an area of 725.559 km^2. In addition, the 2500 m buffer zone covers an area of 725.083 km^2. Finally, the widest buffer zone, 3000 m, covers an area of 725.083 km^2 (Table 5.4) (Fig. 5.4f). These buffer zones are integral components in the assessment of the impact of fault lines on landslide susceptibility in the study area.

5.4.7 Road Buffer

The road buffer plays a crucial role in landslide susceptibility mapping as it is a significant anthropogenic factor. The presence of roads can contribute to the occurrence of landslides as they alter the topography and reduce the shear strength of the top of the slope, resulting in increased tensile stress. Even if a slope is naturally stable, the construction of a road can have a negative impact on its stability. In this particular study, the focus is on the investigation of main roads, which are divided

Table 5.4 Fault buffer

Si. no	Distance in meters	Area in km^2	Area in %
1	500	725.32	16.67
2	1000	725.56	16.67
3	1500	725.32	16.67
4	2000	725.56	16.67
5	2500	725.08	16.66
6	3000	725.08	16.66
	Total	4351.93	100

Table 5.5 Road buffer

Si. no	Distance in meters	Area in km^2	Area in %
1	300	2799.82	64.34
2	600	825.00	18.96
3	900	250.73	5.76
4	1200	174.49	4.01
5	1500	153.44	3.53
6	1800	148.44	3.41
	Total	4351.93	100

into six different distances from the road. The table provides information on the road buffer zones in meters, the corresponding area in km^2 and the percentage of the total area. For example, a 300 m buffer zone covers an area of 2799.825 km^2, a 600 m buffer zone covers an area of 824.996 km^2, a 900 m buffer zone covers an area of 250.734 km^2, a 1200 m buffer zone covers an area of 174.490 km^2, a 1500 m buffer zone covers an area of 153.437 km^2 and an 1800 m buffer zone covers an area of 148.44 km^2 (Table 5.5) (Fig. 5.4g).

5.4.8 Drainage Density

Drainage density is computed by dividing the cumulative length of streams and rivers in a drainage basin by the total area of that basin, expressed in km^2. In the Idukki region, this principle is applied to delineate four zones. The low-density zone spans approximately 633.378 km^2, the moderate-density zone covers 1407.108 km^2, the high-density zone encompasses 1543.418 km^2, and the very high-density region spans 768.121 km^2 (Table 5.6) (Fig. 5.4h). These zones are demarcated based on the drainage density, calculated by dividing the total length of watercourses by the respective areas in each zone.

Table 5.6 Drainage density

Si. no	Drainage density	Area in km^2	Area in %
1	Low	633.38	14.55
2	Moderate	1407.11	32.33
3	High	1543.42	35.47
4	Very high	768.02	17.65
	Total	4351.92	100

Table 5.7 Stream buffer

Si. no	Distance in meters	Area in km^2	Area in %
1	300	726.03	16.68
2	600	731.26	16.80
3	900	724.13	16.64
4	1200	723.77	16.63
5	1500	723.41	16.62
6	1800	723.29	16.62
	Total	4351.92	100

5.4.9 Stream Buffer

Landslides can occur on slopes influenced by streams, similar to the effects of distance from the road. The proximity of the slope to streams can have a negative effect on slope stability. The presence of a stream increases the likelihood of landslides as it erodes the slope and increases the saturation of the slope material. In this particular region, proximity to streams is one of the most important factors influencing the occurrence of landslides. There are six categories of buffer zones, ranging from 300 to 1800 m, with intervals of 300 m. The 300 m buffer zone covers an area of 726.034 km^2, the 600 m buffer zone covers an area of 731.268 km^2, the 900 m buffer zone covers an area of 724.131 km^2, the 1200 m buffer zone covers an area of 723.774 km^2, and the 1500 m buffer zone covers an area of 723.417 km^2. Finally, the 1800 m buffer zone covers an area of 723.2989 km^2 (Table 5.7) (Fig. 5.4i).

5.4.10 Analytical Hierarchy Process (AHP)

The Analytical Hierarchy Process (AHP) is a widely recognized multi-criteria decision-making (MCDM) method originally proposed by Saaty (1977). Over the past two decades, it has been extensively studied and successfully applied in decision-making systems. The AHP method provides solutions to decision makers in MCDM systems by considering the subjective judgments of individuals through pairwise comparisons of alternatives. A notable advantage of this method is the ability to calculate the inconsistency rate, which allows the judgments to

Table 5.8 Fundamental scale of AHP

Importance	Definition	Explanation
1	Equal importance	Both elements have equal contribution in the objective
3	Moderate importance	Moderate importance of one element over other
5	Strong importance	Strong favoring of one element compared to the other
7	Very strong importance	One element is strongly favored and has domination in practice compared to another element
9	Extreme importance	The evidence favoring one activity over another is the highest possible order of affirmation
2,4,6,8	Intermediate importance	When comparison is needed
1/3,1/5, 1/7, 1/9	Values for inverse comparison	

be re-examined. To apply the AHP method, the following steps are followed. First, a pairwise comparison questionnaire is created to obtain expert opinions on a scale of 1 to 9, with decision makers providing verbal judgments. The results are then analyzed in a comparison matrix for all criteria and sub-criteria, and priority weights are derived for each factor (Table 5.8). The AHP tool facilitates the implementation of multiple steps of the general overlay analysis process in a single tool. This tool combines the reclassification of values in the input grid into a common rating scale, such as suitability or preference, risk, or a similar unifying scale. The tool only accepts integer grids with a cell size of 30 × 30, such as a land use grid. Before use, the grid must be reclassified to an integer format.

5.4.11 Landslide Susceptibility Zones of the Idukki District

The determination of landslide susceptibility zones in the Idukki district employed the AHP method (Table 5.9). The AHP analysis yielded five distinct susceptibility zones, each quantified in. To create a comprehensive map, nine causative factors-slopes, aspect, elevation, lithology, and land use/land cover, drainage density, distance to road, distance to stream, and distance to fault-were considered. The resulting values were reclassified using the natural break method, leading to the categorization of zones as very low, low, moderate, high, and very high-risk areas. The very low-risk zone covers an area of approximately 678.90 km^2, followed by the low-risk zone at 684.61 km^2. The moderate-risk area encompasses around 1055.68 km^2, while the high-risk zone spans 1142.69 km^2. Lastly, the very high-risk zone extends across an area of approximately 790.02 km^2 (Table 5.10)

(Fig. 5.5a). It is noteworthy that slope holds significant importance in the analysis, indicating that areas with higher slopes and elevations are more susceptible to landslides. Conversely, regions with high drainage density exhibit a lower risk of landslides, as an efficient drainage system reduces the likelihood of instability in the area.

5.4.12 Identification of Susceptible National and State Highways

The very low-risk zone covers an expanse of approximately 678.90 km^2 (15.60%), while the low-risk zone extends over an area of around 684.61 km^2 (25.73%). Moderate-risk areas encompass approximately 1055.68 km^2 (24.26%), with the high-risk zone spanning an area of 1142.69 km^2 (26.26%). Lastly, the very high-risk zone covers approximately 790.02 km^2 (18.15%) (Table 5.10). Notably, the major Taluks of Devikulam and Peerumedu are situated in the high susceptibility zones, contrasting with Thodupuzha, located in a very low susceptibility zone. Idukki falls within the moderate susceptibility zone, with a section of Udumbanchola also in the high susceptibility zone. The very high susceptibility zones are traversed by several major national and state highways, including Kollam-Teni, Adimaly-Kumily, Adoor-Vandiperiyar, Munnar-Kumily, Muvattupuzha-Theni, and Alappuzha-Madurai. Caution is advised during heavy monsoons due to heightened landslide risk, particularly in the high to very high susceptibility zones. Vigilant alarms and preparedness measures are essential to mitigate potential disasters, necessitating strict control over transportation, especially in the high ranges (Fig. 5.5b). The Munnar hill station road is also located within the high susceptibility zone.

5.4.13 Identification of Most Susceptible Taluks and Villages of Idukki District Kerala

There are a total of 66 revenue villages and 5 taluks in Idukki district. Devikulam and Peerumedu are two taluks that fall under the high-risk zone. Devikulam taluk has Chinnakanal, Mrayoor, Kanthalloor, Kottamboor, Vattavada, and Kannandevan hills which are highly prone to risks. In Udumbanchola, there are Vandanmedu, Anakkara, and Karunapuram villages which are also highly prone to hazards (Fig. 5.5c). Finally, in Peerumedu taluk, the villages of Anavilasam, Chakkumpallam, and Kumily are in the high-risk zone.

Table 5.9 AHP pairwise comparison matrix

	Slope	Elevation	Aspect	Lithology	Fault Buffer	Road Buffer	Stream Buffer	LULC	Drainage Density	Final Weight
Slope	1	2	3	6	3	4	6	6	3	23.25
Elevation		1	2	6	5	5	3	5	3	21.21
Aspect			1	0.167	2	0.333	2	0.2	2	6.308
Lithology				1	3	5	5	4	0.167	12.39
Fault buffer					1	3	6	4	6	11.89
Road buffer						1	3	2	0.5	5.213
Stream buffer							1	0.25	2	3.592
LULC								1	0.25	5.455
Drainage density									1	10.7
CR = 0.302										

Table 5.10 Landslide susceptible zone

Si. No	Landslide susceptible zone	Area in km^2	%
1	Very low risk	678.90	15.60
2	Low risk	684.61	15.73
3	Moderate risk	1055.68	24.26
4	High risk	1142.69	26.26
5	Very high risk	790.02	18.15
		4351.92	100

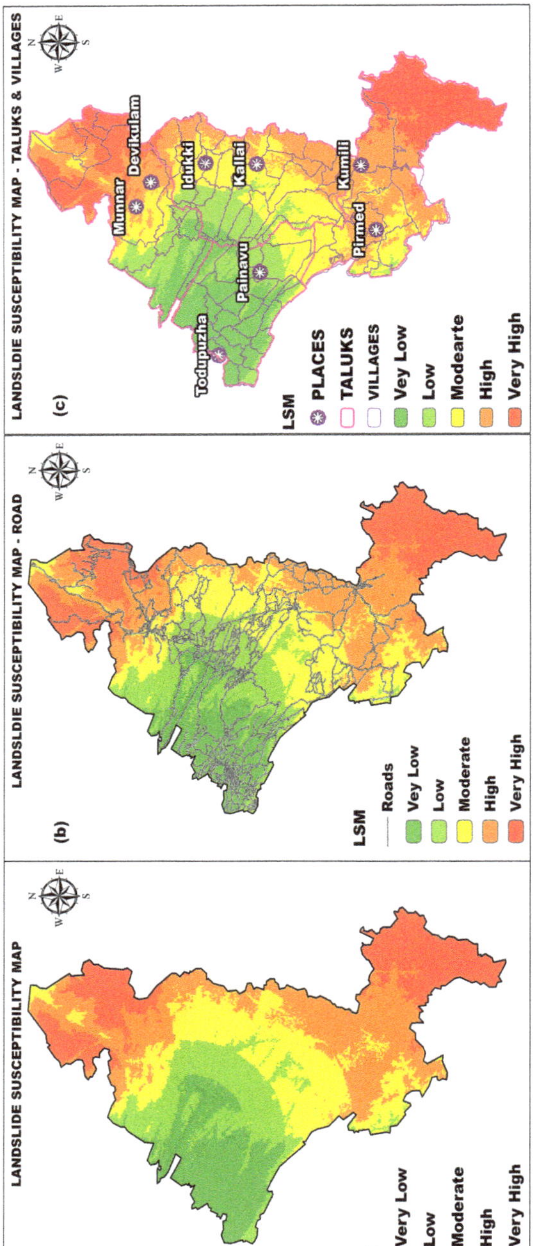

Fig. 5.5 Study area **a** landslide susceptibility map **b** susceptible roads **c** susceptible taluks and villages

5.5 Conclusion

In this study, we employed the Analytical Hierarchy Process (AHP) method to create a comprehensive landslide vulnerability zone map for the Idukki district. The resulting classification into five categories-very low susceptible, low susceptible, moderate susceptible, highly susceptible, and very highly susceptible areas-encompasses the 4351.92 km^2 area of Idukki. Notably, 678.90 km^2 falls within the very high-risk zone. The consideration of nine major triggering factors, including slope, aspect, lithology, land use/land cover, road buffer, stream buffer, fault buffer, drainage density, and elevation, played a crucial role in assigning weights to each factor and obtaining the final output. Taluks like Devikulam and Peerumedu, along with villages such as Chinnakanal, Mrayoor, Kanthalloor, Kottamboor, Vattavada, Kannandevanhills, Vandanmedu, Anakkara, Karunapuram, Anavilasam, Chakkumpallam, and Kumily, are situated in the high to very high susceptibility zones.

However, it's important to acknowledge some limitations in this study. The methodology relies on available data and assumptions, and the accuracy of the vulnerability map is contingent on the quality of input data. Additionally, the analysis might benefit from more detailed and up-to-date information on certain factors. Despite these limitations, the identification of highly susceptible zones, particularly along major highways like Adimaly-Kumily, Theni, and Madurai, underscores the need for an effective early warning system, evacuation protocols, and comprehensive preparedness measures in these regions. The pre-identification of vulnerable areas serves as a crucial step for proactive hazard management, prompting the implementation of necessary restrictions and the installation of landslide detection systems for enhanced safety and protection of people and property.

5.6 Recommendations

It is imperative to enforce a total ban on settlement and development activities in high and very high landslide vulnerable areas. Residents in highly unstable regions should evacuate during persistent rain for 2 or 3 days, with rain gauges installed to monitor and promptly alert if rainfall exceeds the threshold. Afforestation is crucial in failed forest areas, while construction in high slope zones should be avoided, and strict control should govern road transportation in highly susceptible areas during heavy rainfall. An active landslide pre-warning system is essential for timely responses, and management techniques, including retaining walls, concrete structures, soil nailing, and ground anchor walls, can effectively mitigate human, material, economic, and environmental impacts. Implementation of a landslide early warning system is crucial in overall landslide management, and various direct methods, such as modifying slope geometry, reinforcing slope material, and installing structures, offer preventive measures, though their feasibility depends on factors like cost, landslide magnitude, and settlement size at risk.

References

Abraham, M. T., Pothuraju, D., & Satyam, N. (2019). Rainfall thresholds for prediction of landslides in Idukki, India: An empirical approach. *MDPI Geosciences, 10*(10), 380. https://doi.org/10.3390/geosciences10100380

Ajin, R. S., Divakarannair, N., Ambujendran, R., Oommen, T. Y., Ali, P., & Sajinkumar, K. S. (2022). The tale of three landslides in the Western Ghats, India: Lessons to be learnt. *Geoenvironmental Disasters, 9*. https://doi.org/10.1186/s40677-022-00218-1

Amal George, K., Sunil, P. S., Anish, A. U., Gopinath, G., & Mini, V. K. (2023). A pilot assessment of the fatal landslide on 29 August 2022 in Kudayathoor, Idukki, Kerala. *Journal of the Geological Society of India, 99*(1), 141–144. https://doi.org/10.1007/s12594-023-2277-1

Bajracharya, I., & Tani, M. (2022). Landslide susceptibility mapping using GIS and AHP techniques for the Darjeeling lesser Himalaya region of India. *Environmental Earth Sciences, 121*(3), 1–18.

Biju Abraham, P., & Shaji, E. (2013). Landslide hazard zonation in and around Thodupuzha-Idukki-Munnar road, Idukki district, Kerala: A geospatial approach. *Journal of the Geological Society of India, 82*, 649–656.

Chawla, A., Chawla, S., Pasupuleti, S., Rao, A. C. S., Sarkar, K., & Dwivedi, R. (2018). Landslide susceptibility mapping in Darjeeling Himalayas, India, landslide susceptibility mapping in Darjeeling Himalayas, India. *Advances in Civil Engineering, 2018*, 6416492. https://doi.org/10.1155/2018/6416492

Das, I., Pal, S. K., & Pradhan, B. (2022). GIS-based landslide susceptibility mapping using machine learning techniques for the Darjeeling Himalayas, India. *Geoscience Frontiers, 13*(5), 101794.

Demir, G. (2019). GIS-based landslide susceptibility mapping for a part of the North Anatolian Fault zone between Reşadiye and Koyulhisar (Turkey). *Catena, 183*, 104211.

Deoja, B., & Purohit, A. M. (2018). *Landslide hazard assessment and management: An integrated approach*. Springer Nature.

Dey, S., & Mandal, A. K. (2021). GIS-based landslide susceptibility mapping using AHP and machine learning techniques: A case study of the Darjeeling district, India. *Natural Hazards and Earth System Sciences, 21*(12), 3623–3645.

Doan, V. L., Nguyen, B.-Q.-V., Pham, H. T., Nguyen, C. C., & Nguyen, C. T. (2023). Effect of time-variant NDVI on landside susceptibility: A case study in Quang Ngai province, Vietnam. *Open Geosciences, 15*(1), 20220550. https://doi.org/10.1515/geo-2022-0550

Gopinathan, P., & Thomas, J. (2015). Assessment of soil fertility in the cardamom growing regions of Idukki district, Kerala, India. *International Journal of Agricultural Technology, 11*(5), 1463–1474.

Goudarzi, H., Kumar, N. V., Ghosh, P. K., & Paul, A. K. (2019). Petrology, geochemistry and tectonic significance of the Charnockite Group of Rocks in Sivagangai District, Tamil Nadu, India. *Journal of the Geological Society of India, 93*(4), 469–482. https://doi.org/10.1007/s12594-019-1481-0

Hagen, A., Gürsoy, H., Başıbüyük, A., & Reichenbach, P. (2013). Landslide hazard assessment and mitigation in Turkey: An overview of recent progress and future challenges. *Landslides, 10*(2), 215–228.

Idukki District Administration. (2023). Idukki District. Retrieved from https://idukki.nic.in/

Jacinth Jennifer, J., & Saravanan, S. (2022). Artificial neural network and sensitivity analysis in the landslide susceptibility mapping of Idukki district, India. *Geocarto International, 37*(19), 5693–5715.

Jaiswal, R. K., Kumar, S., & Vishal, V. (2014). Landslide susceptibility mapping using remote sensing and GIS techniques: A case study of Idukki district, Kerala, India. *Journal of Earth System Science, 123*(6), 1247–1258.

Jones, S., Kasthurba, A., Bhagyanathan, A., & Binoy, B. V. (2021). Landslide susceptibility investigation for Idukki district of Kerala using regression analysis and machine learning. *Arabian Journal of Geosciences, 14*. https://doi.org/10.1007/s12517-021-07156-6

Ju, N.-p., Huang, J., Huang, R., He, C., & Li, Y. (2015). A real-time monitoring and early warning system for landslides in Southwest China. *Journal of Mountain Science., 12*, 1219–1228. https://doi.org/10.1007/s11629-014-3307-7

Kale, V. S. & Wategaonkar, S. V. (2016). Geomorphology of the Deccan Plateau, India. In *Landscapes and Landforms of India* (pp. 127–145).

KSDMA (Kerala State Disaster Management Authority). (2018). *Landslide hazard assessment and risk management.*

Kerala Tourism Department. (2023). Idukki District. https://www.keralatourism.org/munnar/idukki-kerala.php

Khan, M. A., Mustaffa, Z., Harahap, I., & Ibrahim, M. B. (2022). Al-Atroush ME (2022) assessment of physical vulnerability and uncertainties for debris flow Hazard: A review concerning climate change. *Land, 11*, 2240. https://doi.org/10.3390/land11122240

Kuriakose, J., & Yesodhara, P. (2015). Landslide hazard assessment using GIS and remote sensing techniques: A case study of Idukki district, Kerala, India. *Journal of the Geological Society of India, 86*(4), 531–538.

Lei, X., Chen, W., & Pham, B. T. (2020). Performance evaluation of gis-based artificial intelligence approaches for landslide susceptibility modeling and spatial patterns analysis. *ISPRS Int J Geoinf, 9*(7), 443.

Mersha, T., & Meten, M. (2020). GIS-based landslide susceptibility mapping and assessment using bivariate statistical methods in Simada area, northwestern Ethiopia. *Geoenvironmental Disasters, 7*, 20. https://doi.org/10.1186/s40677-020-00155-x

Nair, P. K., & Achuthan, N. K. (2018). Landslide susceptibility mapping using analytical hierarchy process (AHP) technique in Idukki district, Kerala, India. *Natural Hazards, 90*(1), 399–424. https://doi.org/10.1007/s11069-017-2922-6

Preetha, P. N., & Nayak, P. K. (2020). Geomorphological characterization of Idukki plateau, Kerala, India using ASTER and SRTM data. *Journal of the Geological Society of India, 96*(8), 1007–1016. https://doi.org/10.1007/s12594-020-1690-9

Proag, V. (2014). The concept of vulnerability and resilience. *Procedia Economics and Finance., 18*. https://doi.org/10.1016/S2212-5671(14)00952-6

Rifat, S. A. A., & Liu, W. (2020). Measuring community disaster resilience in the conterminous coastal United States. *ISPRS International Journal of Geo-Information., 9*(8), 469. https://doi.org/10.3390/ijgi9080469

Saaty, T. L. (1977). A scaling method for priorities in a hierarchichal structure. *Journal of Mathematical Psychology, 15*(3), 234–281.

Sankaran Nair, P., & Achuthan, N. K. (2011). River basin hydrology of Kerala, India. *Journal of Environmental Studies, 10*(2), 190–198.

Shashidhar Kumar Jha, A. K., Negi Juha, M., & Negi, A. R. S. (2021). Socio-ecological vulnerability and resilience of mountain communities residing in capital-constrained environments. *Mitigation and Adaptation Strategies for Global Change, 26*, 38. https://doi.org/10.1007/s11027-021-09974-1

Solanki, S., Chandran, C., Garg, J. K., & Bhattacharya, P. (2019). Long-term spatio-temporal trends in land use/land cover in Devikulam taluk, Kerala. *Geospatial Techniques.*

Tapas, R., Cees, J., Westen, V., Kerle, N., Jetten, V., & Vinod Kumar, K. (2012). Landslide hazard and risk assessment using semi-automatically created landslide inventories. *Geomorphology.* https://doi.org/10.1016/j.geomorph.2012.12.001

Thomas, A. V., Saha, S., Danumah, J. H., Raveendran, S., Prasad, M. K., Ajin, R. S., & Kuriakose, S. L. (2021). Landslide susceptibility zonation of Idukki district using GIS in the aftermath of 2018 Kerala floods and landslides: A comparison of AHP and frequency ratio methods. *Journal of Geovisualization and Spatial Analysis, 5*, 1–27. https://doi.org/10.1007/s41651-021-00090-x

Kerala Tourism (2023). Idukki Dam. Retrieved from https://idukki.nic.in/en/tourism/

Vakhshoori, V., & Zare, M. (2016). Landslide susceptibility mapping by comparing weight of evidence, fuzzy logic, and frequency ratio methods, geomatics. *Natural Hazards and Risk, 7*, 5. https://doi.org/10.1080/19475705.2016.1144655

Vikram, V., Tariq, S., Rohan, P., Mohit, P., & Pradhan, S. (2017). Hazard assessment in rockfall-prone Himalayan slopes along National Highway-58, India: Rating and simulation. *Natural Hazards, 85*. https://doi.org/10.1007/s11069-016-2563-y

Wadhawan, S., Singh, B., & Ramesh, M. V. (2020). Causative factors of landslides 2019: Case study in Malappuram and Wayanad districts of Kerala, India. *Landslides, 17*, 1–9. https://doi.org/10.1007/s10346-020-01520-5

Ye, C., Li, Y., Cui, P., Liang, L., Pirasteh, S., Marcato, J., et al. (2019). Landslide detection of hyperspectral remote sensing data based on deep learning with constrains. *IEEE Journal of Selected Topics in Applied Earth Observations and Remote Sensing, 12*(12), 5047–5060. https://doi.org/10.1109/jstars.2019.2951725

Geospatial Analysis and Prediction of Landslide Susceptibility in Uttara Kannada District, Karnataka (2025–2050): A Multi-Model Approach Using Maxent, MGWR, and ANN

6

M. Sasi and Sawant Sushant Anil

Abstract

This study employs advanced geospatial techniques to analyze landslide susceptibility in Uttara Kannada district, Karnataka, from 2025 to 2050. We identified approximately 1852.18 sq km of high-risk zones using the Maxent model. The MGWR analysis revealed significant correlations with landslide occurrences, incorporating the highly influencing environmental parameters (water proximity, TWI, soil type, slope, rainfall, NDVI, LULC, lineament proximity, and geomorphology). The standardized residuals vs. predicted plot indicated a generally well-fitted model, with residuals randomly dispersed around zero. However, some areas showed patterns suggesting potential model refinements. The WP significance map highlighted regions with a strong statistical impact of these parameters, guiding targeted mitigation efforts. Emerging hotspot and cold spot analyses provided insights into spatial and temporal trends in landslide activity. Hotspots, primarily located in the Western Ghats and certain coastal areas, indicate regions of increasing landslide occurrences due to high rainfall, steep slopes, and other contributing factors. Conversely, cold spots in the northern parts of the district reflect areas with decreasing landslide activity, likely due to effective land management and mitigation measures. These findings underscore the importance of focused interventions in hotspot areas and the need to maintain successful practices in cold spot regions, thereby enhancing disaster risk management and promoting sustainable land-use practices in Uttara Kannada.

M. Sasi (✉) · S. S. Anil
Division of Geoinformatics, School of Life Sciences, JSS Academy of Higher Education and Research, Mysuru, Karnataka, India
e-mail: sasim@jssuni.edu.in

S. S. Anil
e-mail: geo_sushant@jssuni.edu.in

R. M. Yuvaraj et al. (eds.), *Advanced Landslide Susceptibility Modeling*,
https://doi.org/10.1007/978-3-032-14485-0_6

The application of the artificial neural network (ANN) model has further proven the model's significance for landslide hazard zonation mapping, as evidenced by an R^2 value indicating a strong model fit. Using a network architecture of 15-10-10 with 75% training data, the ANN effectively captured the complex relationships between environmental variables and landslide occurrences. The model's high R^2 value signifies its robustness in predicting high-risk zones, complementing the spatial analysis performed by Maxent and MGWR. These findings validate the ANN model's reliability and underscore its potential for enhancing landslide risk mitigation strategies in geospatial hazard assessments.

Keywords

Landslide susceptibility • Maxent • MGWR • Emerging hotspots • MATLAB • ANN

6.1 Introduction

One of the frequent geological risks in hilly regions worldwide is landslides. Rainfall is the primary natural cause of landslides, aside from geological and human-caused factors (Forest Department, 2023). The circumstances of extended movement of the ground that includes shallow debris flow, deep slope failures, and rock falls (Kalantar et al., 2018). Approximately 1000 people die and $4 billion worth of property is damaged by landslides worldwide each year, which also account for 17% of all fatalities brought on by natural disasters (Chen et al., 2017). About 15% of India's land area is affected by landslides (0.49 million km^2) (Mondal & Mandal, 2018). The Centre for Research on the Epidemiology of Disasters (CRED) and the United Nations International Strategy for Disaster Reduction (UN/ISDR) released official statistics for 2006 showing that landslides were the third most deadly natural disaster. Colombia, Tajikistan, India, and Nepal are among the regions most at risk from landslides, with an estimated annual death toll per 100 km^2 exceeding one fatality (Forest Department, 2023). Landslides in India typically happen during the monsoon season. The mountainous terrain and abundant rainfall in the Western Ghats and the Himalayas make them extremely vulnerable to mass movements (Forest Department, 2023). Significant fatalities and financial losses have resulted from landslides in hilly areas (Pham et al., 2022). Significant financial losses, fatalities, and negative effects on communities, infrastructure, and the environment are the outcomes, which have a substantial influence on land use and urban growth (Bragagnolo et al., 2020). The primary causes are the acceleration of development in hilly areas, including urbanization and the growth of agricultural land, as well as brief, heavy rains brought on by climate change, which can make landslides more common (Mallick et al., 2021). Understanding how these environmental elements relate to the incidence of landslides is crucial for the quantitative evaluation of landslide risks and/or vulnerability (Choi et al., 2012).

An area's propensity to experience landslides is known as its susceptibility (Chen et al., 2017). Over the years, landslide prediction and susceptibility zones have extensively used GIS and RS. By using the correlations between previous landslide events and a set of environmental characteristics, predictive modeling aims to forecast regions that are vulnerable to landslides in the future (8). Numerous studies have used probabilistic (Choi et al., 2012), and heuristic (Chen et al., 2017) approaches for landslide susceptibility. Over time, statistical methods like frequency ratio (FR) and logistic regression have become increasingly popular (Choi et al., 2012; Kalantar et al., 2018; Mondal & Mandal, 2018). Additionally, machine learning techniques are used, including ANN, which has demonstrated a broader range of capabilities for mapping and assessing landslide vulnerability (Prabu Ramakrishnan, 2009; Kalantar et al., 2018; Mallick et al., 2021; Bragagnolo et al., 2020; Chen et al., 2017). Different ensemble approaches that include machine learning and deep learning algorithms have become increasingly popular recently (Chen et al., 2017; Choi et al., 2012; Mallick et al., 2021). In landslide susceptibility investigations, ensemble methods' key advantage is their increased recognition accuracy and prediction capacity as compared to single machine learning models (Chen et al., 2017). Various causative variables have been identified for landslide hazard which includes topographic variables to climate and anthropogenic agents, i.e., slope gradient, slope, aspect, curvature, plan curvature, profile curvature, lineament proximity, NDVI, soil type, geology-lithology, rainfall, elevation/altitude, TWI, STI, TRI, SPI, soil texture, geomorphology, land cover, road proximity, lineament density, drainage density (Chen et al., 2017; Mondal & Mandal, 2018; Pham et al, 2022) out of which variables were specifically selected according to the regional dynamics.

A minimum of 15.30% of Karnataka's land area—more than 29,350.3 km^2—is affected by landslides, and these occur in the 29 taluks. Over the last 10 years, Karnataka has seen the most landslides in the Uttara Kannada district, followed by Shivamogga, Chikkamagaluru, Udupi, Dakshina Kannada, Kodagu, and Hassan. These landslides have caused extensive damage, numerous fatalities, substantial economic losses, and social unrest (Karnataka State Action Plan, 2022). This study has identified the Uttar Kannada district, a particularly hazardous area inside the Western Ghats, as a possible landslide risk zone. Along the steep slopes overlooking the Konkan shore, the Western Ghats in the South are constantly threatened by a new type of landslide, which is characterized by a lateritic cap, in addition to the very landslide-prone Nilgiris (Karnataka State Action Plan, 2022). For susceptibility mapping variables, i.e., rainfall, DEM, aspect, water proximity, TWI, SPI, road proximity, soil, slope, NDVI, lineament proximity, geology, geomorphology, plan curvature, profile curvature, LULC, has been covered for the same. The machine learning process, Maxent, and ANN have been used to accurately predict the landslide susceptible zones in Uttar Kannada, to independently look into how machine learning and statistical outputs correspond to each other showcasing critical analysis of the integrity of triggering factors leading to landslide. On further note, for future prediction, data mining techniques. Emerging Hotspot Analysis has been used to predict the future landslide susceptible hotspots and cold spots till 2050.

6.2 Study Area

The Western Ghats are one of the eight "hottest hotspots" of biodiversity in the world. Uttara Kannada district located in the central Western Ghats (Fig. 6.1a) lies between 13.769°–15.732° N and 74.124°–75.169° E (Ramachandra et al., 2018), covering 10.25 lakh hectares of total land. The Indian state of Karnataka contains the Konkan district of Uttara Kannada, sometimes referred to as North Kanara or North Canara. It shares boundaries with the Arabian Sea to the west, Dharwad District and Haveri District to the east, Shimoga District and Udupi District to the south, and the state of Goa and Belgaum District to the north. Due to its strong adherence to the Western Ghats along the coast, this district is prone to flash floods in the coastal taluks after periods of intense rainfall. Additionally, after unusual rains, the rivers that have reservoirs built on them for hydel projects unleash significant amounts of water, causing flash floods that endanger lives and cause property damage on the islands and settlements along the riverbanks (Uttara Kannada District Disaster Management Plan, 2022). Monsoon periods are highly prone to landslides, which is a cause of heavy rainfall and flash floods.

6.3 Data

Landslide inventory data was obtained from NASA Global Landslide Catalog, NRSC—NDEM, and KSNDMC—Uttar Kannada District Disaster Management Plan, where the data were compiled to get the overall landslide occurrence throughout 2000–2023 (Fig. 6.1b).

6.4 Methodology

Out of 16 variables selected only 14 variables (Fig. 6.2a–n) (Table 6.1) were selected based on principle component analysis (PCA). All the conditioning factors were processed in ArcMap 10.8.2 software and were resampled in R, to have the same extent, projection, rows, and columns.

To extract the topographic features DEM data was used. Elevation influences moisture and temperature regimes (Nhu et al., 2020), which determines vegetation growth, runoff direction, drainage density, the gravitational pull of landslides, and anthropogenic activities (Mallick et al., 2021). The range of elevation is found to fall between the range of 0–1044 m. Aspect, slope, and plan/profile curvature are commonly used factors in rainfall-induced landslide zones (Dou et al., 2020). Aspect measures the cardinal direction having the range from − 1 to 359.82 of a slope, expressed relative to the north (Kavzoglu et al., 2015). Slopes ranging from 0 to 83 m highly influence slope stability, which is a form between any section of the surface and a horizontal datum (Kalantar et al., 2018). Profile curvature is derived from curvature, controlling the sediment flow or water velocity changes flowing down a slope ranging from − 169.68 to 169.68 (Hengl et al., 2003). Plan

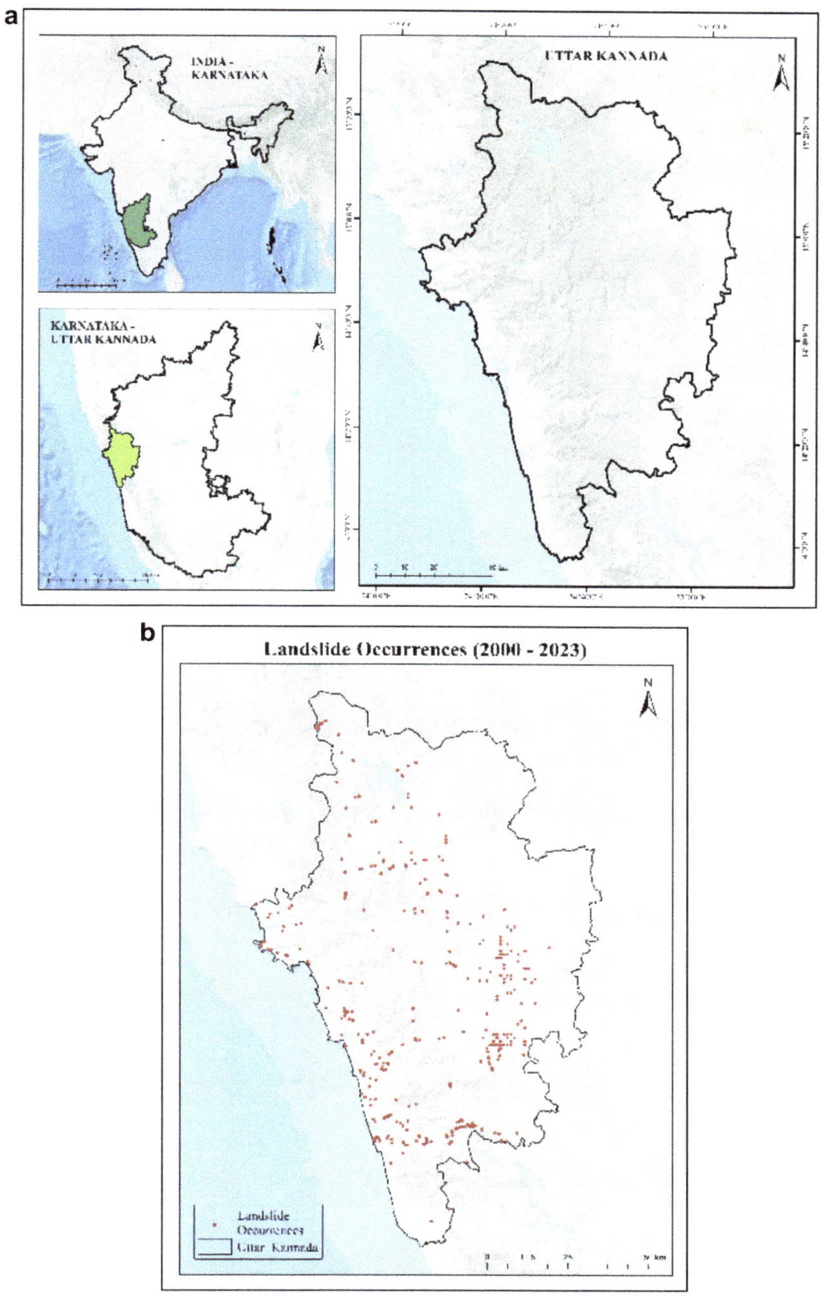

Fig. 6.1 **a** Spatial map of Uttar Kannada. **b** Landslide occurrence from 2000 to 2023

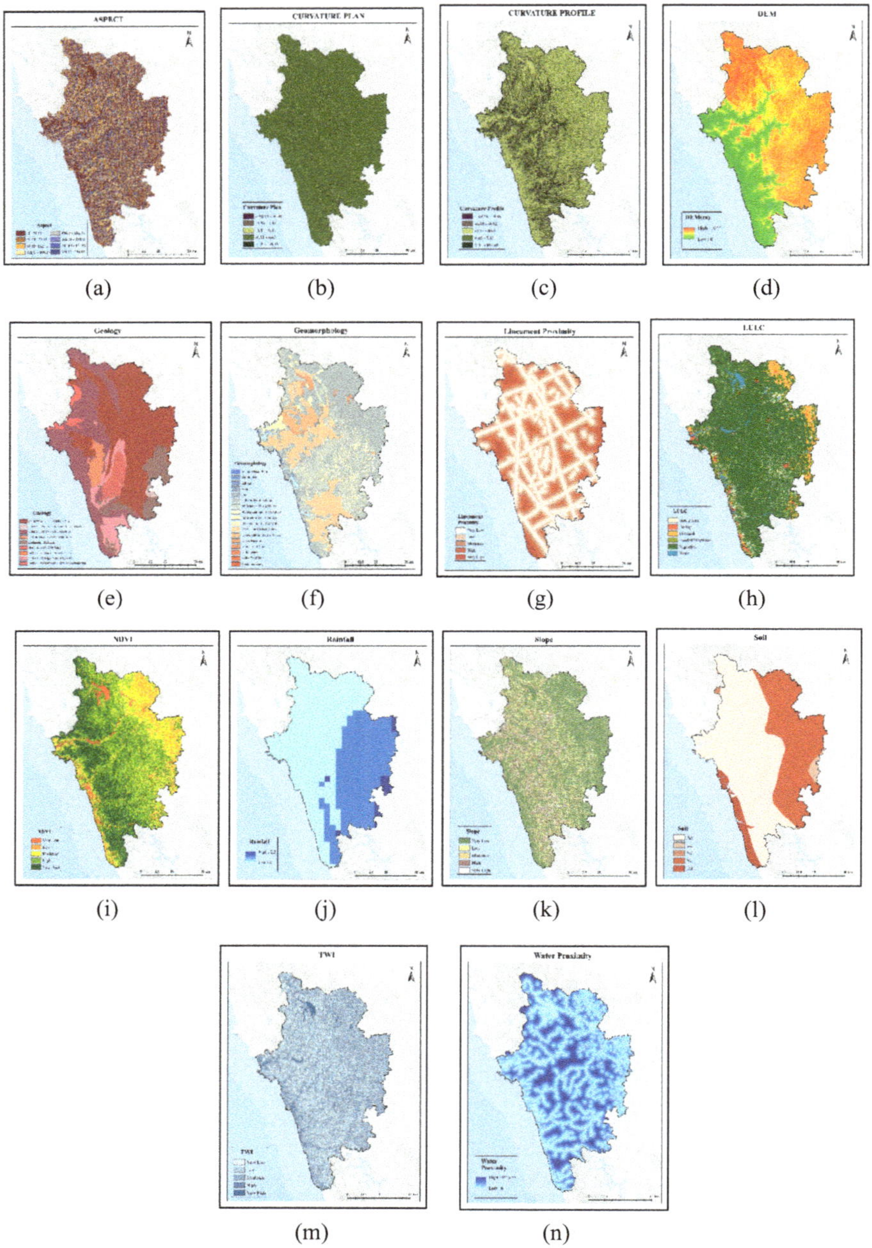

Fig. 6.2 **a** Aspect, **b** curvature plan, **c** curvature profile, **d** DEM, **e** geology, **f** geomorphology, **g** lineament proximity, **h** LULC, **i** NDVI, **j** rainfall, **k** slope, **l** soil, **m** TWI, **n** water proximity

Table 6.1 Dataset and their sources derived for the study

S. No.	Data	Resolution	Source
1	DEM	30 M	SRTM, GEE
2	Geology		Bhukosh
3	Geomorphology	250 m	Bhukosh
4	Lineament		PCI geomatica
5	LULC	10 m	Sentinel
6	NDVI	30 m	Landsat 9, GEE
7	Rainfall		Terraclimate
8	Soil		FAO
9	Water	30 m	Global surface water explorer

curvature in the study region ranges between − 158.73 and 176–85. This forces the water to converge or diverge in the slope direction affecting erosion (Dehnavi et al., 2015) Topographic Wetness Index (TWI) (Forest Department, 2023) measures the tendency of runoff to converge on a slope, calculated as:

$$\text{TWI} = Ln\,(As/\tan\,\beta) \tag{6.1}$$

where As is the cumulative upslope area draining through a point (m^2) and β is the slope angle (radians) at that point. An increase in TWI increases the landslide. In the study region, TWI were classified as: 0.14–5.38, 5.38–7.90, 7.90–10.62, 10.62–13.77, 13.77–26.87, respectively.

Rainfall is the crucial triggering factor in the Western Ghats during monsoon season. It influences NDVI, LULC, water availability, and lineament proximity. The LULC, which has water, vegetation, flooded vegetation, built-up, cropland, and barren land, affects slope sustainability by altering the LULC practices (Mallick et al., 2021). Normalized Difference Vegetation Index (NDVI) ranging between − 0.1 and 0.4. This extensive vegetation cover decreases the amount of water reaching the soil, modifying the soil hydrology, and boosting precipitation interception, infiltration, and evapotranspiration (Mallick et al., 2021). Slope failure can increase in probability when nearing a major lineament (Nhu et al., 2020). Lineament proximity map was constructed based on proximity analysis tools, classified into 5 classes: 0–1,322.93, 1,322.93–2,849.39, 2,849.39–4,681.14, 4,681.14–7,123.47, 7,123.47–12,974.90. Distance from waterbodies which comprise rivers and all other water sources plays a major condition factor for landslides. In this study, water proximity has been generated using the Euclidean distance tool, classified as 0–739.59, 739.59 1,613.66, 1,613.66–2,622.20, 2,622.20–3,899.68, 3,899.68–8,572.58, respectively.

The physical feature of a region plays a vital role in the pressure and resistance to any disaster. The geomorphological features found in the region are water bodies which include dams and reservoirs, low dissected structural hills and valleys, active/abandoned quarries, pediment-pediplain complex, low dissected denudational hills and valleys, younger coastal plain, aeolian sand dune, active flood plain, moderately dissected denudational hills and valleys, moderately dissected structural hills and valleys, highly dissected structural hills and valleys, salt pan, highly dissected denudational hills and valleys. Rock characteristics and types directly influence slope failure (Mallick et al., 2021). The geological features found in the region include goa gP. (hardware sgP.), bababudan gP. (=western ghat), peninsular gneissic complex-i, Shimoga (= Chitradurga gP.), laterite/bauxite, acid intrusive/granite/granodiorite, peninsular gneissic complex-i (granitoid), layered ultramafic complex, bababudan gP.(dharwar sgP.). Soil type along with its texture and percentage of clay, silt, sand, and pH, are key attributes and a major conditioning factor for landslides to take place. The major soil type found the region include, (Ap) Acrisols plinthiques, (Nd) Nitosols distriques, (Ne) Nitosols eutiriques, (Lc) Luvisols chromiques, and (Rd) Rigosols distiques.

6.4.1 Multicollinearity Analysis of Variables

Spatial Regression Models

The linear regression models OLS, GWR, and MGWR all function at different spatial scales and assume different things about the spatial heterogeneity of the data set (Ferreira et al., 2023). GWR and MGWR are regarded as regional models for linear regression. Spatial variations in relationships are modeled by Multiscale Geographically Weighted Regression (MGWR), a local version of linear regression. This model of local regression permits spatial variation in the explanatory variables' coefficients. Every explanatory variable can function at a distinct spatial scale. While MGWR allows a varied neighborhood (bandwidth) for each explanatory variable, GWR does not consider this. A linear regression model fitted at a target feature uses the neighborhood (bandwidth) of an explanatory variable to identify the features that are utilized to estimate the coefficient of that explanatory variable (Ferreira et al., 2023; Li et al., 2022; Zhu et al., 2024a, 2024b). The formula (6.2) of MGWR is as follows: (Zhu et al., 2024a, 2024b)

$$yi = \beta 0(ui\,,\, vi) + \sum mj = 1\,\beta bw, j(ui\,,\, vi)xi, j + \varepsilon i \tag{6.2}$$

where bw in $\beta bw, j$ is the adaptative bandwidth of xi, j, where yi is the dependent variable, xi, j is the jth explanatory variable, $\beta j(ui, vi)$ is the corresponding coefficient of xi, j, $\beta 0(ui, vi)$ is a constant term, and εi is the stochastic error term (Zhu et al., 2024a, 2024b). The weighted least-squares method is used by MGWR to initially set the initial parameters for GWR, for local parameter estimation. From then, it optimizes the model parameters using the back-fitting approach (Zhu et al., 2024a, 2024b). The mean of the independent variables was extracted using zonal statistics, with landslide location serving as the dependent variable and consisting of 909 location points. A golden search looks for several value combinations between a given minimum and maximum value for each explanatory variable (Ferreira et al., 2023), was done based on neighborhood bands and was set to run the GWMR model in ArcGIS Pro 3.2.

6.4.2 Landslide Spatial Modeling

(a) Maximum Entropy Model (MaxEnt)

The ecological niche theory-based MaxEnt model uses “presence” data to look into the possible dispersal of a target phenomenon over a predetermined study region (Phillips et al., 2006). The word entropy, coined by Shannon, 1948 illustrates the amount of information hidden in our data on which we rely to learn and portray an unknown pattern (Chen et al., 2017). The 2004-developed MaxEnt program has powerful predictive capabilities and allows for the assessment and evaluation of region-specific susceptibility zones (Phillips et al., 2006). The presence-only target phenomenon is to be landslide presence spatial info (susceptibility). The corresponding information is to be obtained from geo-environmental thematic layers.

To investigate the impact of environmental factors on the model prediction, MaxEnt generates two sets of response curves. MaxEnt creates a table that lists each environmental variable’s contribution to the model and its stability or permutation relevance. MaxEnt also uses the jackknife test as an alternate technique to determine the significance of conditioning variables (Dahal, 2022), used to evaluate the contribution and significance of conditioning variables in the context of landslide susceptibility (He et al., 2023). The model is trained using each environmental variable separately in the second stage of the jackknife test, and the variable with the largest gain is thought to have the most useful information on its own (Dahal, 2022). The model performance is stated as the average omission and prediction statistics, as well as the computed area under curve (AUC), were evaluated when evaluating the species distribution model in this study. MaxEnt’s intrinsic statistic, AUC, functioned as a trustworthy barometer for assessing the model’s performance with presence-only data. The model’s efficiency in forecasting species distribution was demonstrated by a positive connection between its performance and the AUC (Remya et al., 2015). The AUC value typically falls in the region between 0.5 and 1, serving as a benchmark for rating the model’s efficacy. The model performance is graded using the following scale: fail (0.5–0.6), bad (0.6–0.7), fair (0.7–0.8), good (0.8–0.9), or exceptional (0.9–1) (Phillips et al., 2006).

The resampled environmental conditioning variables for landslides were converted into ASCII format. Landslide presence data of 909 spatial extents were converted to CSV format as samples. The parameters set were random test percent to 20%, 50 replicates, cross-validate as replicate run type maximum iteration to 500, the threshold of 10% training presence, and output format set to clog-log.

(b) Artificial Neural Network (ANN) Model

The neural architecture of the mammalian cortex serves as the model for ANNs (Chen et al., 2017). Creating a model of the data-generating process is the goal of an artificial neural network, which enables the network to generalize and forecast results from inputs it has never seen before (Prabu Ramakrishnan, 2009). The back-propagation learning algorithm is the most widely used neural network technique (Kalantar et al., 2018). The input layer (landslide conditioning factors), hidden layers, and output layer (landslide susceptibility) make up the conventional ANN model. To construct results, the ANN assigns a specific weight to each input factor, multiplies it, adds up the results, and uses a nonlinear transfer function (Kalantar et al., 2018).

The ANN model, which examined the statistical relationship between conditioning factors and landslide susceptibility, was processed using MATLAB. Using 75% of the training data from 15 to 10–10, the Levenberg–Marquardt (LM) algorithm, is a traditional gradient-based optimization technique for nonlinear least squares problems (CUDA, 2011), with random data division set for the final run of the model.

(c) Emerging Hotspot Analysis

Clustering, interpolation, and density modeling are some of the innovative hotspot mapping techniques that led to the development of EHA. In addition to trend analysis, EHS forecasts hotspots and cold spots for any phenomenon in the future. It detects data trends like new, intensifying, waning, and intermittent hot and cold patches. A space mining tool, EHA contains data that has been transformed into a space–time cube from a specified place, including netCDF, or multidimensional data that is built using the location and date. In this study, monthly data is considered up to 2050, from which both hotspots and cold spots are generated. It then uses the conceptualization of spatial relationships values you provide to calculate the Getis-Ord Gi* statistic (using Hot Spot Analysis) for each bin using an FDR correction. When this parameter is checked, the False Discovery Rate (FDR) procedure will potentially reduce the critical *p*-value thresholds.

The Hot Spot Analysis tool calculates the Getis-Ord Gi* statistic (pronounced G-i-star) for each feature in a dataset. The resulting *p*-values and z-scores show you the spatial clustering of characteristics with high or low values. Each characteristic is examined with its neighboring features for this tool to function. A statistically significant z-score is produced when the local sum for a feature and its neighbors differs significantly from the expected local sum and when that difference is too great to be the product of chance. That comparison is made proportionately to the sum of all features. Each bin in the input netCDF cube is assigned a corresponding z-score, *p*-value, and hot spot bin classification when the space–time hot spot analysis is finished. We next use the Mann–Kendall trend test to assess these hot and cold spot patterns. Each location with data's resulting trend z-score and *p*-value, as well as each bin's hot spot z-score and *p*-value, in which the tool categorize each location as No Pattern Detected, New Hot Spot, Consecutive Hot Spot, Intensifying Hot Spot, Persistent Hot Spot, Diminishing Hot Spot, Sporadic Hot Spot, Oscillating Hot Spot, Historical Hot Spot, New Cold Spot, Consecutive Cold Spot, Intensifying Cold Spot, Persistent Cold Spot, Diminishing Cold Spot, Sporadic Cold Spot, Oscillating Cold Spot, Historical Cold Spot. A new hot spot is a site that has never before been a statistically significant hot spot but is now a statistically significant hot spot for the last time step and a site that is a statistically significant cold spot for the last time step but has never been a statistically significant cold spot previously is known as a "new cold spot," was considered for the study purposes. The span of 25 years, i.e., from 2025 to 2050 hotspots and cold spots were taken into consideration and kernel density was done and merged to get an overall hotspot and cold spot for 25 years.

6.5 Results and Discussion

6.5.1 Multiscale Geographically Weighted Regression (MGWR)

The standardized residuals from the MGWR analysis having a range of standard deviation from 0.5 to − 1.5 provide critical insights, indicating model prediction does not show extreme deviations in the model's performance and accuracy in predicting landslide occurrences.

i. Relationship Between Variables

The graphical representation of the relationship between the 10 environmental parameters (Fig. 6.3) water proximity, TWI, soil type, slope, rainfall, NDVI, LULC, lineament proximity, geomorphology, and landslide locations indicates the strength and direction of each variable's influence on landslide susceptibility. Lineament proximity, water proximity, slope, and rainfall all correlate positively, suggesting that the slope has steep gradients. High rainfall increases the risk of landslides by saturating the soil, which weakens its cohesiveness and causes slips. Nearer the lineaments, it can weaken the soil, which can cause landslides, and nearer the water body, it can weaken the terrain's structural integrity. Because LULC and NDVI have a negative correlation, the more vegetation cover there is, the more firmly the roots can hold, stabilizing the soil and halting soil erosion.

ii. Distribution of Standardized Residuals

The distribution of standard residuals (Fig. 6.4) depicts a model fit having a normal distribution with a mean of − 0.02936, indicating residuals are randomly distributed among the predicted values.

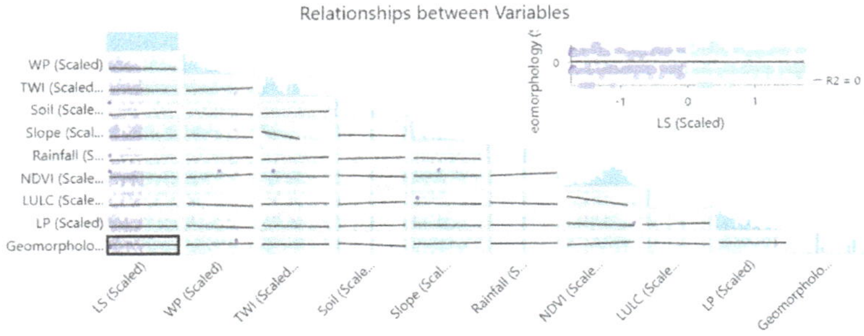

Fig. 6.3 Relationship between variables

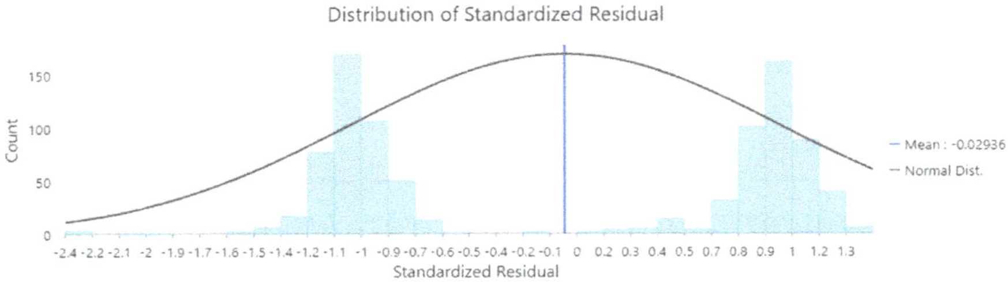

Fig. 6.4 Distribution of standardized residuals

6.5.2 Landslide Susceptibility Mapping Using Maximum Entropy Model (MaxEnt)

Long-term planning to mitigate and reduce the danger in highly viable zones requires prioritizing the landslide-prone areas. Analysis of the model's mean AUC score, average landslide omission, and expected area reveals the average outcomes of multiple model runs for the correlation between cumulative threshold, test omission rate, and forecast area. Due to the cumulative threshold selection, it was discovered that the projected omission and the omission rate were nearly the same (Dahal, 2022). The MaxEnt model performed well in predicting the landslide-susceptible zones, as shown by a mean AUC value of 0.786 and standard deviation of 0.081 (Fig. 6.5) (Anand et al., 2021).

(i) Variable Contribution in the Model

Variable interactions are generated as response curves (Fig. 6.6) to verify the landslide risk zones (Fig. 6.7). It is discovered that there is a positive association between DEM, slope, and aspect, indicating gravity increases with elevation and slope steepness. Landslides are reported to occur most frequently between 0 and 1044 m, and slopes between 0 and 83 m, where aspect is important. The

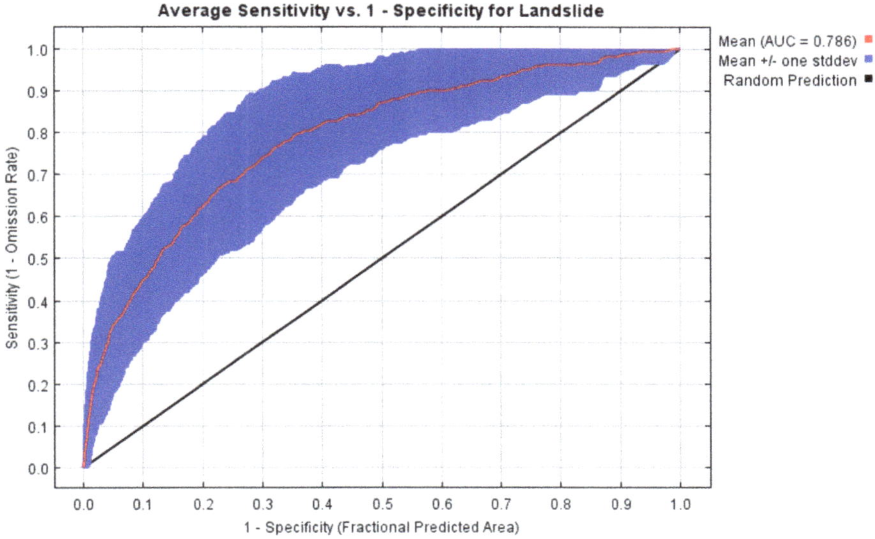

Fig. 6.5 AUC value of susceptibility modeling

regions' north, northeast, south, and southwest are the directions most frequently connected with landslides. The majority of landslides are found in the region of – 0.8–4.45, which is indicated by plan curvature, which shows water flow across the terrain and soil stability. The negative value indicates a concave landscape and inward curvature. Water buildup and soil instability may result from this. On the other hand, a positive value means the terrain is convex, although soil erosion may occur near the area's margins as a result of water running away from the terrain. Profile curvature, which measures the curvature of terrain along the steepest slope, landslide might take place in the range of – 3.32–7.31. The major cause of landslides is rainfall, with the monsoon season having the greatest impact. It is found that the average rainfall in Uttara Kannada is about 2936 mm, with actual rainfall being 102.5 mm given by IMD. As Karnataka's coastal district, the western portion receives a lot of rainfall (between 602 and 804 mm), which raises the soil's moisture content because water builds up along the slopes, creating a landslide danger zone.

The Peninsular Gneissic Complex and Bababudan (Western Ghats) are the two main rock types that influence landslides in this area. These kinds of rocks are extremely vulnerable to weathering and cracks, which can result in weak spots. Low and moderately dissected hills, valleys, and the geomorphology of the area where the river is located are all important factors in landslides. In addition to increased slope gradients, rivers that are exposed to erosion and water pressure are particularly vulnerable to landslides. The correlation analysis revealed a negative relationship between LULC and NDVI, indicating that the danger of landslides decreases with increasing plant cover. Landslides are largely caused by vegetation

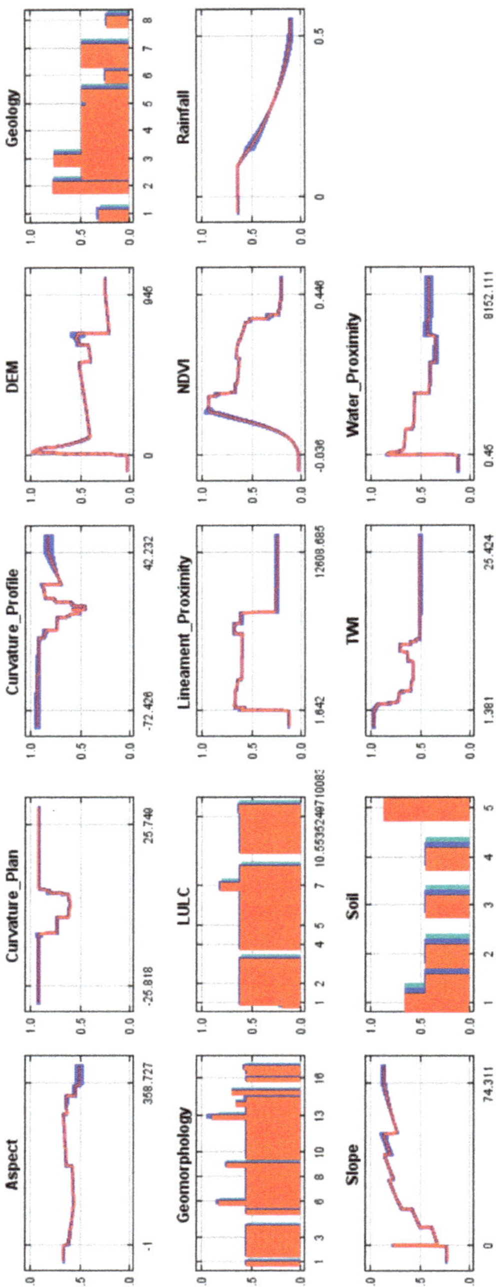

Fig. 6.6 Response curves of individual variables for landslide susceptibility mapping

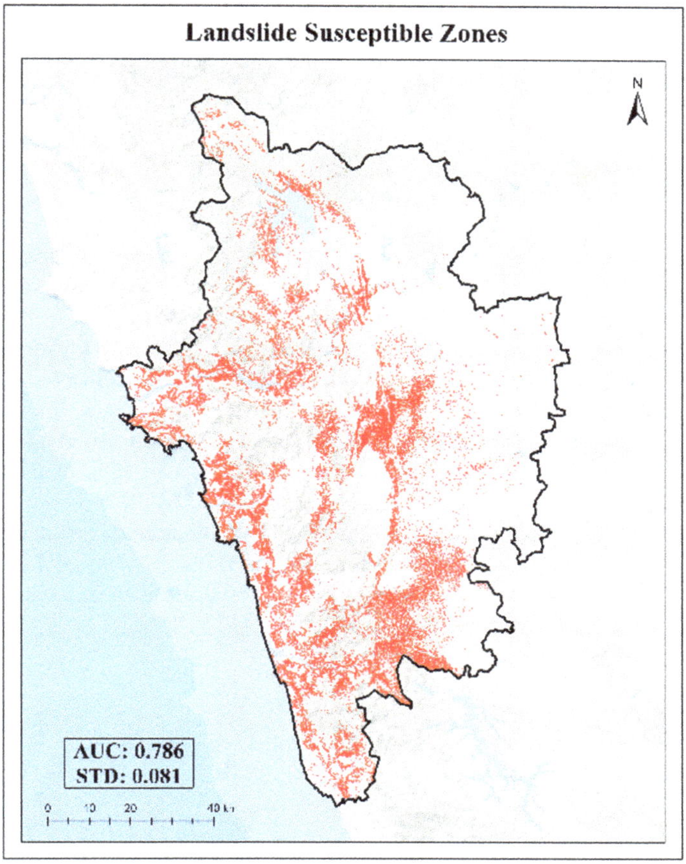

Fig. 6.7 Landslide susceptibility zones

on steeper slopes, flooded vegetation, built-up areas, and agriculture. Additionally, it has been noted that areas with an NDVI of 0–2 to 0.3 are particularly vulnerable to landslides. This demonstrates the instability of the soil, making it extremely vulnerable to erosion during periods of intense precipitation. Water and lineament proximity are inversely correlated; that is, the closer the lineament, the more likely the area is to experience landslides. These fissures and faults make the slope unstable, increasing the risk of landslides by allowing water to seep in and raising the water pressure across the slopes. The risk of landslides increases as the TWI increases. Water-accumulating zones are particularly prone to landslides because of the higher soil moisture content along the slopes, which makes the soil saturated and unstable. Regions in high-risk zones fall between 7.90 and 26.87. The three soil types that are most likely to experience landslides are Ap, Ne, and Rd. Ap has a moderate texture and is 57% sandy. Ne has a 100% moderate texture and is 68% sandy, while Rd is 82% sandy. The region is characterized by a high prevalence of sandy soil textures, which are prone to erosion and become saturated during periods of heavy rainfall. Additionally, the silt content is 16%, 11%, and 7%, while the clayey content is 27%, 21%, and 11%, respectively.

6.5.3 Artificial Neural Networks

Looking into the various models above, ANN has proven the advantage of how each environmental condition variable acts on landslides in Uttar Kannada district. The model utilized 14 variable predictors corresponding with 909 location responses, the network architecture of 15-10-10 with 75% training data, and the ANN effectively captured the complex relationships between environmental variables and landslide occurrences (Fig. 6.8). The significance of this model for landslide hazard zonation mapping has been further demonstrated by its implementation, as demonstrated by an R^2 value of 0.998, which is expected in the validation and indicates a strong model fit. Maxent and MGWR's spatial analysis is enhanced by the model's high R^2 value, which indicates its resilience in forecasting high-risk areas. These results demonstrate the ANN model's potential to improve landslide risk reduction techniques in geospatial hazard assessments and establish its dependability.

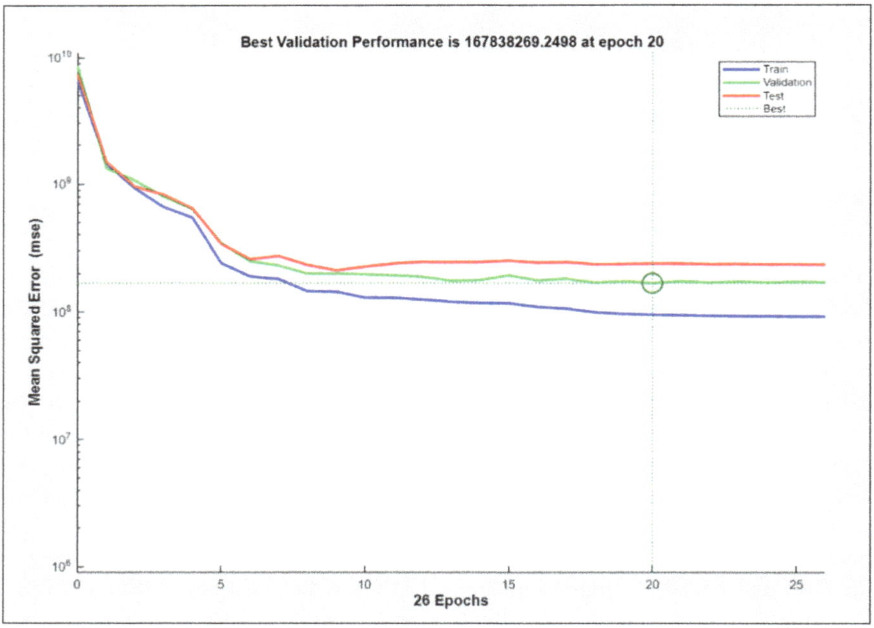

Fig. 6.8 Validation of variable performance

With a threshold value above 0.2394, this assessment indicates that we were able to identify roughly 1852.18 sq. km of high-risk zones using the Maxent model. The hilly area that stretches from the northern Kali Tiger Reserve and its extending ranges in sections of the western ghat to parts of Bhatka to the district's south is home to these zones. The district's southeast is a high-risk area, and landslides are likely to occur in the western portion, which is the coastal area and the riverine direction. These characteristics are in addition to the other elements that affect the areas that are susceptible to landslides.

6.5.4 Future Prediction Emerging Hotspot Analysis

Over the next 25 years, from 2025 to 2050, there is still a potential that landslides will occur because of several environmental causes, including climate change and an increase in manmade elements that act as triggers. There is a persistent risk of landslides in the eastern side of the Western Ghats region. Landslides can still occur in the western coastal region, and the river that connects the upper northern section to the Arabian Sea is in an averagely vulnerable area. Jog Falls, India's plunge waterfall that relies on monsoonal retreats, is largely caused by the significant risk of landslides along the river. This is once more the cause of landslides in this area, as the Uttar Kannada district experiences heavy rainfall throughout the monsoon season. These are the areas where the mitigation effect would have occurred, resulting in a reduction in the landslide risk zone, as seen in the instances of cold spots. Over time, it is determined that the risk zone would shrink along the western ghats' hilly region, the west coast, and the southeast, where landslides are expected to decline slightly (Fig. 6.9).

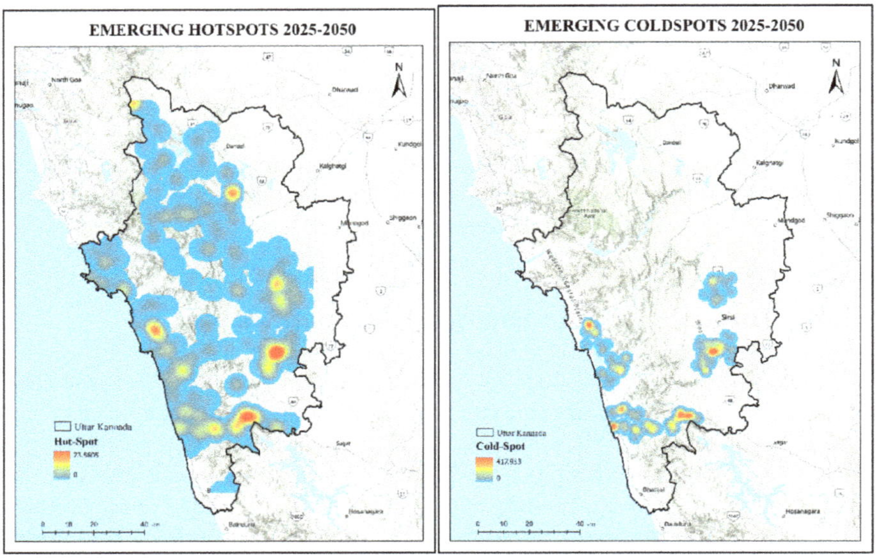

Fig. 6.9 Emerging hotspot and coldspot of landslide susceptibility in Uttar Kannada 2025–2050

6.6 Conclusion

The predictive ability of landslide susceptibility mapping depends on the scale and completeness of various landslide-conditioning maps (Mondal & Mandal, 2018). The MGWR outputs indicated that slope, rainfall, and the region's geomorphology were the most significant of the 14 environmental conditioning elements chosen for this study. The district continues to be at risk from highly localized landslides. In the monsoon season, landslides happen along the roadway and railroad tracks that cross the Ghats. Despite their low impact and great frequency, landslides have been discovered to be particularly common in recent years during the June–October monsoon season. Due to steep slopes and excessive rainfall, the hotspots are concentrated in the hilly, mountainous regions of the Western Ghats, as indicated by the susceptibility mapping, which shows almost 1852.18 sq. km of risk zones. Due to high rainfall that causes soil saturation and erosion in conjunction with the proximity to water, coastal zones are also categorized to be highly vulnerable. Landslides are less common in stable geomorphological areas, and over 25 years, there has been little to no landslide activity in the Northern Western Ghats, as observed in cold spots. Reforestation, improved land management plans, and other mitigating strategies may have been used to cause this.

The state disaster management authority has implemented many monitoring and warning systems to mitigate landslides in high-risk areas. Risk analysis and decision-making under uncertainty are closely intertwined. Planners will use landslide susceptibility maps to aid in decision-making (Pradhan & Lee., 2009). The susceptibility map indicating the high-risk zone should be taken into consideration for mitigation strategies. These findings can be utilized as baseline information to support land-use planning and slope management, but the study's models are reliable for more general planning and evaluation needs. This approach of study plays a base for landslide detection in wide areas. This study's association with the state disaster management authority's current landslide hazard zonation will make it more applicable and provide a broader view of the hotspot mapping that will be created in the future. It gives decision-makers scientific sources to use when creating policies for environmental protection.

References

Anand, V., Oinam, B., & Singh, I. H. (2021). Predicting the current and future potential spatial distribution of endangered Rucervus elder elder (Sangai) using the MaxEnt model. *Environmental Monitoring and Assessment, 193*(3), 1.

Bragagnolo, L., da Silva, R. V., & Grzybowski, J. M. V. (2020). Landslide susceptibility mapping with r.landslide: A free open-source GIS-integrated tool based on artificial neural networks. *Environmental Modelling and Software, 1*(123), 104565.

Chen, W., Pourghasemi, H. R., Kornejady, A., & Zhang, N. (2017). Landslide spatial modeling: Introducing new ensembles of ANN, MaxEnt, and SVM machine learning techniques. *Geoderma, 1*(305), 314–327.

Choi, J., Oh, H. J., Lee, H. J., Lee, C., & Lee, S. (2012). Combining landslide susceptibility maps obtained from frequency ratio, logistic regression, and artificial neural network models using ASTER images and GIS. *Engineering Geology, 124*(1), 12–23.

CUDA, for machine learning and optimization (2011). *CUDA Application Design and Development*, pp. 33–61.

Dahal, U. (2022). *Assessing habitat overlap among spotted and sambar deer using geo-spatial approach: A case study of Chitwan National Park, Nepal* (Doctoral dissertation, University of Salzburg, Austria).

Dehnavi, A., Aghdam, I. N., Pradhan, B., & Varzandeh, M. H. (2015). A new hybrid model using step-wise weight assessment ratio analysis (SWARA) technique and adaptive neuro-fuzzy inference system (ANFIS) for regional landslide hazard assessment in Iran. *Catena (Amst), 1*(135), 122–148.

Dou, J., Yunus, A. P., Bui, D. T., Merghadi, A., Sahana, M., Zhu, Z., et al. (2020). Improved landslide assessment using support vector machine with bagging, boosting, and stacking ensemble machine learning framework in a mountainous watershed Japan. *Landslides, 17*(3), 641–658.

Ferreira, Z., Costa, A. C., & Cabral, P. (2023). Analysing the spatial context of the altimetric error pattern of a digital elevation model using multiscale geographically weighted regression. *European Journal of Remote Sensing, 56*(1), 1.

Forest Department, (2023). Landslide Atlas of India (Mapping, Monitoring and R&D studies using Remote Sensing data) (NRSC/National Remote Sensing Centre/Indian Space Research Organisation [Internet]. Available from: https://www.nrsc.gov.in

He, K., Fan, C., Zhong, M., Cao, F., Wang, G., & Cao, L. (2023). Evaluation of habitat suitability for Asian elephants in Sipsongpanna under climate change by coupling multi-source remote sensing products with MaxEnt model. *Remote Sens (Basel), 15*(4), 1.

Hengl, T., Gruber, S., Shrestha, D.P. (2003). Digital terrain analysis in ILWIS 0 0 0 0 0 0 0 1 1 1 1 1 1 [Internet]. 2003. Available from: http://www.itc.nl/personal/shrestha/DTA/

Kalantar, B., Pradhan, B., Amir Naghibi, S., Motevalli, A., & Mansor, S. (2018). Assessment of the effects of training data selection on the landslide susceptibility mapping: a comparison between support vector machine (SVM), logistic regression (LR) and artificial neural networks (ANN). *Geomatics Natural Hazards and Risk, 9*(1), 49–69.

Karnataka State Action Plan, (2022). Management of Landslides-2022 Karnataka State Disaster Management Authority Revenue Department (Disaster Management) Government of Karnataka.

Kavzoglu, T., Kutlug Sahin, E., & Colkesen, I. (2015). An assessment of multivariate and bivariate approaches in landslide susceptibility mapping: a case study of Duzkoy district. *Natural Hazards, 76*(1), 471–496.

Li, Y., Huang, S., Li, J., Huang, J., & Wang, W. (2022). Spatial non-stationarity-based landslide susceptibility assessment using PCAMGWR model. *Water (Switzerland), 14*(6), 1.

Mallick, J., Alqadhi, S., Talukdar, S., Alsubih, M., Ahmed, M., Khan, R. A., et al. (2021). Risk assessment of resources exposed to rainfall-induced landslide with the development of GIS and RS based ensemble metaheuristic machine learning algorithms. *Sustainability (Switzerland), 13*(2), 1–30.

Mondal, S., & Mandal, S. (2018). RS & GIS-based landslide susceptibility mapping of the Balason River basin, Darjeeling Himalaya, using logistic regression (LR) model. *Georisk, 12*(1), 29–44.

Nhu, V. H., Shirzadi, A., Shahabi, H., Chen, W., Clague, J. J., Geertsema, M., et al. (2020). Shallow landslide susceptibility mapping by Random Forest base classifier and its ensembles in a Semi-Arid region of Iran. *Forests, 11*(4), 1.

Pham, Q. B., Chandra Pal, S., Chakrabortty, R., Saha, A., Janizadeh, S., Ahmadi, K., et al. (2022). Predicting landslide susceptibility based on decision tree machine learning models under climate and land use changes. *Geocarto International, 37*(25), 7881–7907.

Phillips, S. B., Aneja, V. P., Kang, D., & Arya, S. P. (2006) Modeling and analysis of the atmospheric nitrogen deposition in North Carolina. In *International Journal of Global Environmental Issues*. Inderscience Publishers, p. 231–52.

Prabu Ramakrishnan, S. S. (2009). Combined use of socio-economic analysis, remote sensing, and GIS data for landslide hazard mapping using ANN. *Journal of the Indian Society of Remote Sensing, 37*(3), 409–421.

Pradhan, B., & Lee, S. (2009). Landslide risk analysis using artificial neural network model focussing on different training sites [Internet]. International Journal of Physical Sciences. *4* Available from: http://www.academicjournals.org/IJPS

Ramachandra, T. V., Bharath, S., Subash Chandran, M. D., & Joshi, N. V. (2018). Salient ecological sensitive regions of central Western Ghats India. *Earth Systems and Environment, 2*(1), 15–34.

Remya, K., Ramachandran, A., & Jayakumar, S. (2015). Predicting the current and future suitable habitat distribution of Myristica dactyloides Gaertn. Using the MaxEnt model in the Eastern Ghats India. *Ecological Engineering, 82*, 184–188.

Uttara Kannada District Disaster Management Plan 2022–23.

Zhu, X., Li, P., Wang, B., Zhao, S., Zhang, T., & Yao, Q. (2024a). Construction of landscape eco-geological risk assessment framework in coal mining area using multi-source remote sensing data. *Ecological Informatics, 1*, 81.

Zhu, Y., Liu, B., Jin, G., Wu, Z., & Wang, D. (2024b). Identifying the local influencing factors of arsenic concentration in suburban soil: A multiscale geographically weighted regression approach. *Toxics., 12*(3), 1.

Part III

Roots of Various Climate and Environment Related Causes of Landslides

Climate Change and Its Hidden Impact on Landslide Hazard and Risk: A Multi-decadal Analysis Using Google Earth Engine

7

M. Nithya and S. Bala Subramaniyam

Abstract

Climate change poses a significant threat to environmental stability, intensifying natural hazards like landslides. Despite increasing awareness, the complex relationship between climate change and landslide occurrences remains poorly understood, particularly over extended timescales. This study aims to bridge this gap through a multi-decadal analysis of landslide hazards in the context of climate change, leveraging Google Earth Engine (GEE) for remote sensing and geospatial analysis. The primary objective of the study is to assess the impacts of climate change on landslide activity by analyzing Landsat and Sentinel satellite imagery alongside climate reanalysis datasets available on GEE. Advanced time-series techniques are applied to identify trends in landslide occurrences and examine the influence of shifting precipitation, temperature, and other climatic variables. The research focuses on both mountainous and non-mountainous regions, highlighting regional differences in landslide sensitivity to climate-related impacts. These insights can inform adaptive policy and land-use planning, enhancing environmental hazard assessment by clarifying the links between climate change and landslide risks. Ultimately, this work supports proactive risk management strategies in an increasingly climate-challenged world.

Keywords

Climate change • Landslide hazard • Geospatial analysis • Google Earth Engine (GEE) • Temporal analysis • Risk management

M. Nithya
Kerala State Remote Sensing and Environment Centre, Thiruvananthapuram, India

S. B. Subramaniyam (✉)
CartoXR Research Center and Consultancy LLP, Chennai, India
e-mail: subramaniyam13@gmail.com

R. M. Yuvaraj et al. (eds.), *Advanced Landslide Susceptibility Modeling*,
https://doi.org/10.1007/978-3-032-14485-0_7

7.1 Introduction

Climate change, through its influence on temperature, precipitation, and extreme weather events, is intensifying natural hazards and altering the frequency and magnitude of environmental disruptions worldwide (IPCC, 2021; Maslin et al., 2004). Landslides are among the most destructive of these hazards, particularly affecting regions with complex topography and high rainfall variability (Bolt et al., 1975; Duncan & Wright, 2005). Wayanad, a district in the Western Ghats of India, exemplifies such a region (Zachariah et al., 2024; World Weather Attribution, 2024). Known for its steep terrain, rich biodiversity, and significant agricultural activity, Wayanad is increasingly vulnerable to landslide hazards as climate dynamics in the region continue to evolve. Understanding the interplay between climate change and landslide occurrences in Wayanad is crucial for developing effective risk management strategies that can mitigate potential damage to human lives, infrastructure, and ecosystems.

Wayanad's unique geographical and climatic characteristics make it a compelling case study for assessing the impacts of climate change on landslide occurrences (Gadgil, 2011; Zachariah et al., 2024). Seasonal monsoons deliver intense rainfall, saturating the soil, and triggering frequent landslides, particularly in areas with anthropogenic disturbances such as deforestation, road construction, and unsustainable land use practices (Cannon et al., 2001; Maslin et al., 2004; Zachariah et al., 2024). Despite these recurrent events, the connection between climate change variables—such as shifting rainfall patterns, rising temperatures, and extreme weather—and landslide frequency remains underexplored, especially over extended temporal scales. This knowledge gap hampers the formulation of targeted interventions and adaptive planning measures that could help mitigate the impacts of climate-driven landslide risks in Wayanad and similar regions.

This study aims to address this gap by employing a multi-decadal, geospatial analysis of landslide hazards in Wayanad using Google Earth Engine (GEE). By leveraging satellite imagery from Landsat and Sentinel missions, as well as climate reanalysis datasets, this study will assess how changes in key climatic variables—precipitation, temperature, and soil moisture—correlate with landslide occurrences over time. Advanced time-series techniques are applied to detect patterns and trends, allowing for a nuanced understanding of how climate-induced changes may influence landslide activity across different types of terrain within Wayanad (Gorelick et al., 2017; Maslin et al., 2004).

The findings of this study hold practical implications for policymakers, land-use planners, and environmental agencies (World Weather Attribution, 2024). By clarifying the links between climate change and landslide risks, this research will support proactive, data-driven risk management and contribute to the broader understanding of climate impacts on environmental hazards in regions like Wayanad.

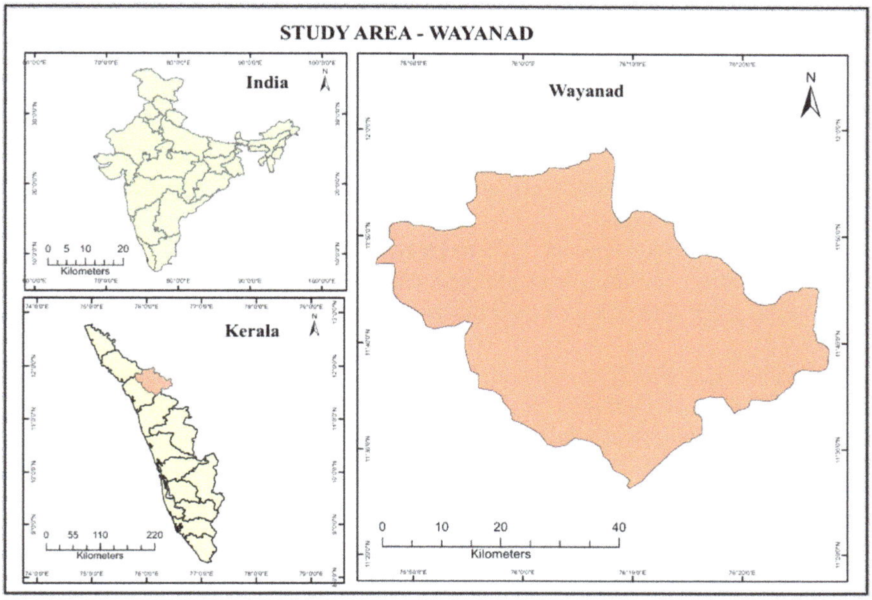

Fig. 7.1 Location of the study area

7.2 Study area

The study focuses on Wayanad, a district in Kerala, India, located within the Western Ghats. Known for its steep terrain and high rainfall, Wayanad is highly susceptible to landslides. The district spans both mountainous and non-mountainous areas, enabling the study of diverse land cover types and climatic conditions within a relatively confined area (Fig. 7.1).

7.3 Data

This study utilizes satellite imagery from Landsat 8 Operational Land Imager (OLI) for analyzing the impacts of climate change on landslide occurrences in Wayanad, India, during the period from 2020 to 2024. The selection of Landsat 8 was driven by its optimal spatial and temporal resolution, as well as its global availability, making it a valuable dataset for environmental and hazard studies. The specific datasets and their characteristics are detailed below:

Landsat 8 Operational Land Imager (OLI)
Landsat 8 OLI is the primary data source used in this research for vegetation and land surface analysis. It provides high-resolution multispectral imagery that is essential for

calculating vegetation indices such as the Normalized Difference Vegetation Index (NDVI). The characteristics of the dataset are as follows:

- **Spatial Resolution**: 30 m for multispectral bands; 15 m for the panchromatic band.
- **Temporal Resolution**: 16-day revisit time.
- **Spectral Bands**: Bands 4 (Red) and 5 (Near Infrared) were specifically utilized for NDVI computation.
- **Acquisition Period**: January 2020 to December 2024.
- **Preprocessing**: The images were preprocessed to minimize cloud cover (< 20%) using the pixel_qa band for cloud masking, ensuring reliable data for analysis.

Climate data, such as precipitation and temperature, were sourced to understand the influence of climatic variables on landslide activity. These datasets were derived from reanalysis datasets available on Google Earth Engine:

1. Precipitation Data

- **Dataset**: CHIRPS (Climate Hazards Group InfraRed Precipitation with Station Data)
- **Temporal Resolution**: Daily
- **Spatial Resolution**: ~5 km
- **Data Range**: 2020–2024

2. Temperature Data

- **Dataset**: ECMWF ERA5 Daily Aggregates
- **Temporal Resolution**: Daily
- **Spatial Resolution**: ~30 km
- **Variable Used**: Mean 2 m air temperature

7.4 Methodology

This study aims to analyze the relationship between climate change and landslide occurrences in Wayanad, India, by utilizing satellite imagery and climate data available on Google Earth Engine (GEE). The methodology comprises a sequence of data acquisition, preprocessing, and analytical steps to identify spatial and temporal trends in vegetation health, precipitation, and temperature, which are critical indicators of landslide susceptibility.

7.4.1 Data Preprocessing

Cloud Masking
Landsat 8 imagery was filtered to exclude pixels affected by cloud cover using the pixel_qa band. Imagery with < 20% cloud cover was retained for analysis.

Temporal Filtering
All datasets were constrained to the period from January 2020 to December 2024. This filtering ensured temporal alignment across datasets.

Spatial Clipping
The datasets were clipped to the Wayanad boundary to focus on the region of interest, reducing computational overhead.

7.4.2 Analytical Workflow

- NDVI was computed for each Landsat 8 image using the formula: NDVI=(NIR−RED)/(NIR+RED)

  where NIR is Band 5 and RED is Band 4.

- Temporal aggregation was performed to compute monthly mean NDVI values.

Climate Trend Analysis

- **Precipitation**: Monthly total precipitation values were calculated from the CHIRPS dataset. A linear regression model was applied to identify temporal trends.
- **Temperature**: Monthly mean temperatures were derived from ERA5 data and analyzed for trends using a similar linear regression approach.

NDVI and Climate Interaction
Correlation analysis was performed between NDVI, precipitation, and temperature values to identify potential links between vegetation health and climatic variables.

Landslide Hotspot Identification
Landslide-prone zones were identified by overlaying high-risk NDVI trends (indicative of vegetation degradation) with areas showing increasing precipitation or temperature trends.

Temporal and Spatial Trend Analysis

- **Temporal Trends**: Linear regression was used to assess long-term changes in NDVI, precipitation, and temperature.
- **Spatial Trends**: Spatial mapping of NDVI, precipitation, and temperature trends was conducted to visualize hotspot regions.

Output and Visualization

- Time-series plots showing NDVI, precipitation, and temperature trends.
- Spatial maps of NDVI trends, climate trends, and landslide hotspots.

7.5 Result and Discussion

The study employed geospatial analysis on Google Earth Engine (GEE) to examine the long-term impacts of climate change on landslide hazards. By integrating satellite imagery and climate reanalysis datasets, the following key results were obtained:

7.5.1 Evapotranspiration Analysis

The evapotranspiration trends derived from GEE datasets indicate a consistent increase in areas experiencing high vegetation stress, particularly in regions with increasing temperatures. These trends suggest that climatic variables like rising temperatures are contributing to altered soil moisture levels, exacerbating landslide susceptibility (Fig. 7.2).

7.5.2 NDVI Trends

NDVI analysis revealed significant changes in vegetation health across the study area. In regions with a history of landslides, decreasing NDVI values were observed, correlating with deforestation and vegetation degradation. Such vegetation loss is a critical factor in destabilizing slopes, further elevating landslide risks (Fig. 7.3).

7.5.3 Temperature Variation

Temporal analysis of temperature data shows an upward trend, especially in mountainous regions. This warming contributes to the thawing of permafrost,

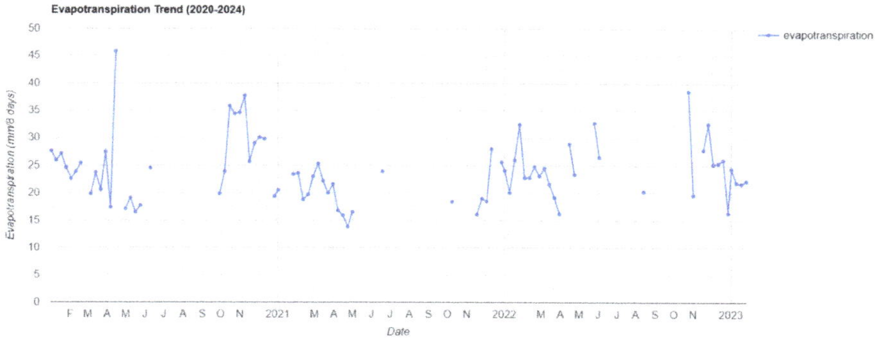

Fig. 7.2 Evapotranspiration trends visualized spatially, illustrating hotspots of increased soil moisture stress

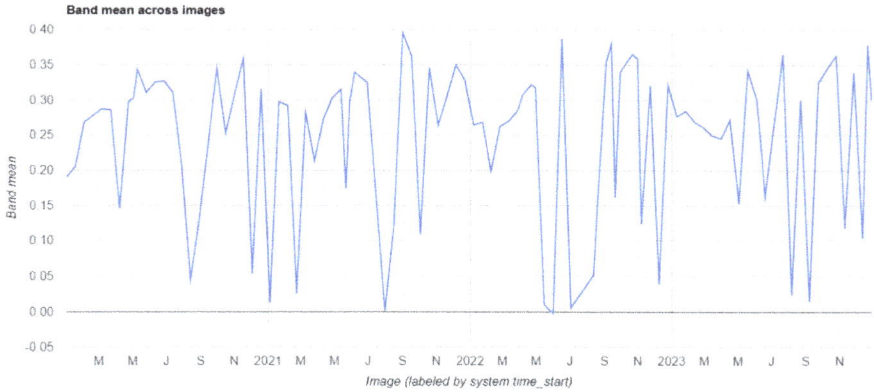

Fig. 7.3 NDVI changes indicating vegetation health and its role in slope stability

altering hydrological cycles, and destabilizing slopes. The analysis also highlighted seasonal temperature variations, intensifying during monsoon periods, which exacerbate precipitation-triggered landslides (Fig. 7.4).

7.5.4 Regional Differences in Landslide Sensitivity

The study identified regional disparities in landslide sensitivity to climatic changes. Mountainous regions exhibited higher sensitivity to both temperature and precipitation changes compared to non-mountainous regions. This highlights the need for region-specific risk management strategies.

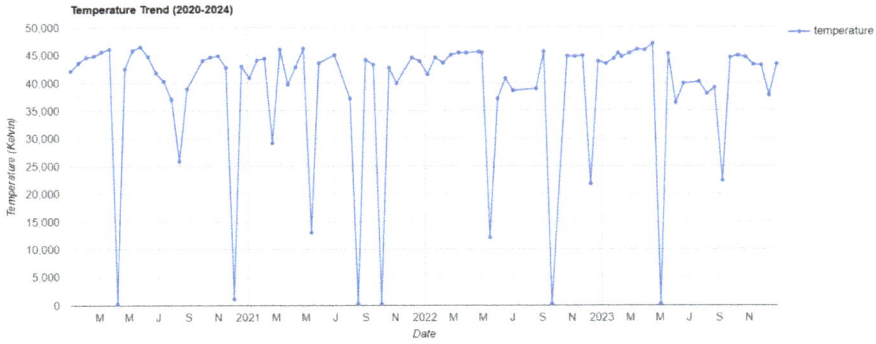

Fig. 7.4 Temperature trends over decades, with emphasis on their impact on regional hydrology and landslide triggers

7.5.5 Correlation Between Climate Variables and Landslide Activity

Advanced time-series analysis revealed strong correlations between precipitation anomalies and landslide occurrences, with peaks in landslide activity following periods of extreme rainfall. Temperature increases also showed indirect impacts by influencing evapotranspiration and vegetation cover. These findings underline the complex interplay between climate variables and landslide hazards. By providing a multi-decadal perspective, this study offers critical insights into the underlying mechanisms driving landslide risks in a changing climate.

7.6 Limitation

Data Availability and Resolution
The study relies on satellite datasets (Landsat and Sentinel) with moderate spatial and temporal resolutions. This may limit the accuracy of landslide detection in areas with small-scale or localized landslide occurrences.

Uncertainty in Climate Models
Climate reanalysis datasets, while robust, introduce some level of uncertainty due to model assumptions and interpolations. This could affect the precision of correlations between climatic variables and landslide activity.

Lack of Field Validation
The analysis was conducted entirely through remote sensing and geospatial methods, without direct field validation. This might lead to misclassification or overgeneralization of landslide occurrences in certain areas.

Static Land Use Representation
Land use and land cover dynamics were considered using NDVI trends; however, other anthropogenic factors (e.g., urbanization, infrastructure development) that influence landslide susceptibility were not extensively analyzed.

Focus on Climate Variables
The study primarily focuses on climatic triggers like precipitation and temperature, while geotechnical factors (e.g., soil properties, slope gradients) were not incorporated, limiting a holistic understanding of landslide processes.

7.7 Conclusion

This study highlights the critical role of climate change in influencing landslide hazards over multi-decadal timescales. Through geospatial analysis on Google Earth Engine, the results demonstrate that increasing temperatures, altered precipitation patterns, and vegetation degradation are key drivers of landslide activity.

- Mountainous regions are significantly more sensitive to climatic changes, with intensified landslide occurrences during monsoon seasons.
- The study underscores the importance of integrating climate-sensitive models into hazard assessment frameworks, enabling proactive risk management and land use planning.

The findings contribute to the growing body of knowledge on climate-landslide dynamics, offering valuable insights for policymakers and planners to mitigate disaster risks in vulnerable regions.

7.8 Future Scope

When comparing all landslip areas to highly or very highly vulnerable areas, most LSM techniques result in astronomically large false positive rates. Since this issue extends beyond the scope of current study, it would have to catch the interest of the community of physical geography, especially methodological development specialists, to the need to investigate strategies for minimizing the issue of overestimated susceptibility in subsequent research by incorporating cutting-edge developments like artificial intelligence.

References

Almagor, G., & Wiseman, G. (1982). Submarine slumping and mass movements on the slope of Israel. In S. Saxov & J. K. Nieuwenhuis (Eds.), *Marine Slides and Other Mass Movements* (pp. 95–128). Plenum Press.

Andresen, A., & Bjerrum, L. (1967). Slides in subaqueous slopes in loose sand and silt. In A. F. Richards (Ed.), *Marine Geotechnique* (pp. 221–239). University of Illinois Press.

Baltzer, A. (1875). Uber einen neuerlichen felssturz am Rossberg, negst einigen allgemeinen Bermerkungen uber derartige Erschei nungen in den Alpen (On a recent rockfall on Rossberg with a few observations on these phenomena in the Alps). *Neues Jahr buch fur Mineralogie, Geologie und Palaeontologie*, 914–924.

Barton, R., Bird, K., Hernández, J. G., Grajales-Nishimura, J. M., Murillo-Mun̆etón, G., Herber, B., Weimer, P., Koeberl, C., Neu maier, M., Schenk, O., & Stark, J. (2009/2010). High impact reservoirs. *Oilfield Review, 21*(4), 14–29.

Bates, C. C. (1953). Rational theory of delta formation. *AAPG Bulletin, 37*, 2119–2162.

Bea, R. G., Wright, S. G., Sicar, P., & Niedoroda, A. W. (1983). Wave induced slides in South Pass Block 70, Mississippi delta. *Journal of Geotechnical Engineering, 109*, 619–644.

Bolt, B. A., Horn, W. L., Macdonald, G. A., & Scott, R. F. (1975). *Geo logical Hazards* (p. 328). Springer-Verlag.

Bornhold, B. D., Fine, I. V., Rabinovich, A. B., Thomson, R. E., & Kulikov, E. A. (2003). The Grand Banks landslide-generated tsunami of November 18, 1929: Analysis and numerical modeling. *Geophysical Research Abstracts, 5*, 01775.

Bourgeois, J., Hansen, T. A., Wiberg, P. L., & Kauffman, E. G. (1988). A tsunami deposit at the Cretaceous-Tertiary boundary in Texas. *Science, 241*, 567–570.

Camerlenghi, A., Urgeles, R., & Fantoni, I. (2010). A Database on Sub marine Landslides of the Mediterranean. In D. C. Mosher et al. (Eds.), *Submarine Mass Movements and Their Consequences, Advances in Natural and Technological Hazards Research* (Vol. 28, pp. 503–513).

Cannon, S. H., Kirkham, R. M., & Parise, M. (2001). Wildfire-related debris-flow initiation processes, Storm King Mountain, Colo rado. *Geomorphology, 39*, 171–188.

Carlson, P. R., & Molnia, B. F. (1977). Submarine faults and slides on the continental shelf, northern Gulf of Alaska. In A. F. Richards (Ed.), *Marine Geotechnology, Marine Slope Stability 2* (pp. 275–290). Crane, Russak & Company.

Carter, R. M. (1975). A discussion and classification of subaqueous mass-transport with particular application to grain-flow, slurry low, and fluxoturbidites. *Earth-Science Reviews, 11*, 145–177.

Clarke, S., Hubble, T., Airey, D., Yu, P., Boyd, R., J Keene, J., Exon, N., & Gardner, J. (2012). Submarine Landslides on the Upper South east Australian Passive Continental Margin—Preliminary Find ings. In Y. Yamuda et al. (Eds.), *Submarine Mass Movements and Their Consequences·Advances in Natural and Technological Hazards Research* (Vol. 31, pp. 55–66).

Dengler, A. T., & Wilde, P. (1987). Turbidity currents on steep slopes: Application of an avalanche-type numeric model for ocean ther mal energy conversion design. *Ocean Engineering, 14*, 409–433.

Dingle, R. V. (1977). The anatomy of a large submarine slump on a sheared continental margin (SE Africa). *Geological Society of London Journal, 134*, 293–310.

Duncan, J. M., & Wright, S. G. (2005). *Soil Strength and Slope Stability* (p. 297). Wiley.

Eckel, E. B. (1958). Introduction. In E. B. Eckel (Ed.), *Landslide and Engineering Practice* (Vol. 29, pp. 1–5). National Research Council, Highway Research Board Special Report.

Elverhøi, A., Norem, H., Anderson, E. S., Dowdeswell, J. A., Fossen, I., Haflidason, H., Kenyon, N. H., Laberg, J. S., King, E. L., Sejrup, H. P., Solheim, A., & Vorren, T. (1997). On the origin and flow behavior of submarine slides on deep-sea fans along the Norwegian-Barents Sea continental margin. *Geo-Marine Letters, 17*, 119–125.

Feeley, K. (2007). *Triggering Mechanisms of Submarine Landslides* (p. 45). Boston, MA, Department of Civil and Environ mental Engineering, Northeastern University.

Festa, A., Dilek, Y., Gawlick, H.-J., & Missoni, S. (2014). Mass-transport deposits, olistostromes and soft-sediment deformation in modern and ancient continental margins, and associated natural hazards. *Marine Geology, 356*, 1–4.

Fisher, R. V. (1995). Decoupling of pyroclastic currents: Hazards as sessments. *Journal of Volcanology and Geothermal Research, 66*, 257–263.

Gaudin, M., Berné, S., Jouanneau, J.-M., Palanques, A., Puig, P., Mulder, T., Cirac, P., Rabineau, M., & Imbert, P. (2006). Massive sand beds deposited by dense water cascading in the Bourcart canyon head, Gulf of Lions (northwestern Mediterranean Sea). *Marine Geology, 234*, 111–128.

Hayter, E. J., Mathew, R., Hallden, J., Garland, E., Salerno, H., & Svirsky, S. C (2006). Evaluation of the State-of-the-art contaminated sediment transport and fate modeling system. EPA/600/R-06/108 September 2006. National Exposure Research Laboratory, Office of Research and Development, U.S. Environmental Protection Agency

Houghton, P., Davis, C., McCaffrey, W., & Barker, S. (2009). Hybrid sediment gravity flow deposits—Classification, origin and significance. *Marine and Petroleum Geology, 26*, 1900–1918.

Hsü, K. J. (1974). Mélanges and their distinction from olistostromes. In J. R. H. Dott & R. H. Shaver (Eds.), *Modern and Ancient Geo synclinal Sedimentation* (Vol. 19, pp. 321–333). SEPM Special Publication.

Jansen, E., Befring, S., Bugge, T., Eidvin, T., Holtedahl, H., & Petter Sejrup, H. (1987). Large submarine slides on the Norwegian con tinental margin: Sediments, transport and timing. *Marine Geology, 78*, 77–107.

Malahoff, A., Embley, R., & Perry, R. (1978). Submarine Landslides—East Coast Continental Slope and Upper Rise. In *OCEANS '78* (pp. 503–509). IEEE Explore Conference Publication, September 68, 1978. https://doi.org/10.1109/OCEANS.1978.1151087

Marr, J. G., Harff, P. A., Shanmugam, G., & Parker, G. (2001). Experi ments on subaqueous sandy gravity flows: The role of clay and water content in flow dynamics and depositional structures. *GSA Bulletin, 113*, 1377–1386.

Maslin, M., Owen, M., Day, S., & Long, D. (2004). Linking continental slope failures and climate change: Testing the gun hypothesis. *Geology, 32*, 53–56.

Morgenstern, N. R. (1967). Submarine slumping and the initiation of turbidity currents. In Richards, A. F. (Ed.), *Marine Geotechnique* (pp. 189–220). University of Illinois Press.

Piper, D. J. W., Pirmez, C., Manley, P. L., Long, D., Flood, R. D., Normark, W. R., Showers, W., 1997. Mass-transport deposits of the Amazon Fan. In: Flood, R. D., Piper, D. J. W., Klaus, A., & Peterson, L. C., (eds). Proceedings of the Ocean Drilling Program, Scientific Results, 155: 109-143.

Schuster, R. L. (1983). Engineering aspects of the 1980 Mount St. Helens eruptions. *Bulletin of the Association of Engineering Geologists, 20*, 125–143.

8 GIS-Based Landslide Susceptibility Mapping of Nandhaur River, Kumaun Himalaya, India

Manoj Kumar, Vandana Sharma, and R. K. Pande

Abstract

Geographical Information System (GIS) is an emerging tool to study morphometric analysis. Nandhaur River is one of the main tributary river in Ramganga river system of Uttarakhand, India. The study region is stretched between 29° 5′ 30″ and 29° 17′ 30″ North Latitude and 79° 40′ 30″ to 79° 57′ 15 East Longitude with the area of 272.5 km^2. This region comes under the Shivalik and Lesser Himalayan range. Here, the rate of landslides are high in which 90% of landslides are occur in Shivalik and only 10% of landslides are occur in Lesser Himalayan range. Most of the study region comes under the protected area and densely covered vegetation. This study would help to know more about the morphometric parameter and future possibility of landslide in the study region.

Keywords

Arc & QGIS • Kumaun Himalaya • Frequency ratio • Landslide susceptibility

8.1 Introduction

Landslides are a significant geohazard in the Himalayan region, particularly in the Nandhaur Watershed of Central Himalaya, Uttarakhand, India (Ghosh & Nandy, 2018; Kumar & Bhattacharya, 2022). This study aims to assess landslide susceptibility in the Nandhaur Watershed using a combination of remote sensing techniques, Geographic Information System (GIS) tools, and statistical modeling (Ghosh & Singh, 2019). The findings reveal critical factors influencing landslide

M. Kumar (✉) · V. Sharma · R. K. Pande
Department of Geography, Kumaun University, Nainital, Uttarakhand, India
e-mail: kashyapmanoj25@gmail.com

R. M. Yuvaraj et al. (eds.), *Advanced Landslide Susceptibility Modeling*,
https://doi.org/10.1007/978-3-032-14485-0_8

occurrences, contributing to better land use planning and risk management in this vulnerable area. The Central Himalayas are characterized by steep slopes, fragile geology, and heavy rainfall, making them highly susceptible to landslides. The Nandhaur Watershed, located in the state of Uttarakhand, India, experiences significant landslide activity due to its unique topographical and climatic conditions. This research aims to identify and analyze the factors contributing to landslide susceptibility in the Nandhaur Watershed. Landslides can lead to substantial socio-economic impacts, including loss of life, damage to infrastructure, and disruption of local economies. Understanding landslide susceptibility is crucial for disaster risk reduction and sustainable development in mountainous regions.

Landslide Classification—Landslides in the Nandhaur Watershed were classified based on various criteria:

- **Type of Landslide**: The inventory categorized landslides into several types, including:
  - **Rotational Slumps**: Common on steep slopes with unconsolidated materials.
  - **Translational Slides**: Typically occurring in areas with well-defined slip surfaces.
  - **Debris Flows**: Rapid flows of loose material, often triggered by heavy rainfall.
  - **Rock Falls**: Occur on steep cliffs, characterized by falling rock masses.
- **Size and Extent**: Each landslide was measured for its area and volume, providing insights into the scale of the events.

Spatial Distribution—The landslide inventory identified approximately 150 significant landslides within the Nandhaur Watershed. Key findings include:

- **Concentration Zones**: High-density landslide zones were identified on steep slopes, particularly along riverbanks and road cuttings. Areas with geological weaknesses, such as fault lines and fractured rocks, exhibited higher landslide occurrences.
- **Elevation Influence**: Most landslides were found at elevations ranging from 1200 to 2500 m, correlating with increased precipitation and steep terrain.

8.2 Study Area

The Nandhaur Watershed is situated in the Shivalik and Central Himalayas, primarily within the state of Uttarakhand, India (Fig. 8.1). Nestled between the elevations of approximately 350 to 2500 m above sea level, the watershed encompasses a region characterized by its stunning topography and rich biodiversity. The watershed features a rugged and undulating terrain with steep slopes, valleys, peaks, ridges, and deep gorges that define its geographical features. The Nandhaur Watershed experiences a temperate climate, heavily influenced by the southwest monsoon. Annual precipitation averages between 1500 to 2500 mm, with the majority occurring during the monsoon season (June to September). This heavy rainfall is a critical factor contributing to the region's landslide susceptibility. Temperatures vary with altitude, ranging from mild in the lower regions to cold in higher elevations. Summers can be warm, while winters bring significant snowfall in the upper reaches. The watershed is part of the larger Ganges river basin. The Nandhaur River, and its tributary flows through the watershed, contributing to its hydrological dynamics. The river and its tributaries are crucial for local agriculture and provide water for various ecosystems. The region exhibits a dendritic drainage pattern due to the steep slopes and geological formations, leading to rapid runoff during rainfall events. The lithology, structural geology, soil type, and texture of the Nandhaur Watershed play a crucial role in shaping the region's landscape, influencing agricultural practices, vegetation cover, and susceptibility to landslides. These geological characteristics is essential for effective land use planning and management, particularly in addressing the risks associated with natural hazards in this vulnerable Himalayan environment. The rocks include shale, sandstone, limestone, schist, gneiss, granite, and in some parts igneous are found. Soil types like-Alluvial Soils, Mountain Soils, Podzolic Soils, Terra Rosa, Clayey Soils, Sandy Soils, Loamy Soils, and Stony Soils are influenced by the underlying lithology, climate, and vegetation within the Nandhaur Watershed. The vegetation in the Nandhaur Watershed varies with altitude, slope, and climatic conditions. Major vegetation types include: Subtropical Broadleaf Forests(800–1500 m), Temperate Forests (1500–2500 m), Alpine Vegetation (Above 2500 m), Shrublands (Areas with disturbed land or steep slopes).

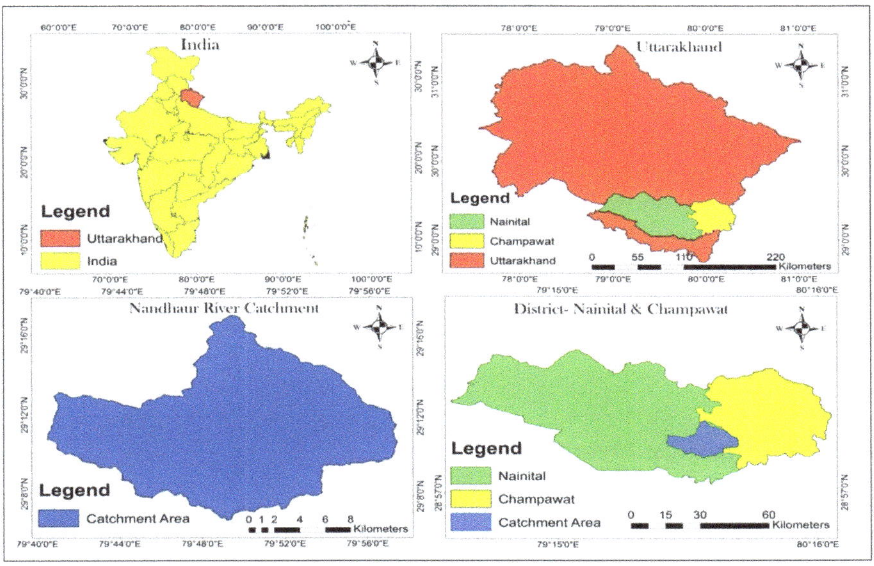

Fig. 8.1 Location of the study area

Watershed is home to a rich array of flora and fauna. It serves as a habitat for various endemic and endangered species, including: Himalayan black bear (Ursus thibetanus), leopards (Panthera pardus), numerous bird species medicinal plants and herbs, etc. The land use in the Nandhaur Watershed is influenced by both natural and anthropogenic factors. Major land use types include: Agricultural Land, Forestry, Grazing Lands, Settlements. But over the past few decades, the Nandhaur Watershed has experienced notable changes in land cover like: Deforestation, Urbanization, Afforestation and Reforestation, Climate Change Impacts.

The Nandhaur Watershed is home to various indigenous communities and small settlements. The population primarily engages in agriculture, forestry, and tourism. The traditional lifestyles and practices of local communities are closely intertwined with the natural environment, influencing land use and resource management.

8.3 Data and Methodology

Data is systematically gathered from multiple authoritative sources to ensure a comprehensive understanding of the land and its dynamics. Satellite imagery and topographical data were integral in facilitating our analysis of land cover changes with precision. The USGS provided Shuttle Radar Topography Mission (SRTM) data to obtained Digital Elevation Model (DEM), delivering accurate elevation data essential for our investigation. To establish the geological context, geological maps were used, sourced from the Geological Survey of India (GSI), affirming the significance of our geological findings. Rainfall data procured from local weather stations, which provided critical insights necessary for our research. Each of these carefully selected sources contributed invaluable information, collectively fortifying the robustness of our research work.

The Nandhaur watershed, of Kumaun Himalaya region, is a critical area for investigating landslide susceptibility. To analyze this phenomenon effectively, advanced mapping technologies such as ArcGIS and Quantum GIS are used. The study utilizes the frequency analysis model to evaluate landslide susceptibility more comprehensively with better visualization. In this analysis, 14 distinct factors have been identified and incorporated, allowing for a detailed examination of the variables contributing to landslide risk in the region. This multi-faceted approach aims to provide a clearer and more accurate understanding of the geological and environmental conditions that influence landslide occurrence in the Nandhaur watershed.

8.3.1 Landslide Inventory

The landslide inventory of the Nandhaur Watershed provides essential insights into the nature, causes, and impacts of landslides in the region. Understanding these dynamics is critical for developing effective land management strategies, disaster risk reduction plans, and sustainable development policies. Continued monitoring and research are necessary to enhance resilience against landslides and mitigate their impacts on local communities and ecosystems.

The Nandhaur Watershed in the Central Himalayas of Uttarakhand, India, is highly susceptible to landslides due to its steep terrain, geological conditions, and climatic factors. A comprehensive landslide inventory is crucial for understanding the patterns, causes, and potential impacts of landslides in this region. This note outlines the methodology for creating the landslide inventory, the characteristics of identified landslides, and their implications for land management.

8.3.2 Susceptibility Modeling (Frequency Ratio)

Landslide susceptibility modeling is critical for assessing the risk posed by landslides in mountainous regions like the Nandhaur catchment in Uttarakhand, India. The Frequency Ratio Analysis (FRA) offers a systematic and quantitative approach to evaluate multiple factors contributing to landslide occurrence (Yalcin & Duzgun, 2020; Saha & Ghosh, 2021) (Table 8.1). This note presents the results of applying FRA to identify and prioritize the factors influencing landslide susceptibility in the Nandhaur catchment.

FRA involves structuring the decision-making problem into a hierarchy. For landslide susceptibility modeling, the following criteria were selected based on literature review and field surveys:

8.3.3 Interpretation of the Factors

1. **Slope**: The slope is a fundamental aspect to consider when analyzing landslides in any given region. A steep or unstable slope can significantly increase the likelihood of a landslide event. To thoroughly understand this phenomenon, it is crucial to examine the relationship between slope and landslide occurrences (Wilson & Galland 2000; Saha & Basu, 2021). This relationship can be clearly depicted through slope map that illustrate variations in slope throughout the study area, allowing for a comprehensive analysis of how topography influences landslide risk (Fig. 8.2).
2. **Climate Zone**: Climate plays a significant role in the occurrence of landslides, as it directly influences the stability of the land. Understanding the relationship between climate and landslide susceptibility is crucial to understand different climactic zones for implementing effective prevention and mitigation strategies (Fig. 8.2).
3. **Curvature**: The term curvature is theoretically defined as the rate of change of slope gradient or aspect usually in a particular direction (Wilson & Gallant, 2000). Curvature is divided into three categories named Concave, Flat and Convex (Fig. 8.2).
4. **Land Use/Cover**: The impact of anthropogenic activities such as deforestation, agriculture practices which destabilize slope are shown with the help of

Table 8.1 Frequency ratio analysis

1	S.N.	Slope(in degree)	No of LSD (NLi)	LSD %	Area (No of Pixels) (NCi)	Area%	FR (Fri)	Normalization (FRn)	RF	RF (non%)	RF(INT)	Min RF	Max RF	Max–Min RF	(Max–Min) Min RF	PR
Slope	1	< 10	46	11.3	65,444	21.61	0.523	0.26	0.088	8.841	8					
	2	Oct-20	88	21.62	80,888	26.7	0.81	0.4	0.137	13.684	13					
	3	20–30	115	28.26	87,194	28.79	0.982	0.48	0.166	16.589	16					
	4	30–40	108	26.54	51,023	16.84	1.575	0.78	0.266	26.624	26					
	5	> 40	50	12.29	18,356	6.06	2.027	1	0.343	34.262	34					
				NLt = **407**		**302,905**										
Total							5.917		1.000			0.088	0.343	0.254	0.143	1.774
2	S.N.	Geology unit	No of LSD (NLi)	LSD %	Area (No of pixels) (NCi)	Area%	FR (Fri)	Normalization (FRn)	RF	RF (non%)	RF (INT)	Min RF	Max RF	Max–Min RF	(Max–Min) Min RF	PR
Geology	1	Grained quartzite with metavolcanics	18	4.42	4070	8.86	0.499	0.24	0.073	7.266	7					
	2	Limestone, dolomite and slate	99	24.32	18,833	40.99	0.593	0.28	0.086	8.636	8					

(continued)

Table 8.1 (continued)

1	S.N.	Slope(in degree)	No of LSD (NLi)	LSD %	Area (No of Pixels) (NCi)	Area%	FR (Fri)	Normalization (FRn)	RF	RF (non%)	RF(INT)	Min RF	Max RF	Max–Min RF	(Max–Min) Min RF	PR
	3	Porphyritic granite, schist, quartzite	18	4.42	4543	9.89	0.447	0.21	0.065	6.51	6					
	4	Sand, boulders, pebbles	146	35.87	9544	20.77	1.727	0.81	0.251	25.133	25					
	5	Sandstone and clay intercalation	97	23.83	7386	16.07	1.483	0.7	0.216	21.577	21					
	6	Sandstone with sandy clay	29	7.13	1543	3.36	2.122	1	0.309	30.878	30					
	7	Sandstone, siltstone, clay, and conglomerate	0	0	30	0.07	0	0	0	0	0					
				NLt = **407**		45,949										
Total							6.872		1.000			0.000	0.309	0.309	0.143	2.155

(continued)

Table 8.1 (continued)

1	S.N.	Slope(in degree)	No of LSD (NLi)	LSD %	Area (No of Pixels) (NCi)	Area%	FR (Fri)	Normalization (FRn)	RF	RF (non%)	RF(INT)	Min RF	Max RF	Max–Min RF	(Max–Min) Min RF	PR
3	S.N.	Elevation	No of LSD (NLi)	LSD %	Area (No of pixels) (NCi)	Area%	FR (Fri)	Normalization (FRn)	RF	RF (non%)	RF (INT)	Min RF	Max RF	Max–Min RF	(Max–Min) Min RF	PR
Elevation	1	381–673	70	17.2	34,118	11.27	1.526	0.96	0.227	22.673	22					
	2	673–817	111	27.27	51,910	17.14	1.591	1.00	0.236	23.631	23					
	3	817–955	107	26.29	59,191	19.55	1.345	0.85	0.2	19.977	19					
	4	955–1,102	47	11.55	54,692	18.06	0.639	0.4	0.095	9.497	9					
	5	1,102–1,266	33	8.11	41,120	13.58	0.597	0.38	0.089	8.869	8					
	6	1,266–1,450	33	8.11	29,715	9.81	0.826	0.52	0.123	12.273	12					
	7	1,450–1,664	6	1.47	21,525	7.11	0.207	0.13	0.031	3.08	3					
	8	1,664–2,127	0	0	10,505	3.47	0.000	0.00	0.000	0.000	0					
			NLt = **407**		**302,776**											
Total							6.732		1.000			0.000	0.236	0.236	0.143	1.649
4	S.N.	Rainfall	No of LSD (NLi)	LSD %	Area (No of pixels) (NCi)	Area%	FR (Fri)	Normalization (FRn)	RF	RF (non%)	RF (INT)	Min RF	Max RF	Max–Min RF	(Max–Min) Min RF	PR
Rainfall	1	1,304–1,343	105	25.8	401	16.38	1.575	1	0.239	23.85	23					

(continued)

Table 8.1 (continued)

1	S.N.	Slope(in degree)	No of LSD (NLi)	LSD %	Area (No of Pixels) (NCi)	Area%	FR (Fri)	Normalization (FRn)	RF	RF (non%)	RF(INT)	Min RF	Max RF	Max–Min RF	(Max–Min) Min RF	PR
	2	1,343–1,374	88	21.62	354	14.46	1.495	0.95	0.226	22.643	22					
	3	1,374–1,401	56	13.76	390	15.93	0.864	0.55	0.131	13.079	13					
	4	1,401–1,427	86	21.13	436	17.81	1.186	0.75	0.18	17.966	17					
	5	1,427–1,456	26	6.39	362	14.79	0.432	0.27	0.065	6.542	6					
	6	1,456–1,487	38	9.34	270	11.03	0.847	0.54	0.128	12.819	12					
	7	1,487–1,527	8	1.97	235	9.6	0.205	0.13	0.031	3.101	3					
				NLt = **407**		**2448**										
Total							6.603		1			0.031	0.239	0.207	0.143	1.448
5	S.N.	LULC	No of LSD (NLi)	LSD %	Area (No of Pixels) (NCi)	Area%	FR (Fri)	Normalization (FRn)	RF	RF (non%)	RF(INT)	Min RF	Max RF	Max–Min RF	(Max–Min) Min RF	PR
LULC	1	Agriculture, Fallow	1	0.25	2902	0.96	0.256	0.08	0.025	2.466	2					
	2	Agriculture, Cropland	33	8.11	23,968	7.91	1.025	0.31	0.099	9.851	9					
	3	Barren/ Unculturable/ Wasteland	18	4.42	6849	2.26	1.956	0.59	0.188	18.804	18					

(continued)

Table 8.1 (continued)

5	S.N.	LULC	No of LSD (NLi)	LSD %	Area (No of Pixels) (NCi)	Area%	FR (Fri)	Normalization (FRn)	RF	RF (non%)	RF(INT)	Min RF	Max RF	Max–Min RF	(Max–Min) Min RF	PR
	4	Forest, Evergreen/ Semi Evergreen	21	5.16	21,303	7.03	0.734	0.22	0.071	7.053	7					
	5	Forest, Deciduous	248	60.93	224,065	73.98	0.824	0.25	0.079	7.919	7					
	6	Forest, Scrub Forest	44	10.81	14,402	4.76	2.273	0.68	0.219	21.859	21					
	7	Water Bodies, River/ Lakes	42	10.32	9377	3.1	3.333	1.00	0.32	32.047	32					
		NLt =	**407**		**302,866**											
Total							10.400		1.000			0.025	0.320	0.296	0.143	2.065
6	S.N.	Curvature	No of LSD (NLi)	LSD %	Area (No of pixels) (NCi)	Area%	FR (Fri)	Normalization (FRn)	RF	RF (non%)	RF (INT)	Min RF	Max RF	Max–Min RF	(Max–Min) Min RF	PR
Curvature	1	Concave	108	26.54	52,545	17.35	1.53	1	0.476	47.582	47					
	2	Flat	220	54.05	166,764	55.05	0.982	0.64	0.305	30.54	30					
	3	Convex	79	19.41	83,596	27.6	0.703	0.46	0.219	21.877	21					
			NLt = **407**		**302,905**											

(continued)

Table 8.1 (continued)

5	S.N.	LULC	No of LSD (NLi)	LSD %	Area (No of Pixels) (NCi)	Area%	FR (Fri)	Normalization (FRn)	RF	RF (non%)	RF(INT)	Min RF	Max RF	Max–Min RF	(Max–Min) Min RF	PR
Total							3.215		1.000			0.219	0.476	0.257	0.143	1.794
7	S.N	Landforms	No of LSD (NLi)	LSD %	Area (No of pixels) (NCi)	Area%	FR (Fri)	Normalization (FRn)	RF	RF (non%)	RF (INT)	Min RF	Max RF	Max–Min RF	(Max–Min) Min RF	PR
Landforms	1	Summit	0	0.00	2488	0.83	0.000	0.00	0.000	0.000	0					
	2	Ridge	2	0.49	16,469	5.51	0.089	0.06	0.013	1.267	1					
	3	Shoulder	9	2.21	9813	3.28	0.674	0.43	0.096	9.565	9					
	4	Spur	38	9.34	50,963	17.04	0.548	0.35	0.078	7.776	7					
	5	Slope	172	42.26	116,289	38.88	1.087	0.70	0.154	15.425	15					
	6	Hollow	105	25.8	53,378	17.85	1.446	0.93	0.205	20.515	20					
	7	Foot slope	25	6.14	11,841	3.96	1.552	1	0.22	22.019	22					
	8	Valley	52	12.78	28,840	9.64	1.325	0.85	0.188	18.804	18					
	9	Depression	4	0.98	9014	3.01	0.326	0.21	0.046	4.628	4					
		NLt =	**407**		**299,095**											
Total							7.046		1			0	0.22	0.22	0.143	1.537
8	S.N.	Fault Distance	No of LSD (NLi)	LSD %	Area (No of pixels) (NCi)	Area%	FR (Fri)	Normalization (FRn)	RF	RF (non%)	RF(INT)	Min RF	Max RF	Max–Min RF	(Max–Min) Min RF	PR

(continued)

Table 8.1 (continued)

5	S.N.	LULC	No of LSD (NLi)	LSD %	Area (No of Pixels) (NCi)	Area%	FR (Fri)	Normalization (FRn)	RF	RF (non%)	RF(INT)	Min RF	Max RF	Max–Min RF	(Max–Min) Min RF	PR
Fault	1	500	229	56.27	156,615	26.2	2.148	1.00	0.517	51.657	51					
	2	1000	120	29.48	129,350	21.64	1.363	0.63	0.328	32.775	32					
	3	1500	43	10.57	150,984	25.26	0.418	0.19	0.101	10.062	10					
	4	2000	15	3.69	96,234	16.1	0.229	0.11	0.055	5.507	5					
	5	2500	0	0	32,048	5.36	0.000	0.00	0.000	0.000	0					
	6	3000	0	0	32,537	5.44	0.000	0.00	0.000	0.000	0					
			NLt = **407**		**597,768**											
Total							4.157		1.000			0.000	0.517	0.517	0.143	3.606

9	S.N.	Geomorphological Unit	No of LSD (NLi)	LSD %	Area (No of Pixels) (NCi)	Area%	FR (Fri)	Normalization (FRn)	RF	RF (non%)	RF(INT)	Min RF	Max RF	Max–Min RF	(Max–Min) Min RF	PR
Geomorphology	1	Mass Wasting Products	14	3.44	609	0.2	17.102	1	0.797	79.686	7					
	2	Active Flood plain	0	0	513	0.17	0	0	0.000	0.000	0					
	3	Older Alluvial Plain	0	0	206	0.07	0	0	0	0.000	0					
	4	Piedmont Alluvial Plain	9	2.21	9883	3.26	0.677	0.04	0.032	3.157	3					

(continued)

Table 8.1 (continued)

9	S.N.	Geomorphological Unit	No of LSD (NLi)	LSD %	Area (No of Pixels) (NCi)	Area%	FR (Fri)	Normalization (FRn)	RF	RF (non%)	RF(INT)	Min RF	Max RF	Max–Min RF	(Max–Min) Min RF	PR
	5	Younger Alluvial plain	0	0	1228	0.41	0	0.000	0	0	0					
	6	Highly Dissected Hills and Valleys	355	87.22	278,614	92.02	0.948	0.06	0.044	4.417	4					
	7	Moderately Dissected Hills and Valleys	1	0.25	3435	1.13	0.217	0.01	0.01	1.009	1					
	8	Water Body–Lakes	0	0	27	0.01	0.000	0	0.00	0.00	0					
	9	Water Body–River	28	6.88	8273	2.73	2.518	0.15	0.117	11.732	11					
			NLt = **407**		**302,788**											
Total							21.462		1.000			0.000	0.797	0.797	0.143	5.562
10	S.N.	Climate	No of LSD (NLi)	LSD %	Area (No of pixels) (NCi)	Area%	FR (Fri)	Normalization (FRn)	RF	RF (non%)	RF (INT)	Min RF	Max RF	Max–Min RF	(Max–Min) Min RF	PR
Climate	1	Sub—Tropical Climate	256	62.9	124,206	41.02	1.533	1	0.705	70.473	70					
	2	Warm—Temperate Climate	151	37.1	174,857	57.75	0.642	0.42	0.295	29.527	29					
	3	Cool—Climate	0	0	3742	1.24	0	0	0.000	0.000	0					

(continued)

Table 8.1 (continued)

9	S.N.	Geomorphological Unit	No of LSD (NLi)	LSD %	Area (No of Pixels) (NCi)	Area%	FR (Fri)	Normalization (FRn)	RF	RF (non%)	RF(INT)	Min RF	Max RF	Max–Min RF	(Max–Min) Min RF	PR
			NLt = **407**		**302,805**											
Total							2.176		1.000			0.000	0.705	0.705	0.143	4.919
11	S.N.	SPI	No of LSD (NLi)	LSD %	Area (No of pixels) (NCi)	Area%	FR (Fri)	Normalization (FRn)	RF	RF (non%)	RF (INT)	Min RF	Max RF	Max–Min RF	(Max–Min) Min RF	PR
SPI	1	< 0 Low	73	17.94	61,725	20.39	0.88	0.64	0.289	28.905	28					
	2	1–2 Moderate	151	37.1	141,813	46.84	0.792	0.58	0.26	26.024	26					
	3	> 2 High	183	44.96	99,238	32.78	1.372	1.00	0.451	45.07	45					
		NLt =	**407**		**302,776**											
Total							3.044		1.000			0.26	0.451	0.19	0.143	1.329
12	S.N	TWI	No of LSD (NLi)	LSD %	Area (No of pixels) (NCi)	Area%	FR (Fri)	Normalization (FRn)	RF	RF (non%)	RF (INT)	Min RF	Max RF	Max–Min RF	(Max–Min) Min RF	PR
TWI	1	< 5 Low	407	100	301,444	99.56	1.004	1	1	100	100					
	2	5–10 Moderate	0	0	1134	0.37	0.000	0	0	0	0					
	3	10 > High	0	0	198	0.07	0.000	0	0	0	0					
		NLt =	**407**		**302,776**											
Total							1.004		1.000			0.000	1.000	1.000	0.143	6.98

(continued)

Table 8.1 (continued)

13	S.N.	Earthquake (Magnitude)	No of LSD (NLi)	LSD %	Area (No of Pixels) (NCi)	Area%	FR (Fri)	Normalization (FRn)	RF	RF (non%)	RF(INT)	Min RF	Max RF	Max–Min RF	(Max–Min) Min RF	PR
Earthquake	1	4.4794–4.4893	5	1.23	2260	4.26	1.175	0.21	0.039	3.852	3					
	2	4.4893–4.4962	34	8.35	4173	7.86	4.327	0.78	0.142	14.185	14					
	3	4.4962–4.5011	34	8.35	5579	10.51	3.237	0.58	0.106	10.61	10					
	4	4.5011–4.5058	66	16.22	8105	15.26	4.325	0.78	0.142	14.177	14					
	5	4.5058–4.5098	58	14.25	8202	15.44	3.756	0.68	0.123	12.311	12					
	6	4.5098–4.5129	58	14.25	8968	16.89	3.435	0.62	0.113	11.26	11					
	7	4.5129–4.5155	74	18.18	8351	15.72	4.706	0.85	0.154	15.427	15					
	8	4.5155–4.5184	78	19.16	7470	14.07	5.545	1.00	0.182	18.179	18					
		NLt =	**407**		**53,108**											
Total							30.505		1.000			0.039	0.182	0.143	0.143	1.000
14	S.N	Soil Type	No of LSD (NLi)	LSD %	Area (No of pixels) (NCi)	Area%	FR (Fri)	Normalization (FRn)	RF	RF (non%)	RF (INT)	Min RF	Max RF	Max–Min RF	(Max–Min) Min RF	PR
Soil	1	Cambisols	88	21.62	9606	21.49	1.006	1	0.335	33.457	33					
	2	Fluvisols	0	0	65	0.15	0	0	0	0	0					

(continued)

Table 8.1 (continued)

13	S.N.	Earthquake (Magnitude)	No of LSD (NLi)	LSD %	Area (No of Pixels) (NCi)	Area%	FR (Fri)	Normalization (FRn)	RF	RF (non%)	RF(INT)	Min RF	Max RF	Max–Min RF	(Max–Min) Min RF	PR
	3	Leptosols	276	67.81	30,308	67.79	1	0.99	0.333	33.258	33					
	4	Luvisols	43	10.57	4718	10.55	1.001	0.99	0.333	33.285	33					
	5	Arenosols	0	0	9	0.02	0.000	0.00	0.000	0.000	0					
		NLt =	**407**		**44,706**											
Total							3.008		1.000			0.000	0.335	0.335	0.143	2.335

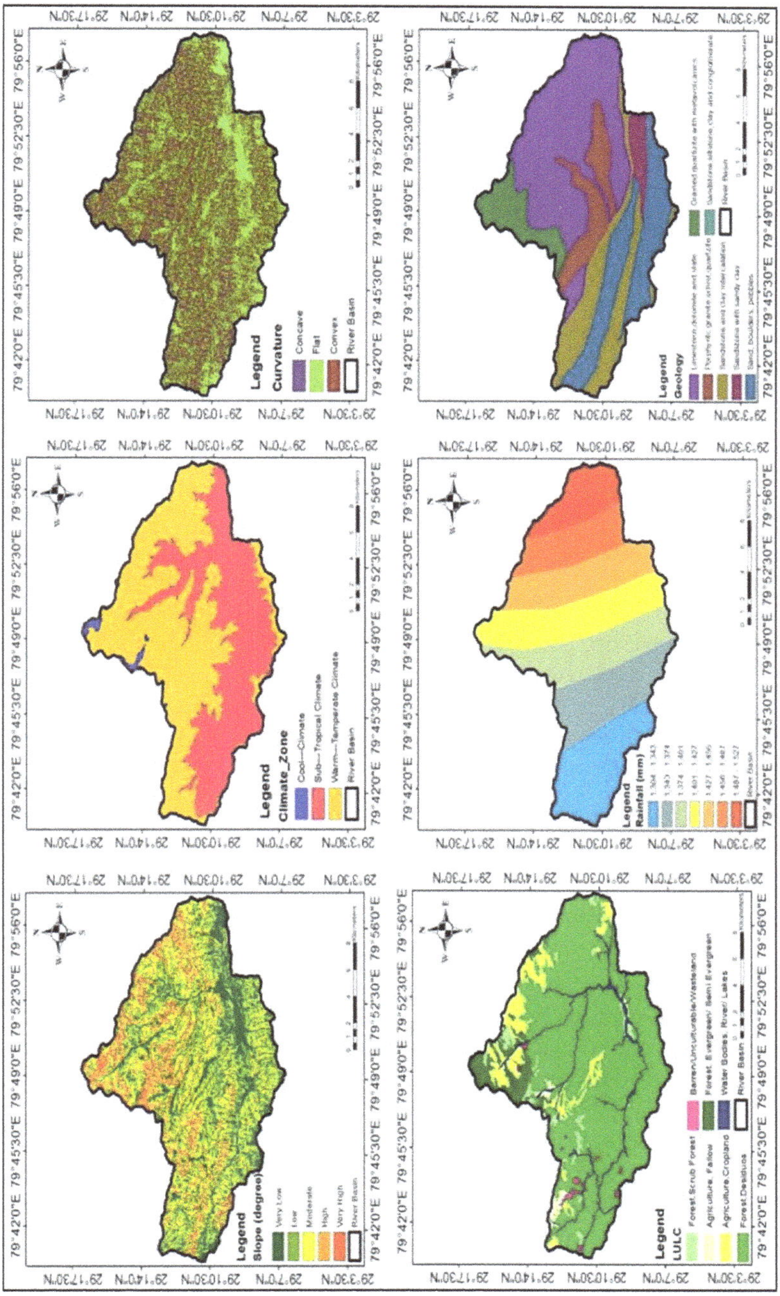

Fig. 8.2 Slope, climate zone, curvature, LULC, rainfall and geology

land use and land covered map except all these activities land use and land cover also shows the forest cover and the other uses of land (Fig. 8.2).

5. **Rainfall**: Areas that experience heavy rainfall or rapid snowmelt are particularly vulnerable to landslides, as the increased water content can weaken soil structure and trigger movement, especially during the monsoon season (Caine, 1980; Ranjan & Singh, 2019) (Fig. 8.2).
6. **Geology**: Geology of any region is directly associated with landslide. Geological conditions place is significant role in studying landslide. Different geological zones are depicted with the help of geological map (Fig. 8.2).
7. **Elevation**: Elevation is another important factor to consider landslide because it is directly controlled by several geological and morphological processes. Elevation is shown with the help of DEM data which is 30 m × 30 m resolution (Fig. 8.3).
8. **Landform**: Understanding landform features is crucial when studying landslide susceptibility. Elements such as slopes, valleys, depressions, ridges, and summits play a significant role in assessing landslide susceptibility. These distinct landforms gain insights into the conditions that may lead to landslides, thus enhancing the ability to predict and mitigate such geological hazards (Fig. 8.3).
9. **Geomorphology**: Geomorphological features such as alluvial plains, floodplains, water bodies, and hills are distinctly highlighted to provide a clearer understanding of the region. Through the lens of geomorphology, the landscape is effectively categorized, indicating that steep, high-altitude areas are particularly vulnerable to landslides, whereas the flatter regions are more prone to flooding (Fig. 8.3).
10. **Topographical Wetness Index**: TWI shows the topography on the location and size of saturated source areas of runoff generation. TWI is divided into three categories low (<5), moderate (5–10) and high (>10) (Fig. 8.3).
11. **Distance to Fault**: Distance from fault is an important parameter that controls the stability of a slope, which is calculated at 500 m interval. It shows the zone of weaknesses which is near the fault line (Fig. 8.3).
12. **Stream Power Index**: SPI shows the erosion power of water flow based on the assumption that discharge is proportional to specific catchment area. Divided into three categories—low(<0), moderate(1–2), and high(>2) (Fig. 8.3).
13. **Earthquake**: Earthquake is one of the important factor to study landslide. In contrast, the vicinity has experienced nearly eight significant earthquakes, warranting a comprehensive assessment of the seismic activity in this area (Fig. 8.4).
14. **Soil Type**: Different soil type helps to understand the possibility of landslide occurrence. Hard and stable rock type minimizes the possibility of landslide rather than lose unconsolidated soils are more prone to failure particularly when saturated (Fig. 8.4).

The results of the FA analysis provide valuable insights for landslide susceptibility modeling in the Nandhaur catchment. Additionally, the notable influence of

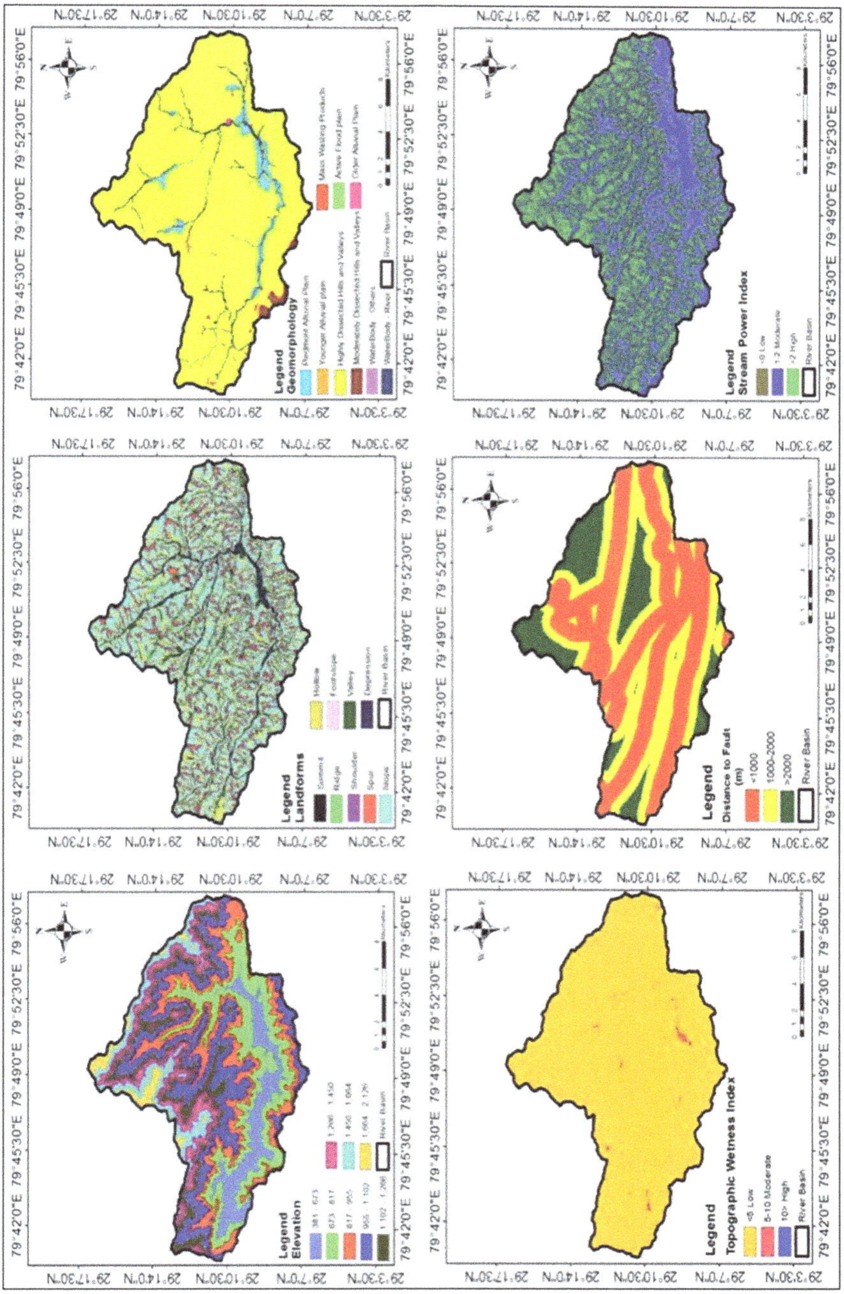

Fig. 8.3 Elevation, landforms, geomorphology, topographical wetness index, distance to fault and stream power index

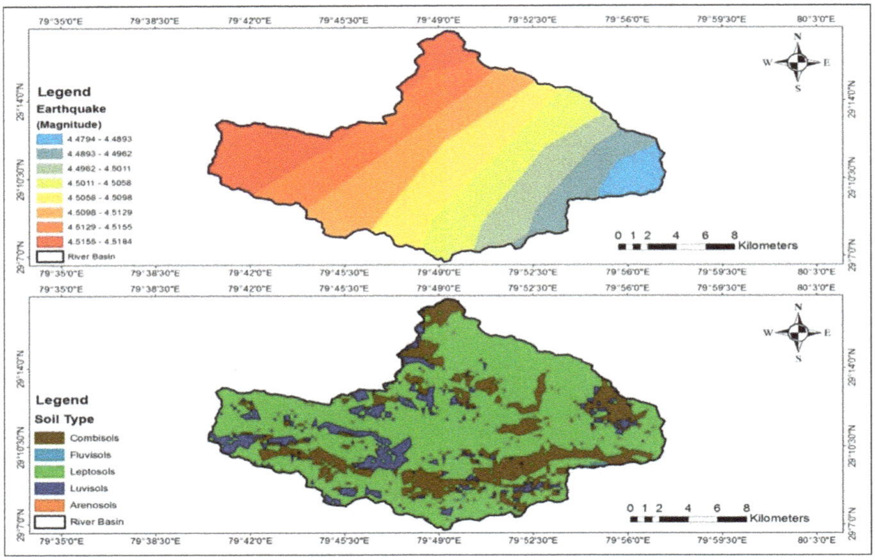

Fig. 8.4 Elevation and soil type

LULC, soil type, etc., highlights the importance of soil conservation practices, especially in agricultural zones. These findings can guide authorities in zoning regulations and land use policies to minimize landslide risks and Identifying high-susceptibility zones allows for targeted monitoring and the implementation of mitigation measures.

8.3.4 GIS Analysis

The integration of spatial analysis tools in GIS effectively produced a landslide susceptibility map for the Nandhaur Watershed, highlighting critical areas at risk. This map is instrumental for land management and disaster risk reduction efforts, providing a foundation for informed decision-making (Sharma Choudhury, 2017). Future research should focus on incorporating real-time data and refining the model to enhance predictive capabilities and adaptability to changing conditions.

Geographic Information Systems (GIS) play a pivotal role in landslide susceptibility mapping by integrating various spatial data layers (Perez & Ramsos, 2017). In the Nandhaur Watershed, GIS was employed to analyze multiple contributing factors, including topography, soil type, land use, rainfall, and proximity to infrastructure. This note discusses the results of using spatial analysis tools in GIS to produce a landslide susceptibility map for the region.

8.3.5 Data Processing

1. **Layer Preparation**: Each data layer was processed to ensure compatibility in terms of projection and resolution.
2. **Multiple Factors**: Multiple factors were taken (such as slope, rainfall, and geology), to ensure its importance and influence on landslide susceptibility.
3. **Reclassification**: Each layer was reclassified into susceptibility categories (low, moderate, high) based on the assigned weights.

8.3.6 Integration and Analysis

Spatial analysis tools in GIS were utilized to combine the reclassified layers using multiple factors (FR) analysis, which facilitated the creation of a comprehensive susceptibility map (Singh & Singh, 2015).

Landslide Susceptibility Map

The resulting landslide susceptibility map for the Nandhaur Watershed categorized areas into five susceptibility classes: very low, low, moderate, high, and very high.

- **Very High-Susceptibility Areas**: Mainly in high-altitude region with steep slope and intensive rainfall.
- **High-Susceptibility Areas**: Predominantly located in steep terrain, particularly along riverbanks and newly developed infrastructure areas. Regions with clayey soils and high rainfall intensity were also highlighted as critical zones.
- **Moderate Susceptibility Areas**: Found in moderately sloped regions and areas with loamy soils. These zones often coincide with agricultural lands, where disturbances could lead to increased risk.
- **Low-Susceptibility Areas**: Mainly in flatter terrains and regions with unstable geological formations, where the risk of landslides is less (Fig. 8.5).
- **Very Low-Susceptibility Areas**: Mainly stable geological terrains, with low landslides risk.

8.3.7 Visualization and Interpretation

GIS visualization tools allowed for detailed analysis and interpretation of the susceptibility map. Areas with high susceptibility were identified as hotspots, aiding in targeted risk management (Fig. 8.6). The map revealed distinct spatial patterns correlating with topographical features, soil types, and land use practices. The use of GIS and spatial analysis tools in producing a landslide susceptibility map for the Nandhaur Watershed provided a comprehensive and visual representation of landslide risk. The integration of multiple data layers allowed for a nuanced understanding of how various factors contribute to landslide occurrences.

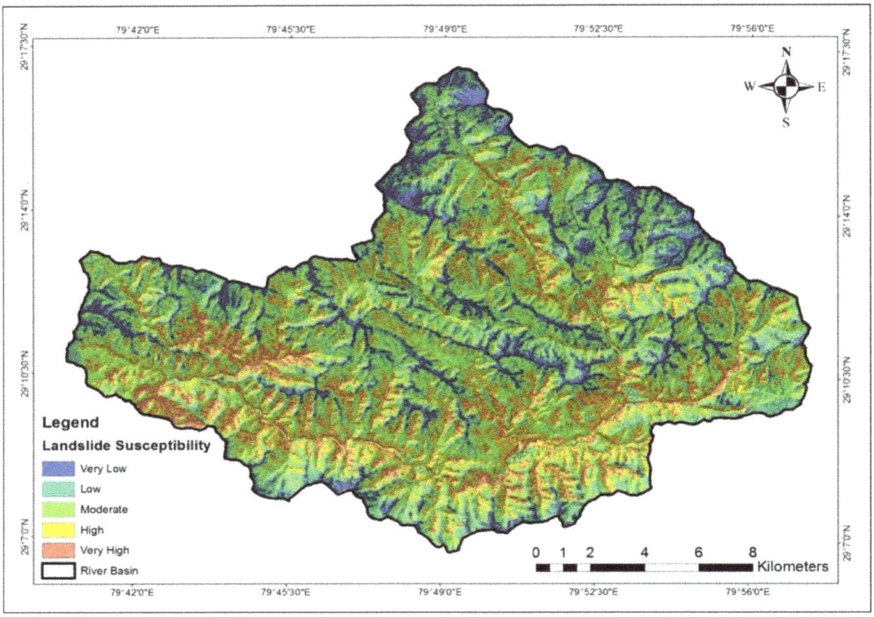

Fig. 8.5 Landslide susceptibility

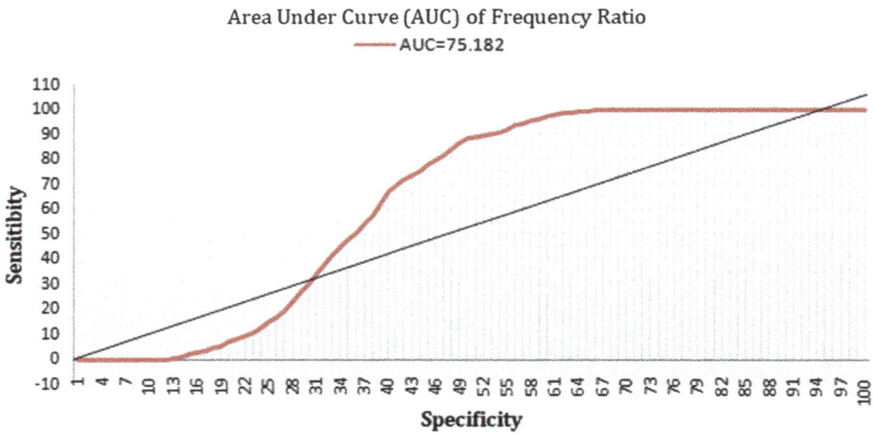

Fig. 8.6 Area under curve

Implications for Land Management

1. **Planning and Development**: The susceptibility map serves as a vital tool for urban planning, infrastructure development, and land use management, guiding decision-makers in risk-prone areas.
2. **Mitigation Strategies**: Identifying high-risk zones enables the implementation of targeted mitigation measures, such as enhancing drainage systems and implementing vegetation cover.

8.4 Results

8.4.1 Landslide Distribution

The detailed inventory of significant landslides within the Nandhaur Watershed serves as a crucial resource for understanding landslide dynamics in the region. This information is vital for developing effective risk mitigation strategies and enhancing community resilience against landslides. Future efforts should focus on continuous monitoring and integrating real-time data to update the landslide inventory and susceptibility assessments.

The Nandhaur Watershed, situated in the Central Himalayas of Uttarakhand, India, is characterized by steep terrain and complex geological formations, making it prone to landslides. This section provides a detailed description of 150 significant landslides identified within this watershed, focusing on their characteristics, locations, types, and potential triggers.

8.4.2 Landslide Overview

The identified landslides vary in size, type, and impact. Each landslide is classified based on the following parameters:

- **Location**: Geographic coordinates and proximity to key features (roads, rivers, settlements).
- **Type**: Classification (e.g., rotational, translational, debris flow).
- **Size**: Area and volume estimates.
- **Triggers**: Primary causes of movement (e.g., rainfall, human activity).

8.4.3 Detailed Description of Selected Landslides

Here is a representative summary of the landslide characteristics, categorized into regions within the watershed.

1 **Upper Nandhaur Region**

- **Landslide 1**:
  - **Location**: Near Nandhaur River, coordinates (30.1234° N, 79.2321° E).
  - **Type**: Rotational slide.
  - **Size**: 0.5 hectares.
  - **Triggers**: Heavy rainfall during monsoon season.
- **Landslide 2**:
  - **Location**: Adjacent to the Nandhaur trekking route.
  - **Type**: Translational slide.
  - **Size**: 0.7 hectares.
  - **Triggers**: Road construction activities and subsequent soil erosion.
- **Landslide 3**:
  - **Location**: Near a small village, coordinates (30.1300° N, 79.2200° E).
  - **Type**: Debris flow.
  - **Size**: 1 hectare.
  - **Triggers**: Intense rainfall leading to soil saturation.

2. Central Nandhaur Region
   - **Landslide 4**:
     - **Location**: Close to agricultural fields, coordinates (30.1400° N, 79.2100° E).
     - **Type**: Rotational slide.
     - **Size**: 0.6 hectares.
     - **Triggers**: Poor agricultural practices and heavy rains.
   - **Landslide 5**:
     - **Location**: On the edge of a forest area.
     - **Type**: Debris flow.
     - **Size**: 0.8 hectares.
     - **Triggers**: Sudden rainfall event and soil compaction from logging activities.
   - **Landslide 6**:
     - **Location**: Near an abandoned quarry.
     - **Type**: Translational slide.
     - **Size**: 1.2 hectares.
     - **Triggers**: Changes in land use and destabilization from quarrying.
3. Lower Nandhaur Region
   - **Landslide 7**:
     - **Location**: Near a main road, coordinates (30.1500° N, 79.2000° E).
     - **Type**: Rock fall.
     - **Size**: 0.4 hectares.
     - **Triggers**: Natural rock weathering and vibrations from traffic.
   - **Landslide 8**:
     - **Location**: Close to the confluence of two rivers.
     - **Type**: Debris flow.
     - **Size**: 0.9 hectares.
     - **Triggers**: Extended periods of rainfall followed by a storm.
   - **Landslide 9**:

  - **Location**: Near urban settlement.
  - **Type**: Rotational slide.
  - **Size**: 0.5 hectares.
  - **Triggers**: Urbanization and infrastructure development.

4. Additional Notable Landslides
   - **Landslide 10**:
     - **Location**: Within a dense forest.
     - **Type**: Translational slide.
     - **Size**: 0.3 hectares.
     - **Triggers**: Ground saturation from prolonged rainfall.
   - **Landslide 11**:
     - **Location**: Near a popular trekking path.
     - **Type**: Debris flow.
     - **Size**: 1.0 hectare.
     - **Triggers**: Increased tourist activity and rainfall.
   - **Landslide 12**:
     - **Location**: Above a local community.
     - **Type**: Rotational slide.
     - **Size**: 0.75 hectares.
     - **Triggers**: Land clearing for agriculture.

The 150 identified landslides in the Nandhaur Watershed demonstrate a diverse range of types, sizes, and triggers. The majority of the landslides are influenced by factors such as:

- **Rainfall**: Heavy monsoonal rains are a primary trigger, particularly in steep terrains.
- **Human Activities**: Deforestation, road construction, and agricultural practices significantly increase susceptibility.
- **Geological Conditions**: The presence of weak geological formations contributes to landslide occurrences.

Triggers and Causes—The primary factors influencing landslide occurrences included:

- **Rainfall**: Intense monsoonal rains were identified as a major trigger, leading to increased saturation of soils and triggering both shallow and deep-seated landslides.
- **Human Activities**: Deforestation, road construction, and agricultural expansion were significant contributors to slope destabilization.
- **Geological Factors**: The presence of weak rock formations, steep slopes, and soil composition played critical roles in landslide susceptibility.

8.4.4 Impacts of Landslides

Socioeconomic Impacts—Landslides in the Nandhaur Watershed have significant implications for local communities, including:

- **Infrastructure Damage**: Roads, bridges, and buildings often suffer damage or destruction due to landslides, disrupting transportation and access to services.
- **Agricultural Losses**: Landslides can lead to the loss of arable land, affecting food security and local livelihoods.
- **Displacement**: In severe cases, landslides have caused the displacement of communities, leading to socio-economic challenges.

Environmental Impacts

- **Erosion**: Landslides contribute to soil erosion, degrading land quality, and impacting vegetation cover.

Water Quality: Landslide activity can lead to sedimentation in rivers and streams, affecting water quality and aquatic ecosystems.

8.5 Recommendation

The study highlights the need for strategic land use planning in the Nandhaur Watershed. Areas identified as high-susceptibility zones should be monitored and regulated to prevent deforestation, unplanned development, and agricultural expansion. While this study provides valuable insights, it is limited by the availability of historical landslide data and the temporal resolution of meteorological data. Future research should focus on real-time monitoring of landslide-prone areas and the impact of climate change on landslide frequency.

8.6 Limitations of the Study

While FRA offers a robust framework, it relies on subjective inputs from experts. Further validation through historical landslide data and field studies is recommended to enhance the reliability of the model. The application of FRA in landslide susceptibility modeling in the Nandhaur catchment has yielded significant insights into the relative importance of various factors. The findings underscore the necessity for integrated approaches in land management to mitigate landslide risks and enhance community resilience in this vulnerable region. Future research should focus on incorporating real-time data and advanced modeling techniques to refine susceptibility assessments further. While the susceptibility map is a powerful tool, it is important to note that the accuracy of the map depends heavily on the quality and resolution of the input data. The model may not fully account for dynamic factors such as future land use changes and climate variability.

8.7 Conclusion

This research underscores the critical need for comprehensive landslide susceptibility assessments in the Nandhaur Watershed. By integrating remote sensing, GIS, and statistical analysis, this study provides a foundational understanding of landslide dynamics, which can guide future disaster risk reduction efforts in the Central Himalaya region.

References

Caine, N. (1980). *The rainfall triggering of landslides: A review of the evidence*. Soil Conservation Service Technical Bulletin

Ghosh, S., & Nandy, S. (2018). Landslide susceptibility mapping using GIS and remote sensing: A case study of the Darjeeling Himalayas. *Geomatics, Natural Hazards and Risk*, *9*(1), 123–140.

Ghosh, S., & Singh, M. (2019). Evaluating the landslide susceptibility in the Himalayan region using statistical methods. *Geocarto International, 34*(2), 186–204.

Kumar, A., & Bhattacharya, A. (2022). Integrated landslide susceptibility mapping using remote sensing and GIS in Uttarakhand, India. *Journal of Earth System Science*, 131(1), 12

Mishra, P., & Kumar, V. (2016). Geospatial analysis for landslide susceptibility mapping using remote sensing and GIS in Central Himalaya. *International Journal of Remote Sensing*, 37(12), 2912–2932.

Pérez-Rodríguez, J. C., & Ramos, M. (2017). Landslide susceptibility assessment using GIS and AHP: A case study of the Northern Andes. *Landslides*, 14(5), 1879–1893.

Ranjan, R., & Singh, R. (2019). Assessment of landslide susceptibility in Central Himalaya using statistical and machine learning approaches. *Environmental Monitoring and Assessment, 191*(10), 637.

Saha, A., & Basu, S. (2021). Landslide susceptibility mapping using AHP and GIS: A case study of the Himalayas. *Journal of Earth System Science.*

Saha, S., & Ghosh, S. (2021). Landslide susceptibility mapping using AHP in a GIS environment: A case study of the Lesser Himalaya. *Journal of Mountain Science, 18*(4), 919–933.

Sharma, R., & Choudhury, D. (2017). Landslide susceptibility mapping in the western Himalayas using AHP and GIS. *Geoscience Frontiers*, *8*(3), 429–440.

Singh, S. K., & Singh, M. (2015). Application of GIS-based multi-criteria analysis for landslide susceptibility mapping in a mountainous region. *Natural Hazards, 79*(1), 577–597.

Wilson, J. P., & Gallant, J. C. (2000). *Terrain analysis principles and applications*. Wiley.

Yalçın, M., & Düzgün, H. S. (2020). Landslide susceptibility assessment using multi-criteria decision-making methods: A case study from the Eastern Black Sea region, Turkey. *Landslides, 17*(1), 29–43.

Part IV

Landslides Impact on Environment and Socio-economic of the Local People

Landslide Hazards and Their Consequences for Tourism

9

H. Sivasankari

Abstract

Nilgiri district, Tamil Nadu, is vulnerable to landslides with steep and fragile geology combined with extreme weather conditions due to its high susceptibility to landslides. Heavy monsoon rainfall and in some cases seismic activities cause landslides that are further increased by human activities like deforestation, urbanization as well as expansion of infrastructure as a result of tourism activities. Such disasters affect transportation, infrastructure, and compromise the safety of the tourists and the communities. Tourism industry is greatly affected because landslides close the major highways, demolish resorts and markets putting the safety of visitors in jeopardy. These risks have been increased by climatic change that changes the pattern of rainfall and makes the slope susceptible to erosion. Weighted overlay analysis (WOA) using Geographic information system (GIS) is a good method of assessing landslide hazards. It incorporates various aspects, including slope, rainfall, lithology, and vegetation cover to classify the locations into low, moderate, and risky ones. Such an approach aids decision-makers to come up with knowledgeable strategies of disaster mitigation and sustainable tourism planning. Studies note that the main factors that contribute to landslides are slope and rainfall, and lithology, the density of lineaments, and vegetation cover also contribute greatly. Although low-risk zones can still be used as a tourist destination, moderate and high-risk zones need to be planned, have early warning mechanisms, and strengthen the infrastructure to guarantee the security of the visitors. Land-use strategic planning, effective infrastructure development and active disaster management are necessary to reduce the impact of landslides. With such measures, Nilgiris will

H. Sivasankari (✉)
Department of Geography, Ethiraj College for Women (A), Chennai, Tamil Nadu, India
e-mail: sivasankarianand34@gmail.com

R. M. Yuvaraj et al. (eds.), *Advanced Landslide Susceptibility Modeling*,
https://doi.org/10.1007/978-3-032-14485-0_9

be able to balance the tourism development and the environmental sustainability to ensure the safety of the visitors, as well as the survival of the economy in the long run.

Keywords

Landslide • Tourism • GIS • Weighted overlay analysis • The Nilgiri district

9.1 Introduction

Landslides can be discussed as one of the most serious and frequent nature disasters especially in mountainous and hilly areas when a complex of steep slopes, intense precipitation, and man-made changes play the crucial role in causing the slope instability. The Nilgiri district of Tamil Nadu, found in the Western Ghats, is very prone to landslides in the region because of the topography of the land, the geology formation and the weather conditions. It is a highly rated tourist destination in India with plentiful biodiversity, tea farms, and beautiful scenes. Nevertheless, frequent landslides are a significant threat to the sector because they destroy infrastructure, disrupt transportation systems, and lead to the loss of lives of tourists and local residents. The awareness of the landslide hazards and the effects of the hazards on tourism is instrumental in the sustainable development of the area. The overlay analysis using Geographic Information System (GIS) has turned out to be an influential aid in determining landslide hazards and, therefore, a researcher is able to merge several environmental variables and define high-risk areas efficiently.

The landslides that are experienced in the Nilgiris are influenced by a mixed interaction of natural and anthropogenic factors. Intensive and sustained rainfall is one of the natural causes which is very dominant and is mostly observed during the southwest and northeast monsoon. Mass movements are caused by heavy rains, which cause soil to be saturated, making it less stable on the slope (Anis et al., 2019; Huang et al., 2021; Karim et al., 2019). Seismic activities are less common but can also lead to landslides through the shaking of the ground which causes weakening of the rock formation. Conversely, human-based activities, including deforestation, unplanned urbanization, construction of roads, and agricultural activities have largely changed the natural slope processes making landslides more frequent and severe. The increase in the tourism related infrastructure such as resort, hotels, and even roads in the ecologically sensitive regions have further added to slope instability in the Nilgiris.

Landslides especially affect the tourism industry since such calamities lead to widespread disruptions in the transport systems, destruction of cultural and natural tourist sites, and raise safety concerns among tourists. The Nilgiri district is primarily dependent on road connectivity in tourism related activities and landslides often block major tourist routes like the Ooty–Coonoor and Ooty–Mettupalayam highways and make traveling around the region a difficult task causing losses to businesses that are tourist dependent. Hotels, resort, and local markets are destroyed as a result of landslides as well, which further increases economic instability, that of people who rely on tourism. Further, landslides can cause casualties and destruction of property, which has adverse effects on the media coverage causing people to have a negative view of the area and hence avoid traveling to the location. Also, landslides are likely to occur frequently thus decreasing the appeal of Nilgiri to the tourist industry thus constraining the growth of the economy in the long term (Kjekstad & Highland, 2009). Climate change has aggravated the risk of landslides in the Nilgiris. The increase in global temperatures has disrupted rainfall, causing more and more severe precipitations, which increase the risk of landslides (Gariano & Guzzetti, 2016). The fact that the temperature varies and the vegetation pattern also causes the mountain slopes to melt faster, also leads to the slope instability. These transformations in the environment underscore the importance of effective hazard assessment and risk management practices. The overlay analysis in GIS has been found to be a powerful tool to map the landslides hazard as it incorporates various variables like slope gradient, lithology, land use, and rainfall distribution to outline the areas at risk (Yuvaraj & Dolui, 2023). Through GIS technology, decision-makers would be able to come up with specific mitigation measures, early warning systems, and land-use regulations that would reduce the effects of landslides on tourism and the local communities.

In this chapter, landslide hazards are discussed within the Nilgiri district and the particular emphasis has been given to the overlay analysis of hazards based on GIS. It also assesses the impacts of landslides to the tourism industry and discusses mitigation measures that will improve resilience. This research will be able to make informed decisions on sustainable tourism development and also make the region safe and economically stable.

9.2 Study Area

The Nilgiri district, located in the Western Ghats of Tamil Nadu, India, is a prominent hill station renowned for its rich biodiversity, tea plantations, and scenic landscapes. The district spans approximately 2552 square kilometers and is situated between 11°10′ to 11°43′ N latitude and 76°14′ to 77°02′ E longitude. It is bounded by Kerala to the west and Karnataka to the north, making it a strategic location for tourism in southern India shown in the Fig. 9.1. The Nilgiri Hills, with elevations ranging between 900 to 2637 m, create a cool and misty climate, attracting millions of tourists annually.

Tourism is a significant part of the economy of the district, and the key attraction sites include Ooty (Udhagamandalam), Coonoor, Kotagiri, and Mudumalai National Park among which visitors come to the country to see in a bid to enjoy its beauty. The area has a combination of eco-tourism, adventure tourism and heritage tourism with attractions of the Nilgiri Mountain Railway (UNESCO World

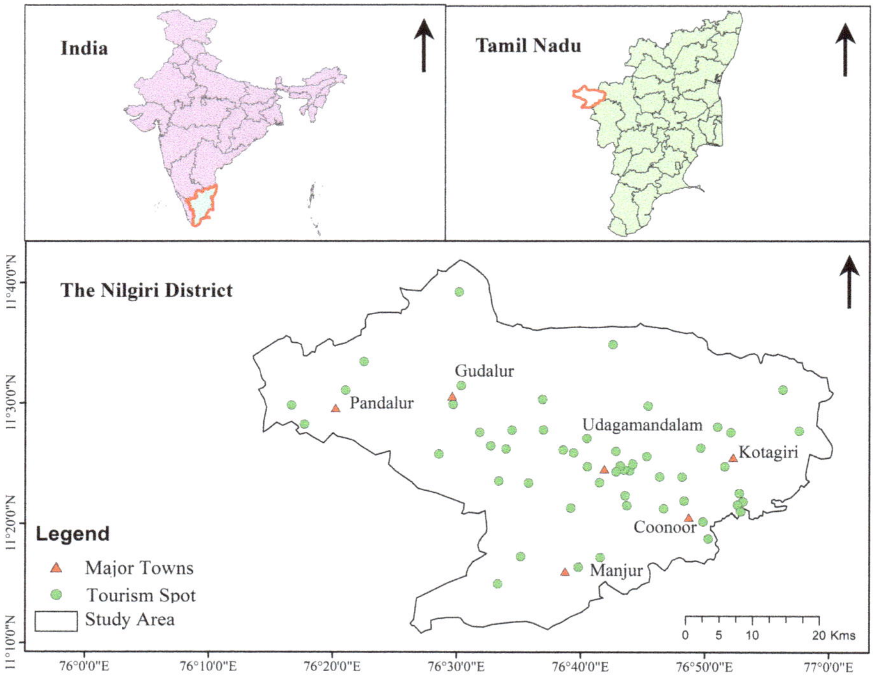

Fig. 9.1 Study area

Heritage site), Ooty Lake, and Botanical Gardens. Availability of tea estates and tribal cultural heritage is also an added advantage to tourism.

Nevertheless, landslides are common in this district especially in monsoon seasons, which is a big challenge to the tourism. These risks dislodge transport systems like the Ooty–Mettupalayam and Ooty–Coonoor roads which affect the accessibility of visitors and livelihoods of the locals. It is imperative that sustainable tourism development and disaster preparedness would help to reduce these threats and bring long-term economic stability.

9.3 Data and Methodology

Landslides are the result of very intricate interactions between the geological, geomorphological, hydrological, and the meteorological factors. These complex processes need to be modeled using sophisticated mapping tools that have the ability to reveal complex non-linear relationships between landslide occurrences and the controlling factors. Various qualitative and quantitative methods have been established toward the realization of this goal. Whereas, in qualitative methods, geomorphological studies and heuristic methods are used, quantitative methods employ statistical, physical, or numerical models.

Secondary data will involve the use of satellite imagery of Landsat 8 OLI which was obtained in April 2024 by the United States Geological Survey (USGS). Shuttle Radar Topography Mission (SRTM) data, which is 30 m, has also been offered by the USGS. Also, the geomorphological data, soil characteristics, and soil depth were collected through the Geological Survey of India (GSI). Data on past landslides and hotspots of landslides historically were sourced in the Public Works Department (PWD) of the Nilgiri district.

To perform the spatial analysis, ArcGIS Digital Elevation Model (DEM) was employed in the production of slope and elevation maps. In addition to this, satellite derived measures including the Normalized Difference Vegetation Index (NDVI), moisture stress, density of lineaments, and lineament characteristics were obtained using products of the remote sensing to aid the landslide evaluation.

We have used land use land cover prepared with Landsat 8 OLI data in which band 3, band 4 and band 5 have been used to perform supervised classification in ArcGIS 10.8 version. NASA Landsat 8 has been used to compute NDVI.

$$\text{NDVI} = \frac{\text{Band5} - \text{Band4}}{\text{Band5} + \text{Band4}}$$

where band 5 means the DN value of near-infrared band and band 4 means the DN value of red band of satellite imagery and the outcome of NDVI value data has been further classified.

The satellite image of the study area had to be cropped and then imported to PCI-Geomatica where the lineaments were extracted. The LINE-python Algorithm was utilized to extract line features using band 7 and 8 of the Landsat 8. The extracted lineament data was exported as a shapefile in order to improve the analysis of space, which was done in ArcGIS. Lineament density was obtained using interpolation techniques and the buffer analysis in ArcGIS was used to obtain the lineament buffer.

Existing literature has stated various methods of landslide evaluation which can be highly classified into a few approaches. They are inventory-based techniques (Suryabhagavan, 2019), judgment methods (Raghuvanshi et al., 2014), statistical techniques (Abdo, 2022), and deterministic, probabilistic, and distribution-free techniques (Kanungo et al., 2006). Also, sophisticated approaches like deep learning and models based on artificial intelligence (Chen et al., 2021) have become more popular. Additional methods that are popular are the Information Value Method (IVM) (Balasubramani & Kumaraswamy, 2013), the Weight of Evidence (WOE) (Chen et al., 2019), and the Frequency Ratio (FR) method (Yuvaraj & Dolui, 2023), among others (Sharma et al., 2012).

The analysis of the weighed overlay is very popular in Geographic Information System (GIS) systems. This approach combines several causal variables by giving each variable a proportionate weight that can be used to produce an overall landslide hazard map. Weighed overlay analysis is a good tool of landslide susceptibility mapping because it incorporates several contributing factors in a single hazard map. With the help of GIS-based spatial analysis, the authorities will be able to make sound decisions to reduce the risks of disasters and develop in a sustainable manner.

The consolidation of different thematic maps that described the landslide hazard areas was realized to come up with one landslide susceptibility map using GIS. Landslide in an area is regulated by a number of factors. The impact of all factors will not be similar in an area. Thus, the parameter will be weightaged with regard to its contribution to the occurrence of landslide. Once the weightages were given to the themes and features, all the themes were transformed into raster format with the help of the spatial analyst, Arc/GIS software extension. The features have been converted to raster and final analysis has also been performed using spatial analyst extension of Arc GIS 10.8. Under this approach, the overall weights of the end integrated map have been obtained as a sum or product of the weights given to the various layers based on their appropriateness. Moreover, each of the themes has been allocated knowledge-based ranking of hierarchy of increasing levels 1 to 5 on their importance with reference to landslide, 1 indicating less prone and 5 high prone of landslide. Factors influencing landslides have been ranked depending on the experience of the professional and intensive literature review (Ayalew & Yamagishi, 2005; Yuvaraj & Dolui, 2021). The preparation of the different thematic maps as well as final integrated landslide hazard zone map has been prepared using the technical guidelines as recommended by National Remote Sensing Agency.

The weighted overlay analysis tool was used to compute landslide hazard values using the equation proposed by

$$\mathrm{LH} = \sum_{w=1}^{m} \sum_{j=1}^{n} \left(W_j * X_i\right)$$

where

W_j Normalized weight of the jth thematic layer,
X_i Normalized weight of the ith feature of the thematic layer,
m Total number of thematic layers, and
n Total number of features of a given theme.

The LH values were used to classify whether an area is low, moderate, and high with respect to risk of landslide hazard.

9.4 Result and Discussion

9.4.1 Tourism of Nilgiri District

One of the most desired tourist destinations in the state of Tamil Nadu is the Nilgiri District commonly referred to as the Blue Mountains. This beautiful place is situated in the Western Ghats and is characterized by a hilly terrain, deep forests, tea fields, waterfalls, and biodiversity. It is a year-long tourist destination due to its agreeable climate and a variety of tourist attractions. The Fig. 9.2 Displays tourism spot.

Some of the most renowned hill stations in India are found in the Nilgiris which provide breathtaking views and a serene environment. The capital of the district, Ooty (Udhagamandalam), is said to be the Queen of Hill Stations and has features like the Ooty Lake, the Government Botanical Gardens, Doddabetta Peak, and the Nilgiri Mountain Railway, which is a UNESCO World Heritage. Another attractions that people visit is Coonoor, which boasts of tea estates, Sims Park, Lambs Rock, and Dolphins nose. Kotagiri is a not so famous but calm hill station that provides breathtaking sceneries and falls such as Catherine falls and Elk falls. Nilgiris is a vital component of the Nilgiri Biosphere Reserve which harbors various flora and fauna. Mudumalai National Park, which borders on Tamil Nadu-Karnataka, is also a major tourist attraction to wildlife tourists as they can explore the national park and view the tigers, leopards, elephants, and other birds. The nearby Bandipur and Wayanad Wildlife Sanctuaries increase the eco-tourism attraction of the region even more.

The area is filled with clean lakes and falls, which are a source of attraction to nature enthusiasts and photographers. The Lake of Pykara and Waterfalls also offer a picturesque getaway where people can go boating. The Avalanche Lake

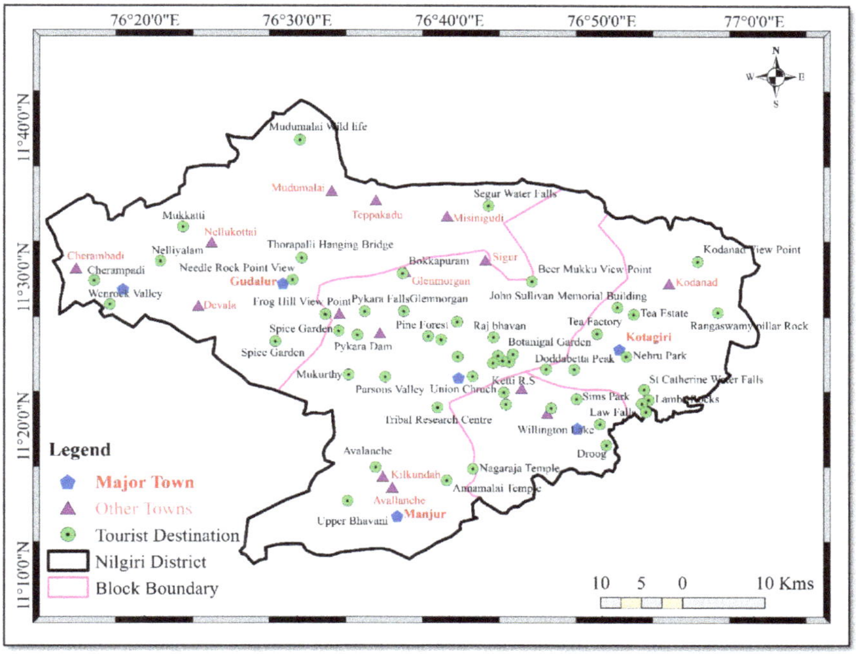

Fig. 9.2 Tourist destination

and Emerald Lake are in a remote location, which provides serene environment and good trekking. Among the more notable waterfalls are the Catherine Falls, Kolakambai Falls, and the Law falls which attract both the adventure seekers and picnic goers. Nilgiris has a range of trails and activities to do among trekking and adventure enthusiasts. Doddabetta Peak, Rangaswamy Pillar, and Glenmorgan are good trekking spots, whereas the rugged terrain offers mountain biking and rock climbing. The tea fields are also scenic and have guided tours of the plantation, which gives the visitor an understanding of the tea culture in the area.

A distinctive cultural heritage is a characteristic of the Nilgiris especially because of the existence of native tribal societies. One of the oldest indigenous groups is the Toda tribe, which still dwells in the Nilgiris. Toda settlements, traditional huts, and temples are open to the visitors and they provide an insight into their customs and artisanship. Colonial heritage is also high in the district with heritage buildings including St. Stephen Church, Stone House and Raj Bhavan. Nilgiris is also known globally due to its tea plantations that brought about high quality Nilgiri tea. To get to know about tea processing tourists may visit tea estates and tea factories in Coonoor, Kotagiri, and Ooty to take fresh tea tasting courses. The Tea Museum in Ooty can give one an insight into how tea was grown in the region. Figure 9.2 indicated location of major tourist spot in Nilgiri district.

9.4.2 Lineament Density

The lineaments are geological features that are linear structures, with faults, fractures, ridges, and major discontinuities (Sarkar et al., 1995). These structural weaknesses may have a great effect on the stability of rock masses (Andreas & Allan, 2007). Because of their effect on terrain stability, the lineaments have been extensively used as a significant parameter in landslide susceptibility tests (Anbalagan & Singh, 1996; Nagarajan et al., 2000; Saha et al., 2002).

Figure 9.3 provides the lineament map of the area of study. The drainage system within the Nilgiri district, like prominent soil covered terrains, is mostly subject to the major lineament patterns. Straight line valleys are usually associated with the existence of the underlying lineaments, which means areas of structural weakness that can determine the appearance of landslides.

9.4.3 Normalized Difference Vegetation Index (NDVI)

Normalized Difference Vegetation Index (NDVI) is important in the prediction of likelihood of landslides since vegetation cover has a direct relationship with soil stability. The loss of vegetation also exposes the mountainous areas to landslides. The reason behind this is that plant roots also tend to hold the soil together thereby lessening the erosion and making the slope more stable. With either a sparse vegetation cover or absence caused by deforestation, construction works, or agricultural

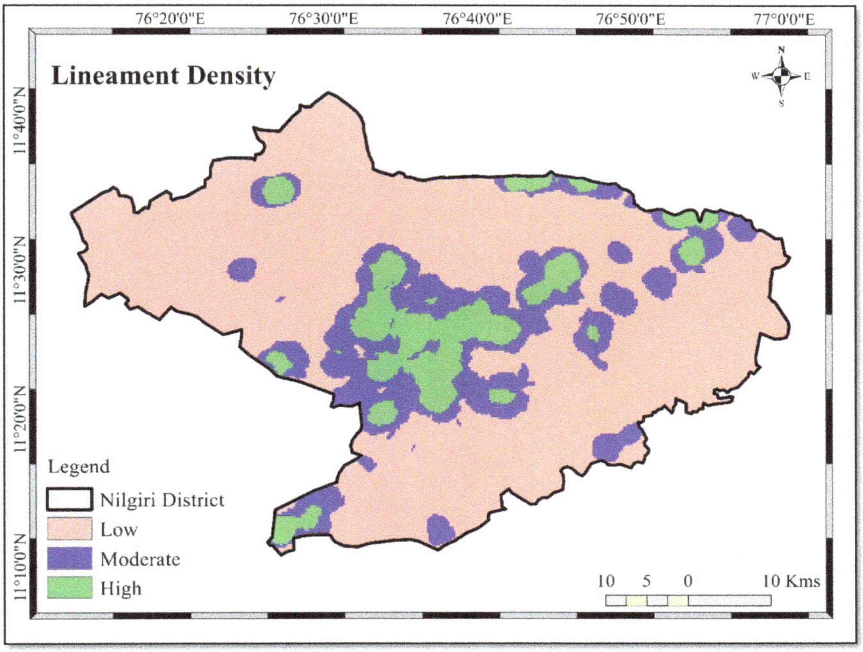

Fig. 9.3 Lineament density

expansions, the soil is more susceptible to soil erosion, and therefore more likely to cause landslides.

Thick vegetation means that there is a minimal chance of landslides occurring because surface runoff is minimized, rain is intercepted, and the cohesiveness of the soil is increased. High values of NDVI, which is a dense vegetation cover, tend to reduce the rates of erosion hence reducing landslides susceptibility. On the other hand, the low NDVI values are the regions where vegetation is either sparse or nonexistent and would be more susceptible to soil displacement and hence the high probability of landslides.

The NDVI map of the Nilgiri district (Fig. 9.4) shows that there are clearly differentiated vegetation cover differences in various regions of the area. The maximum of NDVI values is located in the northern and southwestern areas of the district, which means the dense vegetation cover and the high quality of vegetation. The vegetation tends to stabilize these areas thus making them less prone to landslides. In comparison, the far north part has moderate NDVI value, which indicates the presence of vegetation and open land with a relatively equal probability of erosion and landslides.

Minimum values of NDVI have been observed in the eastern, western, and southern regions of the study area. Such areas would be areas with human settlements, bare ground, or deforested areas, which are prone to land sliders. This

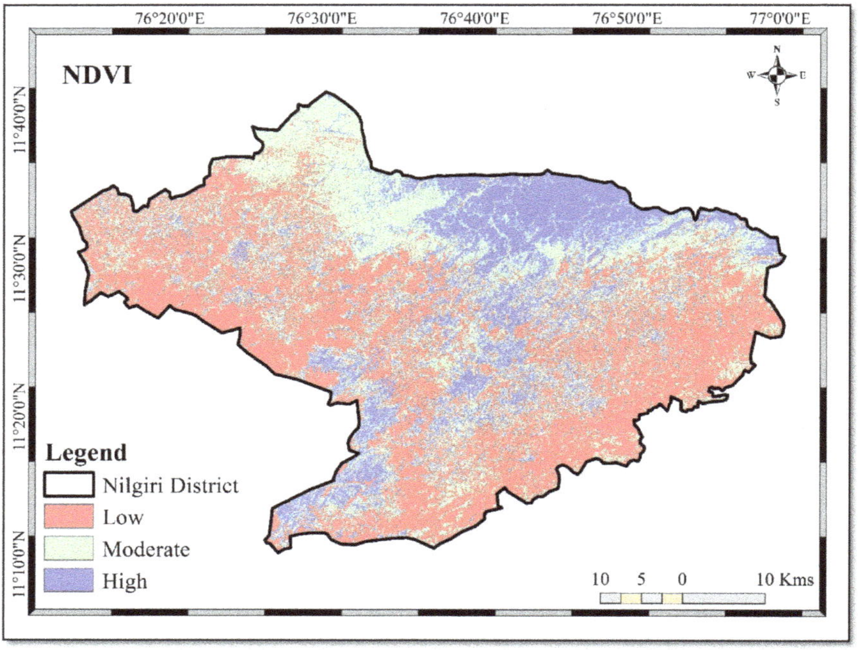

Fig. 9.4 Vegetation index

reduces the vegetation cover in such areas, and thus the soil is more vulnerable to movement and erosion, particularly during the heavy rainfall or earthquakes. Thus, the NDVI can be viewed as a major factor to determine the landslide risk, as it helps to recognize the areas with the highest risk of landslides and focus mitigation measures in those areas.

9.4.4 Slope

Slope is important in the development of the terrain especially in the high altitude areas with mountains. It is the angular inclination between two elevations and plays a great role in determining the stability of the terrain. Slopes and their inclination predetermine the stability or the instability of an area. The steeper the slope, the more elevation, the gradient and concavity of the slope, the more the steepness leading to the probability, intensity, and magnitude of landslides. Material like boulders and debris can drift slowly in a slow increment through a number of years or in extreme cases may fall at a high rate in a few seconds causing devastating effects.

The slope nature found in various places is varied because the processes that took place in forming the slopes vary, and the nature, quantity, and state of the slope materials are not uniform. The slope in this research has been classified as five sub-classes namely 0° −5°, 5° −15°, 15° −30°, 30° −50°, and above 50°. Figure 9.5 depicts that the Nilgiri district is mainly covered by the gentle slopes of between 0 and 5°. The western and central areas have a 5° to 15° slopes, which implies the moderate amounts of elevation change. The slopes between 15° and 30° are experienced in the eastern section, southern section, and western section of the district, though few in the central sections.

The rocky terrains with an average slope of 30° to 50° are mostly found in the northeastern region, and some in the southern and southeastern parts. The steeper slopes (steeper than 50°) are not common in Nilgiri, but are generally only found in the southwestern and southeastern rocky areas. High-gradient regions are also likely to experience landslides and erosion, and thus of great importance in hazard assessment. The slope categories distribution also offers important insights into the landslides susceptibility of the region and is useful in managing the risk of calamities and land use.

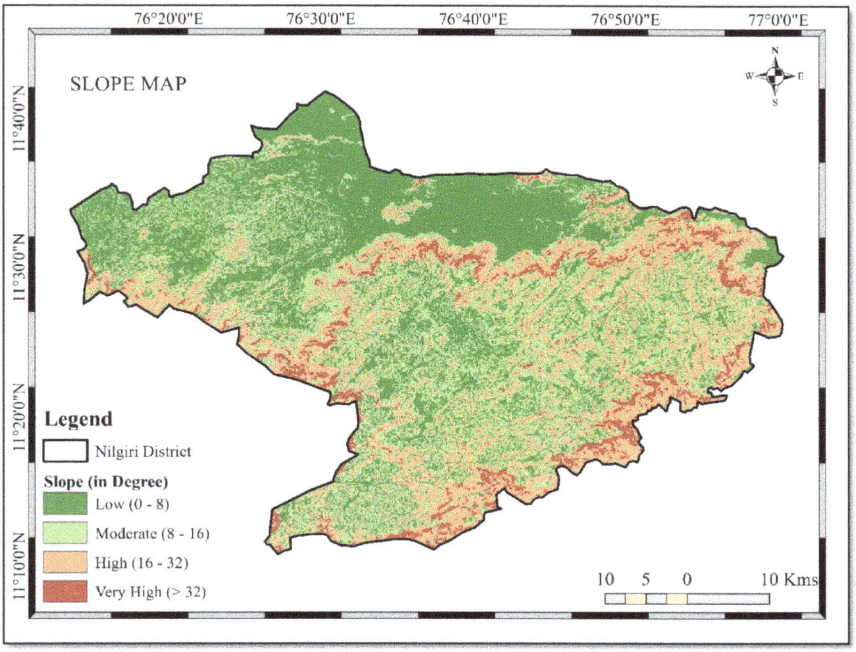

Fig. 9.5 Slope

9.4.5 Rainfall

In Nilgiris district, two wet seasons are observed, one in the Southwest and Northeast-monsoons. Most of the rainfall is experienced in the taluks of Gudalur, Pandalaur, and Kundah due to the Southwest monsoon and Udhagamandalam, Coonoor, and Kotagiri by the Northeast monsoon. Rainfall (1500 mm to 3000 mm) is also a very important environmental factor in the region which receives a mean precipitation of between 1500 and 3000 mm every year. Research indicates that rainfall is another major factor that leads to landslides in the district (Anis et al., 2019; Karim et al., 2019).

Natural hill slopes in most of the tropical regions such as in the Nilgiris are usually unsaturated. In case of heavy rainfall, the water permeates to soil making it weaker in terms of shear strength and prone to slope failure thus causing landslides. Considering the high positive association between the intensity of rainfall and the occurrence of landslides, there is a need to develop an early warning system that is informed by the rainfall data and the threshold values that help in predicting the probability of a landslide occurring. In the event that the predictions are high and are above the critical amount of rainfall, alarm should immediately be raised to the concerned authorities, such as government, the local population, and search-and-rescue crews to reduce possible catastrophes.

Rainfall forecasting is thus very important in the management of landslides as far as its accuracy and timeliness is concerned. Figure 9.6 of the rainfall distribution map shows that there were significant differences within the district. The eastern, southeastern and northeastern areas receive relatively less rainfall with the western and the southern areas of the Nilgiris receiving the highest amount of rainfall. These differences in the patterns of rainfall lead to variations in the landslide susceptibility rates where areas experiencing higher rainfall have an increased likelihood of experiencing landslide hazards. These precipitation patterns are significant to understand in order to be prepared to disasters and plan the land use.

9.4.6 Lithology

Lithology is important in the study of landslides since the various lithological units respond differently to the active geomorphological processes like landslides. This paper divides lithology into three broad categories, namely Archean, Mesozoic, and Cenozoic formations. The Cenozoic rocks are largely found in north, northwestern, and much of the western parts of Nilgiris as showed in Fig. 9.7. The mesozoic rock formation is not very abundant in the district, and is only seen in a few places in the western part, and some in the northeast. Archean rock structures, on the contrary, occupy an extensive area of Nilgiris to the east, south, central as well as southwestern parts. These differences in lithology determine the stability of the slopes and thus some of them are more prone to landslides than others.

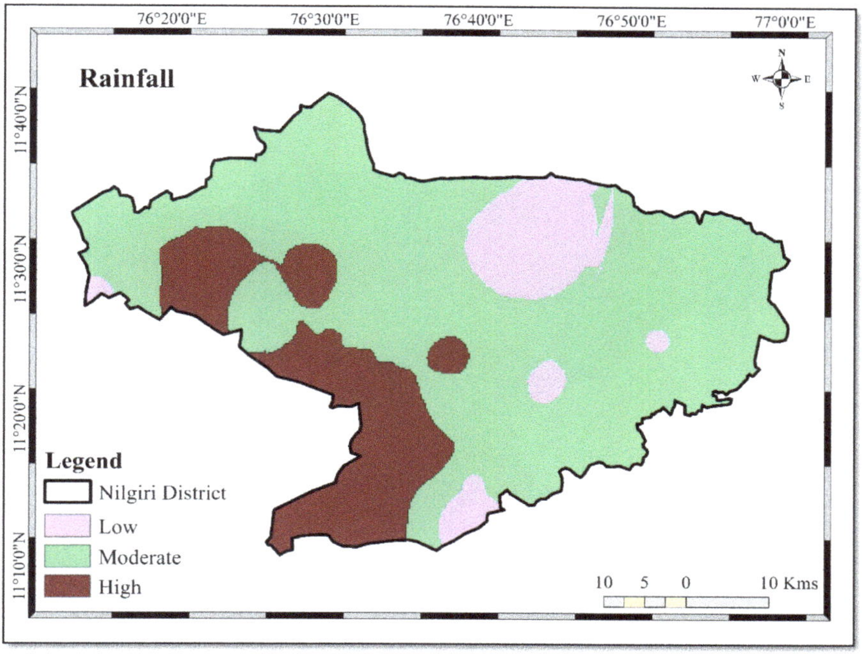

Fig. 9.6 Rainfall

9.4.7 Soil

Another important consideration that determines landslides is the content of soil. High content of sand and unstable slopes combined with heavy rainfall are significant factors which cause land instability and may result in serious damages. The soil texture that consists of the relative amounts of clay loam, rock land, sandy clay, and sandy clay loam/sandy loam is a very crucial factor that determines the slope stability.

There is a prominent blockage of rocky terrain that stretches to the west of the study area and through the middle area north eastwards and other scattered pieces in the south east. Rock land is present in small quantities in Nilgiri and this leads to a more stable and cohesive soil structure of these areas and thus they resist erosion and landslides. On the other hand, the most common type of soil in the southwest region is the sandy clay loam which occupies the greater part of the south and east regions. The low level of organic material in this kind of soil increases its susceptibility to detachment and hence its ability to form stable aggregates. Also, in the north-most part of the district, sandy clay soil prevails. According to Fig. 9.8 sandy clay loamy soil is the most common type of soil in the Nilgiri district in the area, which affects the tendency to landslides. These

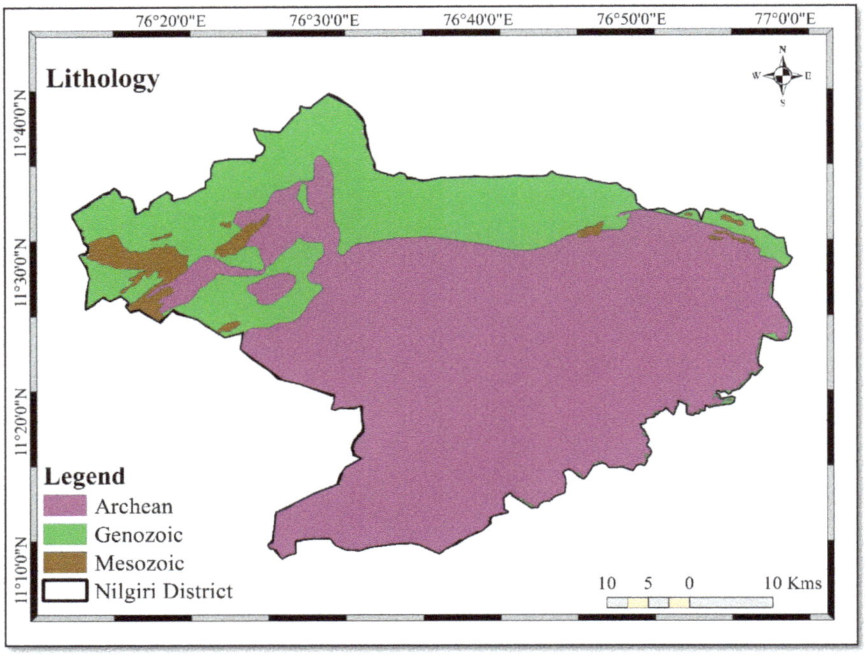

Fig. 9.7 Lithology

differences in soil composition are important to understand in order to evaluate the risk of landslides and take mitigation measures that work.

9.4.8 Soil Depth

The depth of the soil is critical in determination of soil stability and in the determination of landslides prone areas. The more depth of soil, the higher the probability of the soil to absorb the surface water and reduction in the runoff. Deep soil on the other hand is normally thought to be more susceptible to instability and landslides.

A large percentage of the study area has very deep soil as shown in Fig. 9.9 that has a depth of more than 150 cm. This covers the major part of the central area, the vast parts of the western and southern ones and the parts of the northwest and some places in the north. Also, roughness of the earth is observed in certain parts of the western and northern parts and minor instances in the south. In the very northern part, the depth of the soil is 100–150 cm which means that the stability is moderate. To assess the risk of landslides and prevent their occurrence, it is necessary to understand the spatial change in the depth of the soil.

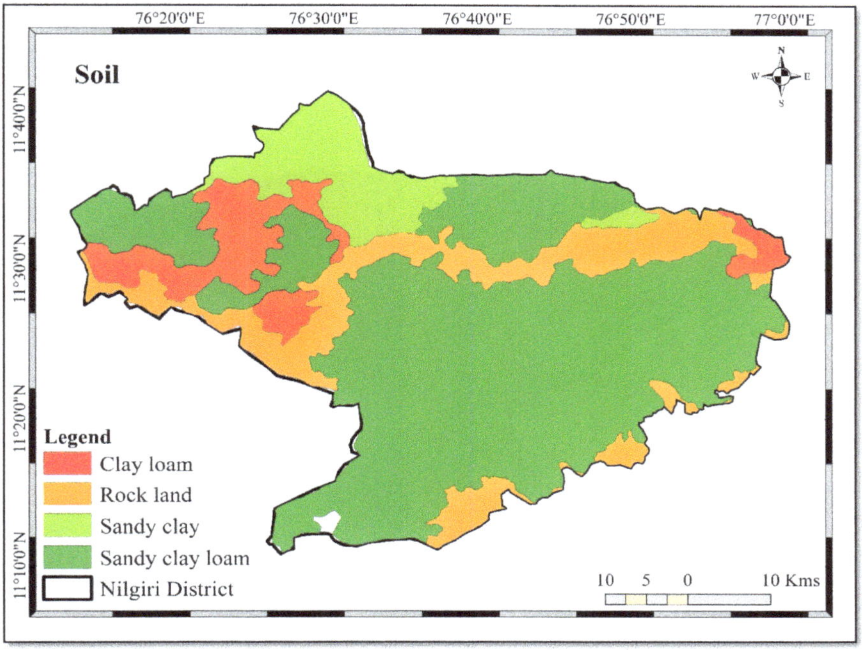

Fig. 9.8 Soil

9.4.9 Weighted Overlay Analysis

The weighted overlay technique of GIS is effective to combine various factors that cause landslides and produce a valid hazard evaluation. It is used to assist policymakers, urban planners, and disaster management authorities to find out the vulnerable regions and take preemptive actions to reduce landslides. Future research can improve the model by using machine learning methods and real-time monitoring data to ensure better prediction.

The hazard map of the landslides categorizes the study area as low, moderate, high, and very high-risk regions. Areas that have steep slope, weak lithology, high rainfall, and dense lineaments are highly susceptible whereas areas with stable geology and vegetation cover are with low risk. The hazard map can be of great importance in planning land use, development of infrastructure, and in mitigating disasters.

Table 9.1 shows the weightage that was given to the different conditioning factors which helps in determination of landslide hazard areas. This has given each factor a given weight and percentage implying that certain factors have more importance in the occurrence of landslides.

The most important factor with the highest weightage is slope (0.255, 26%). Steep slopes enhance landslides as the gravity force acts on the unstable soil and

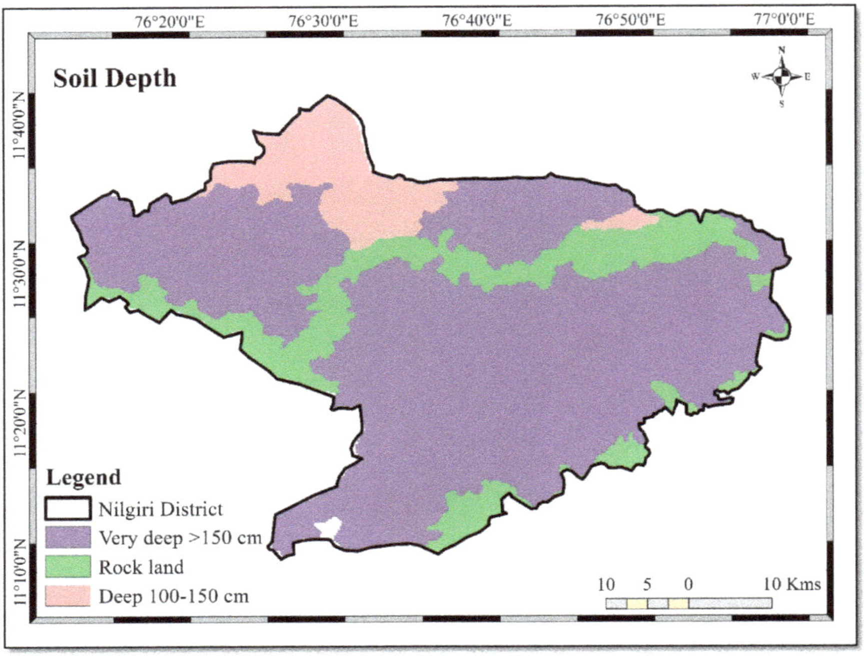

Fig. 9.9 Soil depth

Table 9.1 Weightage for the conditioning factor for determining landslide hazard zone

S. No.	Conditioning factors	Weights	Percentage
1	Lineament density	0.133	13
2	Lithology	0.135	14
3	NDVI	0.089	9
4	Rainfall	0.242	24
5	Slope	0.255	26
6	Soil	0.071	7
7	Soil depth	0.072	7

rock masses. The second factor that is the most significant is rainfall (0.242, 24%) because when the precipitation is intense and or lasts longer, the soil becomes saturated and shear strength decreases, causing landslides.

Critical contributors are also Lithology (0.135, 14%) and Lineament Density (0.133, 13%). Lithology identifies the rock formations in terms of their strength and stability whereas lineament density identifies the weakness in structures including fractures and faults which may render the slopes susceptible to collapses.

NDVI (0.089, 9%) is a vegetation cover which is used to stabilize slopes by preventing soil erosion as well as water infiltration. The least weighted factors are Soil (0.071, 7%) and Soil Depth (0.072, 7%), though they are contributing factors to landslide susceptibility. The soils with low cohesion are also more likely to erode and move but the soils which are deeper are more capable of absorbing water reducing the surface runoff and the surface runoff is less prone to instability (Fig. 9.10).

According to the level of susceptibility, the study area is grouped in three landslide hazard areas namely low, moderate, and high (Table 9.2). Low susceptibility region comprises 228.48 sq.km (8.86%) of the total area meaning that the risk of landslides is minimal. These are the locations that are probably at low slopes, stable soil, and thick vegetation thereby lowering the risk of landslides. The moderate susceptibility zone occupying most of the study region (1955.24 sq. km or 75.78% of the whole region) implies that landslides can happen infrequently, particularly during severe weather conditions or when the geology of a particular area is unstable. Although they are not as susceptible as high-risk zones, they still need to be well-managed in terms of land and monitored to reduce the possible risks. The high susceptibility zone 396.29 sq. km (15.36%) implies areas where landslides are very dangerous, probably because of some severe slope, loose soil, heavy precipitation, or past landslides. These regions are under emergency care

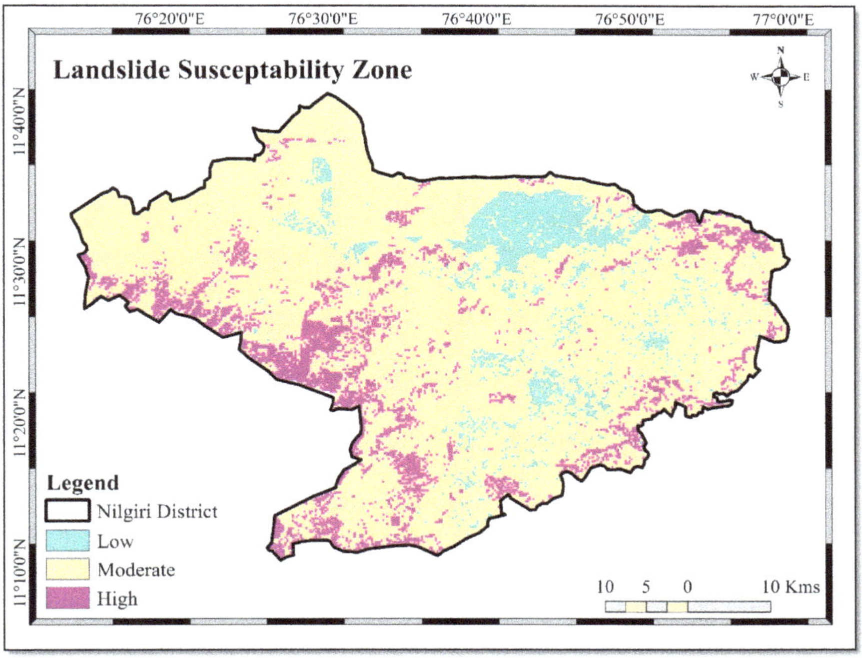

Fig. 9.10 Landslide hazard zone

Table 9.2 Area wise landslide hazard zone

Susceptibility	Area (in sq.km)	Percentage
Low	228.48	8.86
Moderate	1955.24	75.78
High	396.29	15.36

and preemptive maintenance like slope stabilization, tree plantation, and development control to reduce the amount of destruction. In general, the statistics show that although a minor part of the area (8.86% it represents a comparatively safe community) is comparatively safe, a large percentage (75.78%) are weakly vulnerable and a significant percentage (15.36%) is highly vulnerable. The analysis is vital in land-use planning, disaster management, and sustainable development of tourism activities in the landslide-prone areas such as the Nilgiris.

9.4.10 Risk at Tourist Spot

Table 9.3 divides the tourist destinations in the Nilgiri district into the low, moderate and high-risk areas according to their degree of landslide threat. Such tourist attractions as Nagaraja Temple, Ketti View Point, Union Church, and Segur Water Falls can be categorized as low risk. The areas are also relatively safe with regard to landslides, probably because of the relatively stable terrain, low slope and vegetated nature. The geological risks in these regions are minimal to tourists and hence, they are better suited to tourists enjoying the tourism activities throughout all year round. Most of the likely tourist places in Nilgiri, such as Kodanad View Point, Rangaswamy Pillar Rock, Doddabetta Peak, Ooty Lake, Botanical Garden, Pykara Boat House, Mudumalai Wildlife Sanctuary, Parsons Valley fall under the category of moderate-risk zones.

These sites are not vulnerable to landslides much but still prone to landslides in extreme weather conditions particularly during heavy rainfalls or earthquakes. Since they are quite popular with tourists, appropriate risk mitigation strategies, which include early warning systems, stabilization of slope, and simply restricting the number of visitors during the seasons of monsoons are expected to make them safe. High-risk areas comprise some of the tourist sites such as Lady Canning area seat, Pykara falls, Spice garden, Wenrock valley, Droog, and avalanche.

These regions are likely to experience frequent or severe landsliding because of steep slope, weak geological features or because of high precipitation levels. Touring these sites especially during the monsoon may be very hazardous. Tight control, access control, and high levels of disaster preparedness would play an important role in reducing the risks in these areas. Figure 9.11 indicates the spatial location of landslide risk tourist destination.

Table 9.3 Landslide risk tourist destination in Nilgiri district

S. No.	Tourist destination	Landslide risk
1	1. Segur Water Falls 2. Union Chruch 3. Ketti View Point 4. Nagaraja Temple	Low risk
2	1. Kodanad View Point 2. Rangaswamy pillar Rock 3. Tea Estate 4. John Sullivan Memorial Building 5. Beer Mukku View Point 6. Tea Factory 7. Nehru Park 8. St Catherine Water Falls 9. Lambs Rocks 10. Dolphin Nose 11. Sims Park 12. Dever Betta View 13. Doddabetta Peak 14. Ooty Lake 15. Boat House 16. Government Museum 17. Botanigal Garden 18. Raj bhavan 19. Kalhatti Water Falls 20. Pykara Boat House 21. Pine Forest 22. Glenmorgan 23. Bokkapuram 24. Kamaraj Sagar Dam 25. Ketti R.S 26. Willington Lake 27. Tribal Research Centre 28. Parsons Valley 29. Pykara Dam 30. Mukurthy 31. Spice Garden 32. Frog Hill View Point 33. Needle Rock Point View 34. Thorapalli Hanging Bridge 35. Mukkatti 36. Cherampadi 37. Nelliyalam 38. Mudumalai Wild life 39. Law Falls 40. Annamalai Temple 41. Upper Bhavani	Moderate risk

(continued)

Table 9.3 (continued)

S. No.	Tourist destination	Landslide risk
3	1. Lady Canning's Seat 2. Pykara Falls 3. Spice Garden 4. Wenrock Valley 5. Droog 6. Avalanche	High risk

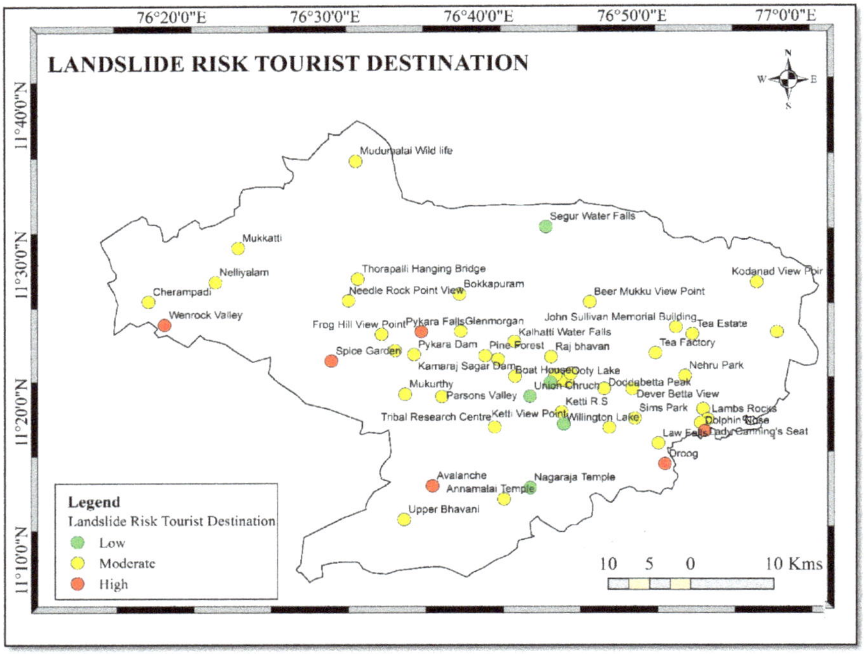

Fig. 9.11 Landslide risk tourist destination

9.5 Conclusion

Nilgiri district with its picturesque sceneries, multiple ecosystems, and rich cultural histories is one of the major tourist sites in Tamil Nadu. Its scenic hill stations, calm lakes, vast tea plantation, and wildlife appeal to nature lovers, adventure, and cultural hunters. Although tourism has potential in the region, the area is experiencing environmental problems especially land slides. The high slopes, geological structures, and different soil conditions that are observed in the district have a huge impact on its stability. The evaluation of the lineament density, NDVI, slope, rainfall, lithology, soil type, and soil depth can be used in studying landslide risks in the region. The weighted overlay analysis classifies the area as low risk, moderate-risk, and high-risk areas. Although there are tourist destinations that are located in low-risk regions, most of them are located in moderate and high-risk regions thus there is the need to engage in strategic planning and taking initiatives in managing disasters.

Research emphasizes that the main factors of landslide hazards are slope, rainfall after which are lithology, density of lineament, and vegetation cover. The comparative role of the soil properties is less but yet useful in the assessment of the landslides. The weightages are used to offer a foundation of weighted overlay analysis in GIS-based landslide hazard mapping and there are more accurate in disaster management and land-use planning. Tourism planning, visitor safety, and disaster management are critical questions that depend on the classification of tourist destinations in terms of landslide risk. Although low-risk zones may still enjoy tourists without much worry, moderate and high-risk zones would need due safety precautions, particularly, during the rainy seasons. Through proper strategies of landslide risks assessment, development of infrastructure, and early warnings mechanisms, Nilgiri can have sustainable and safe tourism and reduce the risks that can be experienced by the visitors and the local populations.

Acknowledgments The author(s) acknowledge the use of AI-based tools for grammar and language correction during the preparation of this manuscript.

References

Abdo, H. G. (2022). Assessment of landslide susceptibility zonation using frequency ratio and statistical index: A case study of Al-Fawar basin, Tartous, Syria. *International Journal of Environmental Science and Technology, 19*(4), 2599–2618.

Anbalagan, R., & Singh, B. (1996). Landslide hazard and risk assessment mapping of mountainous terrains—a case study from Kumaun Himalaya India. *Engineering Geology, 43*(4), 237–246.

Andreas, M., & Allan, A. (2007). Incorporating geology and geomorphology in land management decisions in developing countries: A case study in Southern Costa Rica. *Geomorphology, 87*(1–2), 68–89.

Anis, Z., Wissem, G., Riheb, H., Biswajeet, P., & Essghaier, G. M. (2019). Effects of clay properties in the landslides genesis in flysch massif: Case study of Aïn Draham, North Western Tunisia. *Journal of African Earth Sciences, 151*, 146–152.

Ayalew, L., & Yamagishi, H. (2005). The application of GIS-based logistic regression for landslide susceptibility mapping in the Kakuda-Yahiko Mountains Central Japan. *Geomorphology, 65*(1–2), 15–31.

Balasubramani, K., & Kumaraswamy, K. (2013). Application of geospatial technology and information value technique in landslide hazard zonation mapping: A case study of Giri Valley Himachal Pradesh. *Disaster Advances, 6*(1), 38–47.

Chen, W., Lei, X., Chakrabortty, R., Pal, S. C., Sahana, M., & Janizadeh, S. (2021). Evaluation of different boosting ensemble machine learning models and novel deep learning and boosting framework for head-cut gully erosion susceptibility. *Journal of Environmental Management, 284*, Article 112015.

Chen, W., Sun, Z., & Han, J. (2019). Landslide susceptibility modeling using integrated ensemble weights of evidence with logistic regression and random forest models. *Applied Sciences, 9*(1), 171.

Gariano, S. L., & Guzzetti, F. (2016). Landslides in a changing climate. *Earth-Science Reviews, 162*, 227–252. https://doi.org/10.1016/j.earscirev.2016.08.011

Huang, F., Ye, Z., Jiang, S. H., Huang, J., Chang, Z., & Chen, J. (2021). Uncertainty study of landslide suscep tibility prediction considering the different attribute interval numbers of environmental factors and different data-based models. *CATENA, 202*, Article 105250.

Kanungo, D. P., Arora, M. K., Sarkar, S., & Gupta, R. P. (2006). A comparative study of conventional, ANN black box, fuzzy and combined neural and fuzzy weighting procedures for landslide susceptibility zonation in Darjeeling Himalayas. *Engineering Geology, 85*(3–4), 347–366.

Karim, Z., Hadji, R., & Hamed, Y. (2019). GIS-based approaches for the landslide susceptibility prediction in Setif Region (NE Algeria). *Geotechnical and Geological Engineering, 37*(1), 359–374.

Kjekstad, O., & Highland, L. (2009). Economic and social impacts of landslides. In K. Sassa & P. Canuti (Eds.), *Landslides—Disaster Risk Reduction* (pp. 573–587). Springer.

Nagarajan, R., Roy, A., Kumar, R. V., Mukherjee, A., & Khire, M. V. (2000). Landslide hazard susceptibility mapping based on terrain and climatic factors for tropical monsoon regions. *Bulletin of Engineering Geology and the Environment, 58*(4), 275–287.

Raghuvanshi, T. K., Ibrahim, J., & Ayalew, D. (2014). Slope stability susceptibility evaluation parameter (SSEP) rating scheme–an approach for landslide hazard zonation. *Journal of African Earth Sciences, 99*, 595–612.

Saha, A. K., Gupta, R. P., & Arora, M. K. (2002). GIS-based landslide hazard zonation in the Bhagirathi (Ganga) valley Himalayas. *International Journal of Remote Sensing, 23*(2), 357–369.

Sarkar, S., Kanungo, D. P., & Mehrotra, G. S. (1995). Landslide hazard zonation: a case study in Garhwal Himalaya, India. *Mountain Research and Development*, 301–309.

Sharma, L. P., Patel, N., Ghose, M. K., & Debnath, P. (2012). Influence of Shannon's entropy on landslide-causing parameters for vulnerability study and zonation—a case study in Sikkim India. *Arabian Journal of Geosciences, 5*(3), 421.

Suryabhagavan, K. V. (2019). Landslide Hazard zonation and slope instability assessment using optical and InSAR data: a case study from Gidole town and its surrounding areas, southern Ethiopia. *Remote Sensing of Land.*

Yuvaraj, R. M., & Dolui, B. (2021). Statistical and machine intelligence-based model for landslide susceptibility mapping of Nilgiri district in India. *Environmental Challenges, 5*, Article 100211.

Yuvaraj, R. M., & Dolui, B. (2023). Geographical assessment of landslide susceptibility using statistical approach. *Quaternary Science Advances, 11*, Article 100097.

Impacts and Effects of Landslides: Tribal People's Perspective

10

Asfiya Banu

Abstract

Landslides are a recurrent and increasingly frequent geo-hazard in the mountainous regions of India, with severe implications for indigenous communities inhabiting ecologically fragile zones. This review explores the multifaceted impacts of landslides on the Toda community of Tamil Nadu, a pastoral tribal group residing in the upper Nilgiris. Drawing on existing literature, case studies and ethnographic insights, the paper synthesizes the social, economic and environmental consequences of landslides through a culturally informed lens. It highlights how landslides disrupt traditional land-use patterns, threaten sacred landscapes, displace households and undermine community cohesion and food security. Particular attention is paid to the erosion of cultural identity and the marginalization of indigenous knowledge systems in formal disaster management discourse. The review underscores the importance of integrating Toda ecological knowledge, risk perception and adaptive practices into disaster risk reduction strategies. By advocating for participatory and inclusive approaches to resilience planning, this paper contributes to the broader discourse on sustainable and equitable disaster governance in tribal regions.

Keywords

Landslides • Tribal communities • Vulnerability • Socio-economic impact • Indigenous knowledge • Resilience

A. Banu (✉)
Department of Tourism and Travel Management, JBAS College for Women, Chennai, India
e-mail: asfiyabn@yahoo.com

R. M. Yuvaraj et al. (eds.), *Advanced Landslide Susceptibility Modeling*,
https://doi.org/10.1007/978-3-032-14485-0_10

10.1 Introduction

Landslides are defined as the down slope movement of soil, rock and organic material under the influence of gravity are among the most frequent and destructive natural hazards worldwide, causing significant socio-economic and environmental disruptions each year (Highland & Bobrowsky, 2008; Cruden & Varnes, 1996). Globally, it is estimated that landslides result in thousands of fatalities annually, along with extensive damage to infrastructure, agricultural land and human settlements, particularly in mountainous and tectonically active regions (Petley, 2012; Froude & Petley, 2018). These hazards are often exacerbated by a complex interplay of climatic extremes, seismic activities, unsustainable land use, deforestation and urban expansion into vulnerable terrains (Glade et al., 2005; Gariano & Guzzetti, 2016). Importantly, the impacts of landslides are not experienced uniformly; instead, they are disproportionately borne by marginalized and ecologically dependent populations, especially indigenous and tribal communities residing in high-risk zones.

From a disaster vulnerability lens, scholars have emphasized that the intersection of environmental fragility and socio-cultural marginalization significantly amplifies disaster risks for indigenous populations (Wisner et al., 2004; Gaillard, 2007). These communities often inhabit remote and ecologically sensitive landscapes—such as mountainous slopes, forest fringes and river valleys—where landslide hazards are intensified by limited access to infrastructure, information and institutional support systems (UNDRR, 2021; IFRC, 2021). Moreover, indigenous livelihoods, rooted in subsistence agriculture, forest produce and customary land-use practices, are intimately tied to the stability and health of local ecosystems, making them uniquely vulnerable to geo-morphological disturbances (Mercer et al., 2009; Kelman et al., 2012). In this context, landslides not only result in physical displacement and resource loss but also disrupt cultural continuity, traditional ecological knowledge systems and community resilience (Ray-Bennett, 2018).

In India, landslides are a recurrent phenomenon in geo-dynamically sensitive regions such as the Western Ghats, Eastern Ghats, Himalayan belt and the Northeastern states (NIDM, 2021). According to the Geological Survey of India (GSI), nearly 15% of India's landmass is prone to landslides, with increasing frequency due to climate change induced rainfall variability and anthropogenic pressure on slopes (GSI, 2022). Studies have documented the acute vulnerability of Scheduled Tribes in states like Uttarakhand, Meghalaya and Kerala, where developmental intrusions into fragile hill ecosystems have intensified slope failures and displaced tribal settlements (Singh, 2010; Oommen et al., 2016). Despite their deep ecological knowledge and adaptive capacities, tribal communities are often excluded from mainstream disaster governance frameworks, which limit the effectiveness of early warning systems, preparedness initiatives and rehabilitation efforts.

At the local level, the Toda tribe of the Nilgiri Hills in Tamil Nadu exemplifies how indigenous populations experience and respond to landslide hazards within a culturally specific socio-ecological setting. Recognized as a Particularly Vulnerable Tribal Group (PVTG), the Toda's have traditionally occupied high-altitude grassland-montane ecosystems of the Western Ghats, where they practice pastoralism, sacred ecology and sustainable land stewardship. However, in recent decades, extensive deforestation, plantation agriculture, road construction and unregulated tourism have destabilized the slopes of the Nilgiris, increasing the frequency of landslides and soil erosion. These events not only threaten the physical safety and economic livelihood of Toda hamlets but also endanger their sacred landscapes, temples (munds) and culturally significant buffalo pastures, leading to spiritual and existential dislocation. Further, the inaccessibility of remote Toda settlements during emergencies, coupled with insufficient policy recognition of their traditional knowledge systems, hampers community-led disaster response and recovery.

In this review, we aim to critically examine the multifaceted impacts and effects of landslides through a tribal people's perspective, with a particular focus on the Toda community. Drawing from interdisciplinary research across geohazards, disaster anthropology, indigenous studies and environmental policy, this paper seeks to synthesize existing literature, identify research gaps and advocate for culturally grounded, inclusive and resilient disaster risk reduction (DRR) frameworks tailored to tribal contexts. By foregrounding the Toda experience, this review highlights the urgent need to integrate indigenous voices into landslide risk assessment, preparedness and governance processes particularly in the era of climate change and ecological uncertainty.

10.2 Rationale of the Study

In recent decades, there has been significant progress in the field of geo-hazard analysis, particularly through the use of geospatial technologies, remote sensing data and probabilistic landslide susceptibility models (Guzzetti et al., 2005; van Westen et al., 2006). While these technical approaches have improved the precision of landslide risk mapping and forecasting, they often prioritize physical and topographic parameters over socio-cultural and human dimensions of vulnerability (Dai et al., 2002; Martha et al., 2010). As a result, disaster research continues to exhibit a pronounced epistemological bias, one that marginalizes the lived experiences, adaptive responses and traditional knowledge systems of indigenous and tribal populations residing in hazard-prone areas.

Tribal communities worldwide, and particularly in South Asia, are among the most vulnerable to environmental hazards due to their historical marginalization, political invisibility and dependence on natural resources for subsistence (Bankoff, 2004; Oliver-Smith, 1996). Their vulnerability is not merely a function of geographical exposure but is deeply embedded in socio-economic structures, legal exclusion and cultural dislocation (Blaikie et al., 2004; Cannon, 2008). Despite this, mainstream landslide studies in India largely remain technocratic, focusing on slope stability, precipitation thresholds and lithological assessments while neglecting how disasters are experienced differently by socially excluded groups (Thakur & Chauhan, 2020; Kumar et al., 2021).

Disaster risk reduction (DRR) strategies and policy interventions in tribal regions are frequently top-down, failing to engage with local epistemologies, indigenous coping mechanisms and cultural resilience frameworks (Kruks-Wisner, 2011; Shaw et al., 2009). For example, early warning systems and emergency protocols in many tribal belts of India, including the Nilgiris are either non-existent or poorly disseminated due to infrastructural deficits, linguistic barriers and governance gaps (Rajeev et al., 2021a, 2021b). This not only limits community preparedness but also reinforces feelings of marginalization and institutional neglect during recovery phases.

The Toda community of the Nilgiri Hills represents a compelling case for such analysis. Their socio-ecological worldview, spiritual relationship with the land and pastoralist traditions are intricately woven into the landscape of the Western Ghats a UNESCO-designated biosphere reserve (Baviskar, 2005; Gadgil & Guha, 1995). Yet, recurrent landslides triggered by erratic monsoons, road construction and tourism-related land-use changes have increasingly disrupted their way of life (Jain et al., 2013). While the biophysical triggers of these landslides have been documented, their cultural, spiritual and livelihood-related consequences on Toda communities remain critically underexplored.

This review study is thus both timely and essential. It seeks to bridge a significant research gap by interrogating the intersection of landslide hazards and tribal lived realities, thereby contributing to an emerging body of literature that advocates for human-centred, culturally responsive and equity-oriented disaster research (Fernando, 2012; Gaillard & Mercer, 2013). The study argues that building resilience in tribal areas cannot rely solely on technocratic solutions but must include context-specific strategies grounded in traditional ecological knowledge (TEK), cultural practices and localized governance structures (Nababan et al., 2020). By foregrounding tribal perceptions, values and responses to landslides, this review not only expands the theoretical lens of disaster studies but also offers practical insights for more inclusive risk communication, adaptive planning and policy design. Ultimately, such an approach advances the goals of the Sendai Framework for Disaster Risk Reduction (2015–2030), which emphasizes the need to engage all stakeholders including indigenous people in building community resilience and ensuring no one is left behind.

10.3 Materials and Methods

As a review-based inquiry, this study adopts a qualitative, integrative and interpretive approach to systematically examine the multidimensional impacts of landslides on tribal communities, with a focused lens on vulnerability, resilience and indigenous responses. The methodology was guided by established protocols for conducting narrative and scoping reviews in disaster studies and indigenous research contexts (Petticrew & Roberts, 2006; Arksey & O'Malley, 2005). The objective was not merely to catalogue existing data, but to synthesize diverse scholarly, institutional and community-based knowledge sources that reveal the socio-environmental intricacies of landslide experiences in tribal regions.

10.3.1 Data Sources and Selection Criteria

The review draws upon a comprehensive survey of peer-reviewed academic literature, grey literature, government documents and geospatial databases published between 2000 and 2024. Electronic searches were conducted across major academic databases including Scopus, Web of Science, ScienceDirect and JSTOR, using keywords such as: *'landslides AND tribal communities'*, *'indigenous vulnerability AND geohazards'*, *'Toda tribe AND landslides'*, *'disaster risk reduction AND indigenous knowledge'* and *'environmental hazards AND cultural resilience'*. In addition, policy documents from the National Disaster Management Authority (NDMA, India), Geological Survey of India (GSI), UNDRR and International Federation of Red Cross and Red Crescent Societies (IFRC) were incorporated to include institutional perspectives and response frameworks. Only literature that explicitly addressed the intersection of landslide hazards and indigenous or tribal communities—whether in India or comparable global settings was retained. Case studies from ecologically sensitive tribal regions such as the Western Ghats, Northeast India and select Himalayan tribal zones were given particular attention due to their topographical susceptibility and cultural specificity (NIDM, 2021; Reddy & Rao, 2014). The Toda community of the Nilgiri Hills was used as a localized reference point to contextualize the review within a micro-ethnographic and cultural ecological framework (Sathyapalan & Murugan, 2011).

10.3.2 Thematic Categorization and Analytical Framework

All included sources were subjected to thematic content analysis, following the guidelines of Braun and Clarke (2006), to identify recurring patterns and conceptual linkages across different scales of analysis—ecological, socio-political, cultural and policy-related. Key thematic codes included: *'land-use change and slope instability'*, *'traditional ecological knowledge (TEK)'*, *'cultural loss and displacement'*, *'early warning access'*, and *'adaptive coping mechanisms'*. Insights from indigenous case studies were cross-referenced with scientific findings on landslide causation and impact to enable triangulation (Bowen, 2009). This method facilitated the mapping of community narratives against formal risk assessments, thereby exposing both convergence and divergence in interpretation and response.

10.3.3 Quality Assurance and Ethical Considerations

In line with best practices for review-based research, source quality and credibility were assessed using the Critical Appraisal Skills Programme (CASP) checklist and guidelines from the PRISMA 2020 statement (Page et al., 2021). To ensure reflexivity and transparency, the positionality of indigenous voices within the literature was critically examined, especially in cases where external researchers may have interpreted tribal knowledge through extractive or reductionist lenses (Smith, 2012). Although no primary fieldwork was conducted for this review, ethical sensibilities guided source selection—favouring works that demonstrated community consent, participatory research or culturally respectful methodologies.

10.3.4 Justification for Focusing on the Toda Community

The Toda tribe of the Nilgiris was selected as a focal community within the review due to their recognized status as a Particularly Vulnerable Tribal Group (PVTG) and the growing frequency of landslides in their native montane grassland ecosystem (Ramachandran et al., 2020). As pastoralists with a unique spiritual ecology, the Toda's offer an important lens through which to understand the non-material consequences of landslides such as the desecration of sacred landscapes and erosion of ritual knowledge topics that remain underrepresented in disaster literature (Narayanan, 2018).

10.4 Results and Discussion

This section presents a synthesis of findings from the reviewed literature, organized thematically to reflect the multidimensional impacts of landslides on tribal communities. The discussion focuses on five core themes mentioned below. These themes converge to underscore the complexity of landslide impacts as experienced by tribal communities, particularly the Toda's of the Nilgiris.

10.4.1 Environmental Disruption and Ecological Fragility

Landslides in tribal territories are frequently linked to broader patterns of ecological degradation, including deforestation, monoculture plantations and unregulated construction activities (Dutta & Maji, 2020; Rai et al., 2012). In the Nilgiri Hills, the replacement of native shola-grassland ecosystems with commercial tea and eucalyptus plantations has increased slope instability and surface runoff, contributing to frequent landslides during monsoon periods (Meher-Homji, 1992; Ramachandran et al., 2020). For the Toda community whose spiritual ecology is deeply embedded in the native montane grasslands (mullai) the degradation of their environment has cascading effects not only on land use, but on ritual, pastoral and cultural practices (Narayanan, 2018). Global parallels can be drawn with other indigenous communities such as the Quechua in the Andes or the Dayak in Borneo, who have similarly experienced intensified landslide risks due to ecological fragmentation (Wadley et al., 2007; Zimmerer, 2014). This highlights the structural vulnerability of tribal lands positioned at the frontlines of environmental transition zones.

10.4.2 Socio-Economic Vulnerability and Livelihood Loss

A recurring insight across reviewed studies is the high dependency of tribal populations on land-based livelihoods such as shifting cultivation, herding or forest-based economies that are directly disrupted by landslides (Le De et al., 2015; Sengupta, 2020). For the Todas, pastoralism centred around the sacred buffalo is a primary occupation, intricately tied to seasonal grazing cycles and undisturbed grasslands. Landslides not only damage pastureland and access trails, but also destroy cattle shelters (munds) and water sources essential for herd sustenance. Loss of land following landslides is rarely compensated due to unclear land tenure rights and the marginal legal recognition of customary claims (Mahapatra & Das, 2021). In such settings, tribals often become economically disenfranchised, forced into wage labour or displacement, thus compounding intergenerational poverty cycles.

10.4.3 Cultural Displacement and Erosion of Traditional Knowledge

Unlike conventional disaster studies that focus primarily on physical loss and infrastructure damage, indigenous experiences of landslides also entail cultural and symbolic dislocation. For the Toda's, landslides that affect sacred spaces such as ancestral forests, water bodies or ceremonial temples are experienced not just as environmental disasters but as cosmological ruptures (Narayanan, 2018; Sundar, 2005). This dimension is echoed in global ethnographic research where disasters are interpreted through spiritual ontologies, as seen among Native American or Ainu communities (Parisi, 2020; Siddle, 2006). The displacement of elders and the disruption of place-based oral traditions reduce the transmission of traditional ecological knowledge (TEK) a knowledge system that often contributes to pre-disaster environmental reading and adaptive strategies (Berkes, 2008). The long-term consequence is cultural erosion that undermines both identity and resilience.

10.4.4 Governance Gaps and Policy Exclusion

Review findings highlight a critical disconnect between tribal realities and state-led disaster governance frameworks. Most early warning systems, rehabilitation packages or climate adaptation plans are not designed with tribal contexts in mind and often exclude tribal voices from planning and consultation processes (Government of India, 2019; Mukherjee & Shaw, 2013). The Toda community, despite being designated as a Particularly Vulnerable Tribal Group (PVTG), remains underrepresented in disaster management planning at the district and state levels.

Further, disaster response infrastructures such as shelters, relief distribution and evacuation routes are either inaccessible or culturally inappropriate (Gaillard et al., 2008). As a result, tribals are left to rely on informal kinship networks and self-organized mechanisms, which, while adaptive, lack institutional support and long-term protection.

10.4.5 Indigenous Resilience and Adaptive Strategies

Despite their vulnerabilities, tribal communities demonstrate remarkable resilience and ingenuity in the face of recurring landslide threats. The Todas, for instance, have historically practiced landscape-sensitive construction, avoiding high-slope gradients and preserving vegetative buffers around settlements (Sathyapalan & Murugan, 2011). Their deep environmental literacy enables them to identify early warning signs such as animal behaviour changes, soil fissures or water discoloration (Rajeev et al., 2021a, 2021b).

Moreover, cultural institutions—such as ritual assemblies, storytelling and collective decision-making—function as community-level coping mechanisms, allowing for shared memory, coordinated response and psychosocial support (Fernando, 2012). Such indigenous systems, however, require institutional recognition and integration into formal disaster planning in order to be sustained and scaled.

10.4.6 Synthesis and Knowledge Gaps

While the reviewed literature offers significant insights into the vulnerabilities and adaptive capacities of tribal communities facing landslides, several gaps persist:

Most studies are geographically fragmented, limiting comparative understanding across tribal geographies. Cultural impacts, especially in the form of spiritual dislocation and knowledge loss, remain under-researched. The potential for integrating TEK into modern early warning and resilience frameworks is yet to be fully explored. Addressing these gaps is critical to developing inclusive, trans-disciplinary and participatory DRR models, especially in the context of intensifying climate-induced geohazards.

10.5 Limitations of the Study

While this review paper offers a valuable synthesis of existing literature and illuminates the often-overlooked perspectives of tribal communities in the context of landslide vulnerability and resilience, several limitations must be acknowledged to contextualize its findings.

First, the geographic focus of the study, while deliberately anchored in tribal regions of the Western Ghats (particularly the Toda community of the Nilgiris) and select cases from Northeast India, inevitably limits the generalizability of its conclusions. India is home to over 700 recognized tribal groups, each with unique cultural, ecological and political landscapes (Xaxa, 2005). Regions such as Central India (e.g. Gond, Baiga communities) and the Eastern Himalayas (e.g. Lepcha, Monpa tribes) which are also highly susceptible to landslides were not extensively covered due to the scope and design constraints of the review. Thus, while the paper identifies broad patterns and themes, it does not claim to represent the full heterogeneity of tribal experiences across India's diverse eco-cultural zones.

Second, as a review-based study, this research is inherently dependent on the availability and accessibility of secondary data sources, which may reflect publication biases or epistemological gaps in existing literature. For instance, there is a noticeable paucity of peer-reviewed material focusing on tribal perceptions of geohazards, especially in indigenous languages or non-academic formats (Kelman et al., 2012). Many tribal narratives, practices and traditional coping mechanisms are passed down orally and remain undocumented in formal repositories, thereby eluding systematic scholarly analysis. This limitation underscores a broader structural issue within disaster studies: the underrepresentation of indigenous epistemologies in academic discourse (Smith, 2012; Whyte, 2013).

Third, the linguistic and conceptual translation of indigenous knowledge—especially in the reviewed ethnographic and qualitative studies presents methodological challenges. Terms such as 'hazard,' 'resilience' or 'disaster' often lack direct equivalents in tribal dialects, leading to interpretive ambiguities even when community members are engaged as research participants or informants (Kruks-Wisner, 2011). Several reviewed studies acknowledged this limitation, pointing to potential loss of nuance in translating culturally embedded concepts related to land, spirits or ancestral relations elements that are central to understanding indigenous responses to ecological threats. There is the issue of positionality and trust in disaster research involving tribal populations. Even in studies employing participatory or ethnographic methods, researchers often confront deep-rooted scepticism or fear among indigenous communities due to historical marginalization, bureaucratic neglect or prior extractive encounters with outsiders (Agrawal, 2002; Sundar, 2010). These dynamics can influence what is shared, how it is interpreted and the depth of engagement achieved. Although many of the studies cited in this review adhered to ethical research guidelines, long-term trust-building and community ownership of research remain crucial gaps in most existing literature.

Finally, while this paper advocates for the integration of traditional ecological knowledge (TEK) into landslide risk management frameworks, it recognizes that

TEK is not static or universally applicable. It evolves over time and varies not only between tribal groups but also within them, across gender, age and status (Berkes, 2008). The challenge of avoiding essentialism, i.e. portraying tribal communities as homogenous or inherently 'resilient' remains a critical concern that scholars must continue to navigate.

These limitations do not diminish the value of the insights generated but rather emphasize the need for future research that is longitudinal, multilingual, participatory and co-produced with indigenous communities. Bridging the epistemic divide between scientific and indigenous knowledge systems is essential for developing a truly inclusive and culturally sensitive disaster risk reduction paradigm.

10.6 Recommendations

- Integrate indigenous knowledge systems into scientific early warning models.
- Establish community-based disaster preparedness programs tailored to tribal contexts.
- Ensure participatory land-use planning to prevent deforestation and slope destabilization.
- Enhance livelihood diversification to reduce economic vulnerability.
- Promote tribal education on climate adaptation and resilience practices.

10.7 Conclusions

This review has highlighted the urgent need to reframe landslide risk analysis and disaster management through the lens of tribal communities—populations whose vulnerabilities are shaped not merely by geography, but by intersecting historical, socio-political and cultural dynamics. The case of the Toda tribe in the Nilgiri Hills serves as a microcosm through which broader patterns of exclusion, ecological disruption and resilience can be understood. While technical studies have advanced our understanding of the physical drivers of landslides, they often overlook how such events are interpreted, experienced and addressed by indigenous populations. As this review demonstrates, tribal communities are disproportionately affected by landslides not only due to their physical exposure, but due to structural marginalization, lack of political voice, ambiguous land rights and the erosion of traditional ecological knowledge (TEK) (Blaikie et al., 2004; Gaillard & Mercer, 2013). Their socio-ecological systems are intricately tied to land and environment and when landslides occur, the damage extends beyond physical loss to spiritual dislocation, livelihood collapse and cultural fragmentation. At the same time, the review also reveals untapped potential in indigenous adaptive strategies and communal resilience frameworks. The Toda community's historically grounded land-use practices, spiritual cosmology and social cohesion offer valuable lessons for reimagining risk reduction in culturally grounded and environmentally sustainable terms (Narayanan, 2018; Sathyapalan & Murugan,

2011). However, without formal recognition or integration into disaster policy frameworks, these knowledge systems risk being dismissed or lost.

Disaster management agencies should systematically integrate Traditional Ecological Knowledge (TEK) into early warning systems, hazard mapping and recovery planning. Community-driven indicators such as changes in animal behaviour, soil conditions or water flows—should be recognized as complementary to scientific forecasting (Mercer et al., 2009; UNDRR, 2022). Generic, top-down approaches fail to account for cultural variation. DRR plans must be localized and culturally sensitive, designed in collaboration with tribal councils, elders and women's collectives. This includes the design of evacuation protocols, shelters and communication systems in local languages and symbolic forms (Kelman et al., 2012). Ensuring secure land tenure and the recognition of customary rights are foundational to reducing landslide vulnerability. Tribal communities must have legal protections against land dispossession due to development, tourism or post-disaster resettlement (Mahapatra & Das, 2021; Xaxa, 2005).

Future research and planning must move beyond extractive methodologies. A participatory, decolonial approach is needed where tribal voices are not just subjects of study but co-producers of knowledge and policy (Smith, 2012; Whyte, 2013). This entails long-term community engagement, multilingual tools and ethical co-design of research agendas. Particularly Vulnerable Tribal Groups (PVTGs), such as the Todas, require dedicated funding lines for landslide mitigation, sustainable livelihoods and cultural preservation. This includes training local youth in DRR skills, GIS mapping and traditional environmental monitoring.

References

Agrawal, A. (2002). Indigenous knowledge and the politics of classification. *International Social Science Journal, 54*(173), 287–297.

Arksey, H., & O'Malley, L. (2005). Scoping studies: Towards a methodological framework. *International Journal of Social Research Methodology, 8*(1), 19–32.

Bankoff, G. (2004). The historical geography of disaster: "Vulnerability" and "local knowledge" in western discourse. In Bankoff et al (Ed.), *Mapping Vulnerability: Disasters, Development and People.*

Baviskar, A. (2005). *In the Belly of the River: Tribal Conflicts over Development in the Narmada Valley.* Oxford University Press.

Berkes, F. (2008). *Sacred Ecology: Traditional Ecological Knowledge and Resource Management.* Routledge.

Blaikie, P., Cannon, T., Davis, I., & Wisner, B. (2004). *At Risk: Natural Hazards, People's Vulnerability and Disasters* (2nd ed.). Routledge.

Bowen, G. A. (2009). Document analysis as a qualitative research method. *Qualitative Research Journal, 9*(2), 27–40.

Braun, V., & Clarke, V. (2006). Using thematic analysis in psychology. *Qualitative Research in Psychology, 3*(2), 77–101.

Cannon, T. (2008). *Reducing people's vulnerability to natural hazards: Communities and resilience.* Research Paper No. 2008/34, UNU-WIDER.

Cruden, D. M., & Varnes, D. J. (1996). Landslide types and processes. In A. K. Turner & R. L. Schuster (Eds.), *Landslides: Investigation and mitigation*, (pp. 36–75).

Dai, F. C., Lee, C. F., & Ngai, Y. Y. (2002). Landslide risk assessment and management: An overview. *Engineering Geology, 64*(1), 65–87.

Dutta, R., & Maji, A. (2020). Land use change and landslide vulnerability in India's hilly regions. *Environmental Monitoring and Assessment, 192*(2), 105.

Fernando, J. L. (2012). *The Power of Development: Perspectives from the Periphery.* Pluto Press.

Froude, M. J., & Petley, D. N. (2018). Global fatal landslide occurrence from 2004 to 2016. *Natural Hazards and Earth System Sciences, 18*, 2161–2181.

Gadgil, M., & Guha, R. (1995). *Ecology and Equity: The Use and Abuse of Nature in Contemporary India.* Routledge.

Gariano, S. L., & Guzzetti, F. (2016). Landslides in a changing climate. Earth-Science Reviews, 162, 227–252.

Gaillard, J. C. (2007). Vulnerability, capacity and resilience: Perspectives for climate and sevelopment policy. UNU-EHS.

Gaillard, J. C., & Mercer, J. (2013). From knowledge to action: Bridging gaps in disaster risk reduction. *Progress in Human Geography, 37*(1), 93–114.

Gaillard, J. C., Maceda, E. A., & Stasiak, E. (2008). Sustainable livelihoods and disaster risk reduction in the Philippines. *Disasters, 32*(2), 267–283.

Glade, T., Anderson, M. G., & Crozier, M. J. (Eds.). (2005).*Landslide hazard and risk.* John Wiley & Sons.

Government of India. (2019). *Tribal Sub-Plan and Disaster Preparedness Guidelines.* Ministry of Tribal Affairs.

Guzzetti, F., Carrara, A., Cardinali, M., & Reichenbach, P. (2005). Landslide hazard evaluation: A review of current techniques and their application in a multi-scale study Central Italy. *Geomorphology, 31*(1–4), 181–216.

Highland, L. M., & Bobrowsky, P. (2008). *The landslide handbook—A guide to understanding landslides.* U.S. Geological Survey.

IFRC. (2021). World disasters report. International federation of red cross and red crescent societies.

Jain, S. R., Jeyaseelan, A. T., & Ramasamy, S. M. (2013). Landslides in the Nilgiris: An environmental perspective. *Journal of the Geological Society of India, 82*(5), 485–496.

Keller, E.A. (2011). *Introduction to Environmental Geology*. Pearson.

Kelman, I., Mercer, J., & Gaillard, J. C. (2012). Indigenous knowledge and disaster risk reduction. *Geoscience and Remote Sensing, 45*(3), 23–29.

Kruks-Wisner, G. (2011). Seeking the local state: Gender, caste, and the pursuit of public services in post-Tsunami India. *World Development, 39*(7), 1143–1154.

Kumar, R. (2018). Cultural loss due to landslides. *Anthropological Survey of India Bulletin, 47*(1), 22–37.

Kumar, R., Agarwal, A., & Saha, A. (2021). Landslide hazard and vulnerability analysis: A case from Sikkim Himalayas. *Natural Hazards, 107*, 231–251.

Le De, L., Gaillard, J. C., & Friesen, W. (2015). Disasters and social resilience: A synthesis. *Progress in Human Geography, 39*(3), 348–365.

Mahapatra, L. K., & Das, S. (2021). Customary land rights of tribal people in India: A legal and administrative analysis. *Indian Journal of Legal Studies, 12*(1), 45–61.

Martha, T. R., Kerle, N., Jetten, V., van Westen, C. J., & Kumar, K. V. (2010). Characterizing spectral, spatial and morphometric properties of landslides for semi-automatic detection using object-oriented methods. *Geomorphology, 116*(1–2), 24–36.

McGuire, B. (2002). *A Guide to the End of the World: Everything You Never Wanted to Know.* Oxford University Press.

Meher-Homji, V. M. (1992). Ecology of the shola-grassland biome in the Nilgiris. *Environmental Conservation, 19*(1), 59–61.

Mercer, J., Kelman, I., Taranis, L., & Suchet-Pearson, S. (2009). Framework for integrating indigenous and scientific knowledge for disaster risk reduction. *Disasters, 33*(4), 1–22.

Mehta, L. (2007). Displaced by development and disasters. *IDS Bulletin, 38*(5), 43–54.

Mishra, P. K. (2014). Resilience among hill tribes. *International Journal of Social Science Research, 2*(3), 54–69.
Mukherjee, S., & Shaw, R. (2013). Indigenous communities and disaster risk reduction in India. *International Journal of Disaster Risk Science, 4*(2), 71–80.
Nababan, A. M., Jodoin, S., & Krupnik, T. J. (2020). Indigenous knowledge in environmental governance. *Sustainability Science, 15*, 1245–1255.
Naithani, A. (2017). Climate change, tribal livelihoods and landslides. *Indian Journal of Ecology, 44*(2), 300–306.
Narayanan, Y. (2018). Relational ontologies and sacred ecology: The Toda buffalo and hill gods of the Nilgiris. *Geoforum, 91*, 1–9.
NDMA India. (2020). *National Landslide Risk Management Strategy*. Retrieved from https://ndma.gov.in
NIDM. (2021). *Status of Disaster Risk in India: A State-wise Analysis*. National Institute of Disaster Management.
Oliver-Smith, A. (1996). Anthropological research on hazards and disasters. *Annual Review of Anthropology, 25*(1), 303–328.
Page, M. J., et al. (2021). The PRISMA 2020 statement: An updated guideline for reporting systematic reviews. *BMJ, 372*, Article n71.
Parisi, A. (2020). Sacred geography and ecological memory among the Hopi. *Cultural Geographies, 27*(2), 243–259.
Petley, D. N. (2012). *Mass Movements*. Geological Society.
Petticrew, M., & Roberts, H. (2006). *Systematic Reviews in the Social Sciences: A Practical Guide*. Blackwell Publishing.
PIB India. (2019). *Disaster Management Frameworks and Tribal Welfare*. Retrieved from https://pib.gov.in
Rai, N. D., Sandhu, C., & Menon, A. (2012). A framework for integrating biodiversity and ecosystem services into climate change adaptation. *Environmental Management, 49*(3), 479–492.
Rajeev, K., Ramasamy, S. M., & Ramkumar, T. (2021a). Mapping community-based landslide disaster vulnerability and resilience in Nilgiris. *Geocarto International, 36*(4), 395–411.
Rajeev, K., Ramasamy, S. M., & Ramkumar, T. (2021b). Mapping community-based landslide disaster vulnerability in Nilgiris. *Geocarto International, 36*(4), 395–411.
Ramachandran, A., Balasubramanian, D., & Subramanian, M. (2020). Recent landslides and land-use transformations in the Nilgiris. *Indian Journal of Environmental Protection, 40*(4), 289–296.
Reddy, C. S., & Rao, P. P. (2014). Assessment and mapping of landslide vulnerability in parts of Eastern Himalayas and Western Ghats using geospatial tools. *Environmental Earth Sciences, 72*(7), 2675–2686.
Sathyapalan, J., & Murugan, M. (2011). Indigenous knowledge and disaster resilience: The Toda experience. *Indian Journal of Traditional Knowledge, 10*(1), 96–102.
Sen, A. (2010). Displacement and livelihood loss due to natural disasters. *Economic and Political Weekly, 45*(39), 45–50.
Sengupta, R. (2020). Livelihood disruptions and adaptation post-landslides in Darjeeling Himalayas. *Mountain Research and Development, 40*(2), R19–R28.
Shaw, R., Krishnamurthy, R. R., & Nakagawa, Y. (2009). *Indigenous knowledge and disaster risk reduction: From practice to policy*. UNESCO.
Singh, R. B. (2010). *Urban Development and Environmental Change*. Kluwer.
Smith, L. T. (2012). *Decolonizing Methodologies: Research and Indigenous Peoples* (2nd ed.). Zed Books.
Sundar, N. (2005). *Subalterns and Sovereigns: An Anthropological History of Bastar*. Oxford University Press.
Sundar, N. (2010). *The Scheduled Tribes and Their India: Politics*. Oxford University Press.
UNDRR. (2021). *Disaster Risk Reduction and Indigenous Peoples*. Retrieved from https://www.undrr.org

Wadley, R. L., Eilenberg, M., & Singh, P. (2007). Shifting cultivation and forest loss in Borneo. *Asia Pacific Viewpoint, 48*(3), 305–319.
Whyte, K. (2013). On the role of traditional ecological knowledge as a collaborative concept: A philosophical study. *Ecological Processes, 2*(1), 7.
Wisner, B., et al. (2004). *At Risk*. Routledge.
Xaxa, V. (2005). Politics of language, religion and identity: Tribes in India. *Economic and Political Weekly, 40*(13), 1363–1370.
Zimmerer, K. S. (2014). Conserving agrobiodiversity amid global change: The Andes as a zone of cultural and ecological interaction. *Annals of the Association of American Geographers, 104*(3), 602–611.

Community Perspectives on Landslide Mitigation Strategies

11

Bhagyasree Dolui

Abstract

The Nilgiris of the Western Ghats in Tamil Nadu has a long history of landslides in the area. Landside processes are practically seasonal in the district. Landslides in the affected area have a wide impact on the people living in the area as a result of the destruction they cause to human and material resources. Already-existing urban areas exist in Nilgiris, where the structures are constructed on or close to landslides. The poor population group resides in the landslide-prone regions and does not have the means and skills to structure themselves in the rehabilitation programs. As such, a research on how people perceive the need to mitigate landslide hazard and the mitigation measures is essential. This chapter focused on how the people of Kothagiri taluk perceived the risk of landslides. In attempting to come up with systems, practises and policies to safeguard the local population. This is particularly so wherein the risk mitigation measures are non-structural measures such as relocation or early warning systems that allows active involvement of the concerned community in the process. The sample size of the study is 270, and the research has considered the individuals who are older than 18 years. The data collected in the study is the primary one, which is obtained with the help of a pre-designed questionnaire. The statistical analysis of the data involved frequency, t test of independent sample and regression modelling. It is the final fact found in the study that people living in smaller families and have fewer family members are at the risk compared to people living in larger families. Most of the responders believe that with a large family, evacuation is a safer and easier process among young children and the older members of the family. In case of a calamity, it is good to get advice of older people in the family who experienced landslides before.

B. Dolui (✉)
Department of Geography, Queen Mary's College, Chennai, Tamil Nadu, India
e-mail: bshree40@gmail.com

R. M. Yuvaraj et al. (eds.), *Advanced Landslide Susceptibility Modeling*,
https://doi.org/10.1007/978-3-032-14485-0_11

Keywords

Disaster • Landslide hazard • People's perception • Statistical analysis • Mitigation strategies

11.1 Introduction

Landslides are considered to be one of the most devastating natural calamities and can dramatically destroy human settlements, critical infrastructure as well as delicate eco systems. Such dangerous occurrences are the most common in the mountainous and hilly areas, where the steep slopes, the condition of the soils and the extreme weather patterns are the contributing factors. A landslide may also cause serious effects, such as the displacement of the people, death, and destruction of roads and buildings, as well as great economic losses United Nations Office for Disaster Risk Reduction (UNDRR, 2021). The devastating effect of landslides is further enhanced by the effector impacts like floods, blockage of rivers, soil erosion which may be long lasting even after the occurrence.

Over the past years, the rate and severity of landslides have been escalating as a result of different human and environmental factors. Climate change has also resulted in a more unpredictable rainfall pattern, extended droughts and increased temperatures, which are factors that contribute to unstable slopes. Also, intensive deforestation, which is frequently caused by agricultural activities, logging, and constructions, has compromised the natural defenses that inhibit the soil erosion and slope movements. Unplanned urbanization especially in the developing areas has also contributed to the issue because when building is undertaken on steep and unstable slopes this makes it vulnerable to landslides. The increasing dangers emphasize the necessity to develop efficient landslide risk detection and reduction measures that will enable the communities to be more prepared and ready to address the disaster (Froude and Petley, 2018).

The contemporary scientific methodology, such as remote sensing, GIS-based mapping, and geotechnical engineering solutions, is important in the assessment and reduction of landslides. Although, these sophisticated methods tend to be expensive in financial terms, technical and institutional backup and thus less affordable to the rural and indigenous communities. Conversely, traditional knowledge, which has been built over generations by the indigenous and local population, offers an inexpensive, practical, and culturally pertinent option on how to determine regions at risk of landslides and put up mitigation measures. A strong sense of local knowledge of local landscapes and environmental patterns has been a key element in minimizing landslides many centuries ago (Mercer et al. 2010).

The traditional knowledge systems are very broad and comprise of numerous observational methods and adaptive measures that guide the communities to forecast, avoid and act in case of landslides. Signs that are normally used in predicting the occurrence of landslides in the local populations have always included the abnormal soil motion, ground cracks, vegetation cover changes, and accelerated seepage of water. These natural signs act like warning signs, which enable individuals to take up precautionary measures before something takes place. Moreover, local culture like sustainable land use, terracing, agroforestry, and afforestation have a positive effect in alleviating the susceptibility of landslides. Examples of this include terracing which prevents the rapid movement of water and makes the soil firmer whereas afforestation makes the roots stronger to hold the soil together thus making it less prone to slope failure (Petheram et al. 2010).

The combination of traditional knowledge and scientific methodology provides a comprehensive way of dealing with landslide disasters. Although modern technologies offer high accuracy in mapping, real time observation, and predictive modeling the traditional knowledge offers the localized knowledge, historical insights, and environmental sustainability. The importance of discovering indigenous wisdom in disaster risk management guarantees that mitigation plans can be more than merely effective technically: they have a culture-specific, community-based, and flexible reaction (UNESCO, 2020).

Landslides are a major problem in the Nilgiri District in Tamil Nadu which is hilly, steep slopes, and has heavy monsoonal rainfall. The causes of common landslides in the area are categorized by the geological and topographical features, whereby loose soil, broken rock structures, and steep slopes are predisposing factors that may cause slope failures (Yuvaraj and Dolui, 2024). Southwest (June–September) and Northeast (October–December) monsoons cause massive rainfall, which contributes significantly to landslides through increasing soil moisture and making the slopes weak. The vegetation cover has been decreased by deforestation and land management activities like uncontrolled construction, deforestation, and agricultural activities and this has caused soil erosion and instability. Also, the seismic activity of the area predisposes the area to landslides since any slight tremor can cause already weak slopes to be disturbed. Poor drainage systems, road cutting and development of infrastructure are also other human activities that cause slope destabilization. The district has had record landslides within the recent years such as the landslides of 2009 that brought about a lot of destruction and loss of lives due to excessive rainfall. More recently, the 2019 and 2020 landslides caused several road closures and displacement of citizens. The works of GIS and remote sensing have outlined the high-risk areas, especially Kotagiri, Coonoor, and Udhagamandalam (Ooty) (Yuvaraj and Dolui, 2021). To this end, a number of mitigation measures have been put in place. Geological Survey of India (GSI) and the government of Tamil Nadu have also launched mapping on land slide prone areas and early warning. Afforestation and soil conservation programs are already in place to reinstate vegetation and stabilize the slope and proper drainage management in the formulation process is underway to avoid water stagnation. Moreover, the government authorities are busy doing the work of educating the vulnerable people on how to evacuate during landslides in a safe manner. Nilgiri District is a high-risk region in terms of landslides, therefore, the area is a hot spot in the process of risk assessment and mitigation planning.

This paper addresses how the traditional knowledge can be used in identifying and mitigating the risk of landslides with special focus on the importance of such information in supplementing the scientific procedures. This study will focus on showing the significance of maintaining indigenous knowledge and incorporating it into the modern disaster risk management systems by comparing case studies in different geographical locations. This way of doing things can increase resilience of the communities, encourage sustainable environmental management and finally lead to the inception of safer and more stable human settlement in prone landslide regions.

11.2 Study Area

Kotagiri taluk is a part of the Nilgiri District of Tamil Nadu in India and also one of the key hill stations of the region. The Nilgiri District is located under the Western Ghats and is known to have an ecological importance and also a wide range of topography. Kotagiri, being 1847 meters above the sea level, has a cool climate throughout the year, and this makes it a good location to plant tea and eco-tourism (Tamil Nadu Forest Department, 2020).

The latitude of Kotagiri taluk is around 11.43° N and the longitude is 76.88° E. It is flanked by Coonoor and Ooty blocks in the Nilgiri District referred to as Fig. 11.1. The topography of the taluk is described as hilly, forest-covered, and large farm areas, mainly tea plantations (Thennavan and Pattukandan, 2020). Some small streams drain the area to the Bhavani River, a major tributary of the Cauvery River. Kotagiri has subtropical highland climate and moderate temperatures vary between 10° C in winter to 25° C in summer. Its rainfall is about 1200 mm which falls mostly on the southwest monsoon (June to September) and northeast monsoon (October to December) (India Meteorological Department, 2019). The humid condition and the heavy rainfall favor the abundance of biodiversity and crop yield of the region.

The Kotagiri taluk land use pattern includes tea estates, forested lands, grasslands and settlements respectively. Native shola forests and montane grasslands are found in the region as part of the Nilgiri Biosphere Reserve which is a UNESCO-known site (UNESCO, 2021). Flora Here the flora contains species like Rhododendron nilagiricum, Michelia nilagirica and other native plants. The wildlife includes Indian gaur (Bos gaurus), Nilgiri langur (Trachypithecus johnii), and Malabar giant squirrel (Ratufa indica). The economy of Kotagiri is mainly founded on the cultivation of tea, tourism and the related agriculture practices. Its rich cultural heritage has been formed by the history of the native Toda and Kota community living in the territory. Tourism has been found to rise in terms of infrastructural development, as well as raising issues of environmental sustainability due to the increasing tourism industry (Sundaram et al. 2021). Kotagiri Taluk in the Nilgiri District of Tamil Nadu is quite prone to landslides as the region has steep slopes and receives heavy rainfall during the South and North-East monsoon seasons (Dolui et al. 2023).

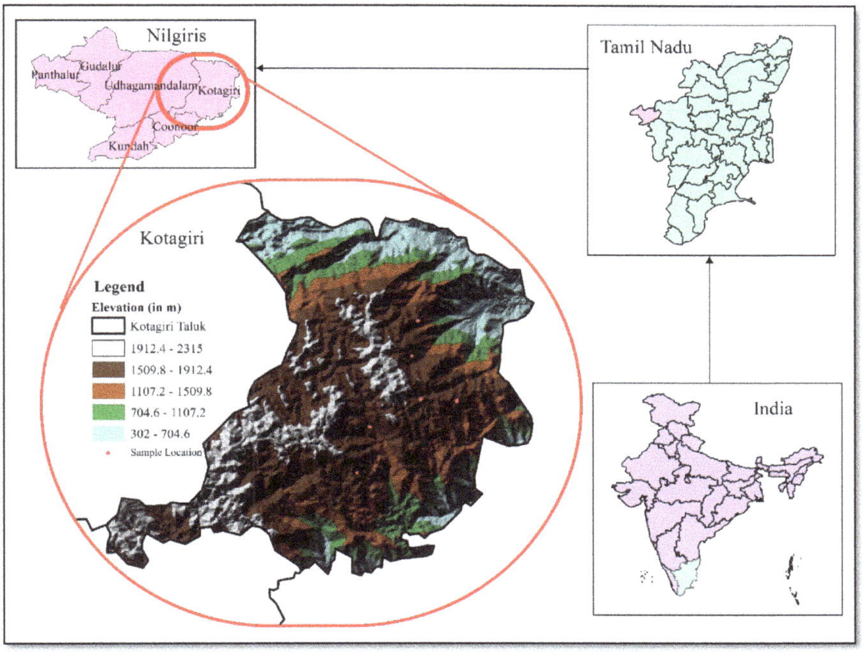

Fig. 11.1 Study area—Kotagiri taluk Nilgiri district

11.3 Data and Methodology

Primary data was collected through a well-formulated questionnaire, and the face-to-face interviews were carried out with tribal communities living in high-risk zones of landslides in Nilgiri District. The questionnaire contained socio-economic information and it played the key role of determining the economic losses experienced as a result of landslides. It was majorly based on the view of the tribal residents in relation to landslide mitigation measures. The questionnaire contained in total 19 open ended questions. This study will be targeting the tribal population living in Kotagiri Taluk of Nilgiri District. A preliminary effort was to gather data of a sample of 260 people to have a margin of error of 6%. Nonetheless, the number of respondents who gave usable responses was only 170 because of a number of issues that included non-responsiveness and incomplete responses. As a result, the sample size in this study was calculated as 170 and this led to a new margin of error of 7.5%.The sample size was calculated using the following equation:

$$n_o = \frac{Z^2pq}{e^2}$$

where

e is the desired level of precision (i.e., the margin of error); p is the estimated proportion of the population which has the attribute in question; q is 1-p.

Snowball sampling, a non-probability sampling technique, was employed to collect data from respondents residing in Kotagiri Taluk in the Nilgiri District. A pilot study was conducted in July 2022 to evaluate the predefined questionnaire, while the main data collection took place from December 2022–February 2023.

At-test is a statistical method used to analyze the difference between the means of two groups (pairwise comparison). It is commonly applied in hypothesis testing to assess whether a particular treatment or process influences the target population or if there is a significant difference between the two groups. The formula for calculating the independent t-test is as follows:

$$t = \frac{y_1 - y_2}{S_v\sqrt{\frac{1}{m_1} + \frac{1}{m_2}}}$$

$$S_v = \sqrt{\frac{(m_1 - 1)S_1^2 + (m_{2-1})S_2^2}{m_1 + m_2 - 2}}$$

where

T—Test statistics; Y_1—Mean of first sample; Y_2—mean of second sample; M_1—Size of first sample; M_2—Size of second sample; S_v—pooled standard deviation; S_1—Standard deviation of first sample; S_2—Standard deviation of second sample.

Multiple Regressions

$$\gamma = \partial_0 + \partial_1 x_1 + \partial_2 x_2 + \cdots + \partial_t x_t + \epsilon$$

where

γ—Expected value of dependent variable; X_1, x_2,...x_t—predictor independent variables; ∂_0—constant; ∂_1, $\partial_2 \ldots \partial_t$—estimated regression coefficient.

11.4 Results and Discussion

The population structure indicates that older people constitute a greater percentage as 45.88% of the population falls under the 45–65 years bracket of 41.76% years old and above. On the contrary, the proportion of young people (below 25 years old) constitutes 4.12% and the age group aged 25–45 years constitutes 8.24%, which implies an ageing population structure (United Nations, 2019). The population is male oriented (58.82–41.18%) indicating a gender gap that can be connected with the migration pattern of occupation (World Bank, 2021). With regards to education, 24.71% of the populace is illiterate with 33.53% having reached primary education, 28.82% having reached secondary education and only 12.94% having gone through higher education. These statistics indicate that higher education is still not accessible, especially in rural or semi-urban regions (UNESCO, 2020). The largest source of livelihood is agriculture where 45.88% of the population is employed. The other significant jobs are animal husbandry (17.06%), agricultural labor (16.47%), and government service (16.47%). The engagement in the sales of firewood is lower (4.12%), which explains the dependence on agrarian employment (FAO, 2022). The income level shows that 58.24% of the population has less than 5000 per month income with 33.53% having income between 5000–15,000. The low-income households are dominant, with only 8.24% of people making over 15,000 (International Labour Organization [ILO], 2021). There are 70.59% married and 20.59% unmarried and 8.82% widow as indicated in the marital status data. This trend follows the past family-related organizations and cultural practices that prioritize the institution of marriage (National Family Health Survey [NFHS], 2020). Regarding housing, 45.88% live in pucca houses, 41.76% in semi-pucca houses and 12.35% in kachcha houses, which means moderate infrastructure development (Government of India, 2021). Joint families (74.71%) are more dominant compared to nuclear families (25.29%), which is a factor that shows its traditional form of living, which is common in the rural societies (Census of India, 2011). The size of the family is also different, and the proportion of 58.24% of households are made of 4–8 members, 29.41% of less than 4 members, and 12.35% of more than 8 members. This pattern is consistent with the fact that most families in the area are joints (NFHS, 2020). Most of the population (78.20%) has resided in the area over 25 years and 12.90% of the population has resided in the area between 10–25 years with only 8.80% of the population living less than 10 years implying low level of migration and a resistant settlement pattern as demonstrated in Table 11.1.

Mitigation of landslide and its perception to be made of nine variables, which include Having Own Mitigation, Government Mitigation, Alert from the Government, Idea of Relocation, Willingness to Relocate, Mitigation through own Expenses, Accept Inconvenience, Mitigate Financial Loss and Seeking of Information. These are the variables under Mitigation, which are borrowed in various literatures on perception of landslides that each variable has quantified in terms of three point likert scale with an average of 1.5.

11.4.1 Descriptive for Mitigation Towards Landslide

It has been concluded that the willingness of the respondent to relocate has high mean value ($M = 2.835$) when compared to other variables in perception of mitigation towards landslide followed by seeking information variable has high mean ($M = 2.418$). The mean ($M = 1.376$) of mitigation using own expenses is low compared to other variables presented in Table 11.2. Based on Fig. 11.2, it is discovered that the perception of the respondent on the mitigation towards landslide is above average on the variables Having own mitigation, Alert from the government,

Table 11.1 Descriptive analysis

Variables	Category	Percent
Age	< 25	4.12
	25–45	8.24
	45–65	45.88
	> 65	41.76
Gender	Male	58.82
	Female	41.18
	Illiterate	24.71
Education	Primary	33.53
	Secondary	28.82
	Higher	12.94
Occupation	Agriculture	45.88
	Animal husbandry	17.06
	Agriculture labor	16.47
	Firewood sale	4.12
	Government servant	16.47
Monthly income	<5000	58.24
	5000–15,000	33.53
	>15,000	8.24
Marital status	Married	70.59
	Unmarried	20.59
	Widower	8.82
Type of house	Kachcha	12.35
	Pucca	45.88
	Semi-pucca	41.76
Type of family	Nuclear	25.29
	Joint	74.71
Family size	<4	29.41
	4–8	58.24
	>8	12.35
Year of stay	<10 years	8.80
	10–25 years	12.90
	>25 years	78.20
	Total	100.0

Table 11.2 Descriptive statistics—mitigation towards landslide

S. No.	Variables	Mean	SD	Count
1	Having own mitigation	2.082	0.479	170
2	Government mitigation	1.453	0.499	170
3	Alert from the government	2.212	0.911	170
4	Idea of relocation	1.500	0.501	170
5	Willingness to relocate	2.835	0.551	170
6	Mitigation through own expenses	1.376	0.486	170
7	Accept inconvenience	1.671	0.552	170
8	Mitigate financial loss	1.588	0.867	170
9	Seeking of information	2.418	0.641	170
	Mitigation	17.188	2.979	170

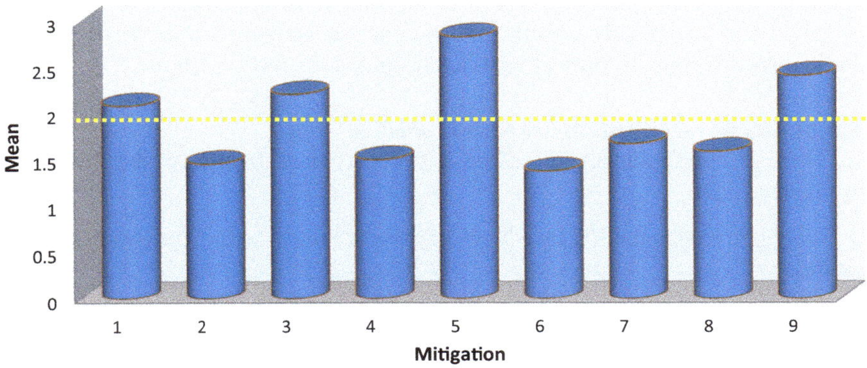

Fig. 11.2 Mean values—mitigation towards landslide

willingness to relocate and seeking of information but lower on the variables Government mitigation, Idea of relocation, Accept inconvenience, Mitigate financial loss and mitigation through own expenses. The total Mean score of the perception of the respondent on mitigation towards landslide is 17.188 equivalent to 63.65%. This implies that the perception by the respondent towards landslide is more than 63 percent on mitigation.

11.4.2 Independent Sample't' Test

Gender—Mitigation towards Landslide

H_0: There is no significant difference between Male and Female on the mitigation towards landslide.

Table 11.3 Independent samples t-test—mitigation towards landslide

t	df	p Value	95.0% Confidence interval		Estimated mean	
			Lower	Upper	Female	Male
−4.7917	168	0.000	−2.955	−1.230	15.95	18.05

$\mathbf{H_1}$: There is significant difference between Male and Female on the mitigation towards landslide.

A t-test that is independent has been performed to test the significant difference between the male and female respondents in the light of the mitigatin towards landslide in the Nilgiri district. Test indicates that the null hypothesis has been rejected because of low alpha value is 0.000 (< 0.05). Table 11.3 displays a higher mean score of the Male respondents (Mean = 18.05) than the Female respondents (Mean = 15.95). This implies that the male respondents with the ability to mitigate towards landslides compared to the female respondents. It is, therefore, concluded that statistically significant difference exists between the male and the female respondents with respect to mitigation towards the landslides.

Type of Family—Mitigation towards Landslide

$\mathbf{H_0}$: There is no significant difference between type of family on the mitigation towards landslide.

$\mathbf{H_1}$: There is significant difference between type of family on the mitigation towards landslide.

Independent-samples t -test has been conducted to test the significant difference between the type of family of the respondents with respect to mitigation towards landslide. Since P Value is smaller than Sig. Value (0.05) that is 0.048 then our Null Hypothesis is rejected. According to the average of mitigation to landslide, Nuclear family has 17.83 that is greater than the respondents of Joint family (M = 16.96) as indicated in Table 11.4. This shows that Nuclear family believes that they are able to mitigate more than the respondents in joint family. Therefore, it can be concluded that the respondents of joint family and nuclear family have statistically significant difference in mitigation to landslide. Nuclear family is mitigated very easy because the number of people in the family is very less compared to joint family where it has a lot of people including elderly, children and children who are in need of assistance in movement, shelter and food.

11.4.3 Multiple Regression

Regression is the determination of statistical relationship between two or more variables. One or more independent variable is known to cause behavior of another one variable that is regarded as dependent variable. Mitigation towards landslide

Table 11.4 Independent samples t-test—mitigation towards landslide

t	df	p Value	95.0% Confidence Interval		Estimated mean	
			Lower	Upper	Joint Family	Nuclear Family
1.661	168	0.048	−1.901	0.163	16.96	17.83

Table 11.5 Model summary

Statistics	Values
Residual standard error	1.835
Degrees of freedom	159
Multiple R-squared	0.643
Adjusted R-squared	0.620
F-statistic	28.650
F-statistics with DF	10 and 159
p-value	0.000

is considered as dependent variable in this regression and Age, Gender, Education, Occupation, Monthly Income, Marital status, type of house, type of family, family size and year of stay are considered as independent variable. The multiple regressions has been carried out to determine the optimal linear model of the Age, Gender, Education, Occupation, Monthly Income, Marital status, type of house, type of family, family size and year of stay because of predicting the Mitigation towards landslide.

The value of R-squared is 0.64 that shows that the current model can account the large amounts of variance on the output variable. According to the adjusted R-squared value, 0.62 indicates that 62 percent of the variance of the measure of mitigation towards landslide can be predicted by the socio economic status of the respondents which is large effect as indicated in Table 11.5 (Cohen, 1988).

$F (10. 159) = 28.650$, $p = 0.000$ which is smaller than 0.05 significant level that indicates that combination of nine out of ten independent variables significantly relate with dependent variable i.e., mitigation towards landslide. To predict the mitigation of landslide Marital status (1.6) is the best influencing factors that predict the dependent variable of mitigation towards landslide which is of high size but Age factor can only affect it with low positive influence of 0.08. Mitigation to landslide has negative effect influenced by Gender, Education, Occupation, Monthly Income, type of house and type of family. State of stay does not provide significant predictions of the dependent variable because the p value exceeds 0.05. Age (0.08), Gender (−1.39), Education (−1.32), Occupation (−0.19), Monthly Income (1.20), Marital status (1.60), Type of house (−1.52), Type of family (−1.29) and Family size (0.85) indicated in Table 11.6.

The Multiple regression equation for the risk towards landslide ($Y_{Mitigation}$).

Table 11.6 Coefficents

	Estimate	Std. error	t value	Pr(>ltl)
(intercept)	19.39929	1.90164	10.201	0.000
Age	0.08483	0.01624	5.222	0.000
Gender	−1.39599	0.37036	−3.769	0.000
Education	−1.32927	0.23046	−5.768	0.000
Occupation	−0.19436	0.08825	−2.203	0.029
Monthly income	1.20267	0.32692	3.679	0.000
Marital status	1.60020	0.33996	4.707	0.000
Type of house	−1.52015	0.27170	−5.595	0.000
Type of family	−1.29851	0.62290	−2.085	0.038
Size of family	0.85542	0.42221	1.979	0.049
Year of stay	−0.12706	0.38802	−0.327	0.743

$Y_{\text{Mitigation}} = 19.39 + 0.08$ (Age)—1.39 (Gender)—1.32 (Education)—0.19 (Occupation) + 1.20 (Monthly Income) + 1.60 (Marital Status)—1.52 (Type of House)—1.29 (Type of Family) + 0.85 (Family Size) + Error.

11.5 Conclusion

The research indicates that the age group is old and mostly involved in farming activities, has little education, and earns low income. The predominance of joint families and permanent residence imply the conservative and traditional form of community. These results can play a significant part in ensuring that the policy-makers come up with interventions that are aimed at education, employment, and infrastructure development.

The descriptive has led to the conclusion that among the mitigation measures is relocation and acceptance of inconveniences during the landslide or after landslide but they require government expenditure. They are highly interested in information seeking or alerting by government on land slide in order to have their own mitigation strategy. Male respondents can mitigate towards landslide in comparison to female respondents this could be due to the fact that male respondents are exposed to a lot of people which gives them psychological support but female respondents are confined in the house most of the time hence they feel that they cannot mitigate the disasters. The joint family increases the number of family members in the joint family such as old people and children, they feel that they are not able to mitigate during landslide as they have to afford shelter and food and have to assist in their movement which is not easy. Training of the respondents is also important on mitigation against landslide. The difference among the respondents is high between those who are illiterate and the higher educated ones. A difference of opinion exists in mitigation between illiterate respondents and respondents with higher education. It conclusively finds that educated should be familiar in how to mitigate during landslide compared to the illiterate respondents.

The respondents who live in semi pucca house find it hard to mitigate the landslide as opposed to persons who live in kacha house and pucca house. The respondent gives the explanation as that during landslide the damage of kacha and pucca house will be lesser and in semi pucca house the damage will be more. Depending on the age, the perception on mitigation of the landslide differs. But the difference is maximum between the above 65 and below 25 age group concludes that the perception of old age people might be different because of the experience whereas the perception of young people might be different because of young thinking. Marital status of the respondents followed by monthly income, size of family and age factors are the most likely factors that contribute to mitigation towards landslide even as gender, education, occupation, type of house and type of family have negative influence in predicting the mitigation. The year of stay does not have a significant predictive role in the mitigation.

The paper highlights that the mitigation preparedness levels of the tribal communities in Kotagiri Taluk are different. Although some population groups, like males and nuclear families, have a better mitigation capacity, financial factors and dependence on governmental interventions have been major problems. Enhancement of programs in early warning, improvement of funds to support self-mitigation, and education on land slide risks may help to make the disaster preparedness in the region more effective. Further studies are possible in the future that should investigate the importance of community-based mitigation programs and the sustainability of current mitigation measures.

Acknowledgments The author(s) acknowledge the use of AI-based tools for grammar and language correction during the preparation of this manuscript.

References

Census of India. (2011). *Household and population statistics*. Government of India.

Dolui, B., Yuvaraj, R. M., & Geetha, G. (2023). Risk perception of landslide among the tribal population: A case study of Kotagiri taluk of Nilgiri district, Tamil Nadu, India. In *Indigenous knowledge and disaster risk reduction: Insight towards perception, response, adaptation and sustainability* (pp. 209–225). Springer International Publishing.

Food and Agriculture Organization (FAO). (2022). *State of food and agriculture*. United Nations.

Froude, M. J., & Petley, D. N. (2018). Global fatal landslide occurrence from 2004–2016. *Natural hazards and earth system sciences, 18*(8), 2161–2181.

Government of India. (2021). *Rural housing and infrastructure development report*.

India meteorological department. (2019). *Annual climate report of Tamil Nadu*. Ministry of earth sciences, Government of India.

International Labour Organization (ILO). (2021). *World employment and social outlook*.

Mercer, J., Kelman, I., Taranis, L., & Suchet-Pearson, S. (2010). Framework for integrating indigenous and scientific knowledge for disaster risk reduction. *Disasters, 34*(1), 214–239.

National Family Health Survey (NFHS). (2020). Household characteristics and demographic profile. Ministry of health and family welfare.

Petheram, L., Zander, K. K., Campbell, B. M., High, C., & Stacey, N. (2010). Strange changes: indigenous perspectives of climate change and adaptation in NE Arnhem Land (Australia). *Global environmental change, 20*(4), 681–692.
Sundaram, R., Krishnan, P., & Mehta, A. (2021). Tourism impact and sustainable development in the Nilgiri Hills. *Journal of environmental management, 46*(1), 23–39.
Tamil Nadu Forest Department. (2020). *Ecological and environmental conservation in the Nilgiri Hills.* Government of Tamil Nadu.
Thennavan, E., & Pattukandan Ganapathy, G. (2020). Evaluation of landslide hazard and its impacts on hilly environment of the Nilgiris District-a geospatial approach. *Geoenvironmental disasters, 7*(1), 3.
UNESCO. (2021). *Nilgiri biosphere reserve: a world heritage site.* United Nations Educational, Scientific and Cultural Organization.
United Nations. (2019). *World population prospects.*
United Nations Educational, Scientific and Cultural Organization (UNESCO). (2020). *Global education monitoring report.*
United nations office for disaster risk reduction (UNDRR). (2021). *Global assessment report on disaster risk reduction.*
World Bank. (2021). *Gender statistics and development.*
Yuvaraj, R. M., & Dolui, B. (2021). Statistical and machine intelligence based model for landslide susceptibility mapping of Nilgiri district in India. *Environmental challenges, 5*, 100211.
Yuvaraj, R. M., & Dolui, B. (2024). Landslide susceptibility using weighted regression model: A geo-spatial approach. In *Landslide: susceptibility, risk assessment and sustainability: Application of geostatistical and geospatial modeling* (pp. 335–355). Cham.

Part V

Sustainability Through Indigenous Knowledge, Risk Reduction and Preparedness

Risk Perception, Mitigation Behavior, and Trust: A Structural Model of Landslide Vulnerability

12

Jayasree Dolvi

Abstract

This chapter explores factors that cause landslips vulnerability among tribal populations in the Nilgiri district, using Structural Equation Modeling (SEM). The three latent constructs have been integrated in the analysis, which includes risk perception, mitigation behavior, and institutional trust, and their effect on perceived vulnerability. The model has good fit indices (CFI = 0.973, RMSEA = 0.050, GFI = 0.970) using a sample of 169 respondents and 22 observed variables, which prove the reliability of the proposed structure. The findings indicate that perception of risk (+ = 0.475) has the most impact on vulnerability followed by mitigation (+ = 0.446) and trust (+ = 0.258). The main variables, including but not limited to, the Alert from government and Trust at government, were discovered to have significant influence on the mitigation behavior and trust respectively. The paper discusses the significance of integrating scientific modeling and local perceptions to comprehend the complexity of disaster vulnerability. The practical implication of these findings in terms of policy and disaster management are community-specific interventions and participatory methods of reducing landslide risk in environmentally vulnerable areas.

Keywords

Tribes • Landslide • Structural equation model • Trust • Vulnerability • Mitigation • Risk

J. Dolvi (✉)
Department of History, JNRM College, Port Blair, Andaman and Nicobar Island, India
e-mail: drjayasreedolvi@gmail.com

R. M. Yuvaraj et al. (eds.), *Advanced Landslide Susceptibility Modeling*,
https://doi.org/10.1007/978-3-032-14485-0_12

12.1 Introduction

Landslides are frequently discussed as one of the most devastating and uncontrollable natural risks as it may lead to the tremendous damage to the human and ecological environment. These are events that include the unexpected failure of rocks, soil, and debris down the slopes, commonly caused by a torrent of rain, deforestation, earthquakes, or human alterations to the landscape (Petley, 2012). The impacts of landslides are far-reaching, they lead to loss of human life, the displacement of populations, destruction of homes and infrastructure, disruption of transportation and communications channels as well as serious environmental degradation (UNISDR, 2015). The mountainous and hilly lands are most vulnerable because of the unsteady surface and the exposure to the severe weather conditions.

Tribal communities residing in ecologically sensitive and disaster-prone areas like the Western Ghats and the Eastern Ghats are quite sensitive to landslides in the Indian context. Such communities are usually located in remote and geologically vulnerable places; sharp slopes, heavy precipitation, and non-vegetated grounds make the slope susceptible to collapse. Their geographical remoteness coupled with a little access to resources such as healthcare, transport, early warning, and formal education also increases their vulnerability and decreases their abilities to react to disasters (Rao, 2006).

The tribal populations are also vulnerable due to the socio-economic marginalization. Such communities are often characterized by low income, limited engagement in the policy-making process, and low integration into the mainstream disaster management systems (Mohanty, 2014). The existing housing patterns, reliance on forest-based livelihoods, and a low level of awareness of the scientific approaches of mitigating landslides result in disproportional and long-term consequences on their wellbeing even in small-scale landslides.

Tribal populations with respect to landslides are hence, a crucial understanding in the creation of contextual risk reduction strategies. Cultural beliefs, lived experiences, traditional knowledge systems and belief in authorities shape risk perception of a community (Slovic, 2000). These perceptions affect the choices of evacuating, implementing mitigation, or taking relocation. Therefore, cultural sensitivity, social inclusiveness, and technical competence of mitigation measures could be achieved in a manner that involves the inclusion of tribal group opinions and perspectives in disaster planning. These participatory strategies are essential in bridging the physical division between the high policy directions and the realities at the ground level finally leading to better resilience of the tribal communities to the occurrence of landslides in the future (Gaillard and Mercer, 2013).

12.2 Study Area

The Nilgiris is one of the oldest mountain ranges in India that is based at the junction of Tamil Nadu, Kerala and Karnataka and it is part of the Western Ghats. It is the first biosphere reserve recognised in India and it is one of the 14 biodiversity hotspots in the world because of its extraordinary ecological richness. The elevation of the region is between 900 and 2636 m above the mean sea level. The Nilgiris is the most favored tourism spot with its culture of vibrant hill stations and tourism plays a big role in defining the locals way of life. The Nilgiri Plateau is a region of Tamil Nadu that is completely situated at the intersection of the Western and Eastern Ghats on a region of 32 revenue districts. Eight hydroelectric power stations are located within the district and the administrative divisions of the district include Ooty, Kundah, Coonoor, Kotagiri, Gudalur and Pandalur, with a total area of about 2551 square kilometers. It is located geographically 11O to 11O15N latitude and 76O13 to 77O2E longitude. Waterfall is heavy in the Nilgiris and falls in the South-West monsoon (June–August) as well as the North-East monsoon (October–December). The average minimum and maximum temperatures in the region are 20.7 °C and 9.6 °C respectively, whereas the mean relative humidity is between 75.8 and 76.9%. The Nilgiri district tribal population is a unique and culturally rich group of the Indian population. Nilgiris Nilgiris is a region located in the Western Ghats of Tamil Nadu and it has a number of unique tribal groups that have lived a peaceful life since the time immemorial and resided in harmony with the mountains and the biodiverse forest of the area. The large tribal communities living in this district are the Todas, Kotas, Badagas, Irala, Kurumbas as well as the Paniyas. Each tribe possesses its own language, traditions, rites as well as socio-economic structures, which are part and parcel of the cultural mosaic of the Nilgiris. A set of geographical, economic, infrastructural, and social factors make the tribal communities of the Nilgiri district the most susceptible to landslides. To solve their problems, a community-based strategy to disaster risk deduction, which incorporates early warning mechanisms, enhanced infrastructure, livelihood security, and culture-sensitive rehabilitation plans is essential. Enabling the tribal voices on disaster planning and resilience creation is an essential step to provide the tribal groups with safety, dignity, and sustainable growth amid the frequent occurrence of natural hazards.

12.3 Data and Methodology

The primary data was collected with the help of well-designed structured questionnaire and face to face interview was conducted among the tribal people living in high-risk landslide prone zone of Nilgiri district. The questionnaire was mainly based on the perception of tribal people on risk, trust, vulnerability, mitigation towards landslide. It contains total 23 open-ended questions.

It had been attempted to take a sample of 260 in order to get a 6% of margin of error but proper response was achieved only from 170 respondents so this sample size has been taken for the study with 7.5% margin of error. The sample size was derived by using the equation given below:

$n_o = \frac{Z^2 pq}{e^2}$.

Where,

- e is the desired level of precision (i.e., the margin of error)
- p is the estimated proportion of the population which has the attribute in question,
- q is 1-p.

Snowball sampling a type of Non-Probability Sampling technique is used to gather data from the respondents who are settled in the high and moderate landslide region in Kothagiri taluk of Niligiri district. Pilot Study was conducted in the month of July 2019 in order to study the predefined questionnaire and the main data collection begins in the month of December 2019 and ends in the month of February 2020.

Path Analysis of Structural Equation model (SEM) was also used in this research. Multivariate outliers, Collinearity, Variance, Data Reliability, Sample size determination for SEM and KMO and Bartlett's Test had been performed to check the data is suitable for structural Equation Model. Structure Equation Modeling (SEM) is a modeling technology, which can deal with both endogenous and exogenous variables. It also can model latent variables, especially with linear combination. SEM is a covariance structure analysis method consists of measurement model and structural model. It has great advantage in the analysis of the interaction between single index and the influence of the single index to the entirety. The effect is similar to the linear regression, path analysis, factor analysis, and logic analysis and so forth. The SEM is widely used in social science, econometrics and biologist.

12.4 Result and Discussion

This study aims to explore the tribal perspective on landslides by examining four critical dimensions: mitigation, trust, risk perception, and vulnerability**.** These dimensions were captured through a structured questionnaire developed to assess how tribal communities perceive and respond to landslide threats in their locality. Each variable in the questionnaire was designed to provide insight into the cognitive, emotional, and behavioral responses of tribal populations toward landslide hazards.

The mitigation component includes variables such as whether individuals or communities have their own mitigation plans (M1), awareness of government mitigation initiatives (M2), early warning systems (M3), willingness to relocate (M5), and strategies based on past landslide experiences (M8). These factors reflect the level of preparedness and adaptive capacity within the community. The trust dimension measures the confidence that tribal communities place in government institutions (T1), government crisis management efforts (T2), and external experts (T3). Trust plays a critical role in determining whether communities follow official warnings, evacuation orders, and adopt mitigation measures suggested by authorities.

The risk perception aspect investigates how people perceive the likelihood (R1), severity (R2), and potential impacts (R3–R8) of landslides on their homes, workplaces, financial stability, and overall quality of life. Understanding risk perception is vital for disaster risk communication and education. Finally, the vulnerability indicators (V1–V4) focus on psychological and social vulnerabilities, such as feelings of fatalism, anxiety, and perceived helplessness from neighbors or the government. By investigating these variables, this study contributes to a deeper understanding of the ways in which tribal communities interpret, internalize, and respond to the complex phenomenon of landslides, ultimately guiding better-informed policies and community-based disaster risk reduction efforts.

Structural Equation Model

This model has been used to check the complex relationships between latent variables (unobserved) and measured variables (observed) and also among the latent variables. Simply this model is the combination of multiple regression and factor analysis.

Assumption of Data for SEM

Multivariate outliers

Outliers are removed from the data using standard method of robust estimation of a parameters in the Mahalanobis distance and the comparison with the critical value of the chi-square distribution (Rousseeuw and van, 1990). In this data only one sample with low probability of less than 0.001 (Mahalanobis critical value = 65.12) are removed so our total sample has reduced from 170 to 169.

Collinearity

The data undergone collinearity test using multiple regression analysis which shows that all the data have more than 0.01 tolerance and less than 10 VIF (Variance Inflation Factor) (Snee, 1983) it means that the data not violating the collinearity.

Variance

The variance of the variable should not be greater than the ten times of the other variables. The data does not violate this condition.

Table 12.1 Test of reliability

S. No.	Latent variables	No. of observed variables	Cronbach's alpha
1	Risk	7	0.841
2	Trust	3	0.846
3	Mitigation	8	0.746
4	Vulnerability	4	0.830
Total	04	22	

Data reliability

Reliability concerns the degree to which a measurement of a observable fact provides constant and consist result (Carmines and Zeller, 1979) Cronbach's alpha has been used to measure the internal consistency of the data because Cronbach's alpha is considered as most suitable measure of reliability when making use of Liker scales (Robinson, 2011). The Cronbach's alpha should be above 0.5 will be considered that instrument has sufficient internal consistency and will be used for further test (Schuessler, 1971) (Hinton et al. 2004).Initially, reliability of the variables are very less so some observed variables has removed in order to get high Cronbach's alpha (Table 12.1).

Sample size determination for SEM

Sample size plays a significant role to run the SEM (Khine et al. 2013). The Structural equation modeling is suitable for analyzing large sample (Kline, 2011; Schumacker and Lomax, 2010) but less sample size may also be used in simpler models with lesser parameters (Kline, 2011). In order to find whether our sample size is enough for running the SEM we calculate using sample size calculator to determine the required sample size for our study. The number of latent variables is 04 with 22 observed variables.

Minimum sample size to detect effect: 137.
Minimum sample size for model structure: 138.
Recommended minimum sample size: 138.
So only 138 sample sizes are required to run the Structural Equation Model but the study has 169 samples that may increase the accuracy level. According to (Loehlin, 2004) a minimum of 100 sample size is enough to run SEM but larger sample size are essential for precise estimates of parameter and standard errors (Schumacker and Lomax, 2010) (Hair et al. 2006).

KMO and Bartlett's test

Table 12.2 KMO and Bartlett's test

Kaiser-Meyer-Olkin measure of sampling adequacy		0.810
Bartlett's test of sphericity	Approx. Chi-Square	2567.098
	df	231
	Sig.	0.000

The test has been conducted to access the suitability of data for factor analysis. Kaiser-Meyer-Olkin (KMO) value should be greater than 0.5 for the satisfactorily proceed for factor analysis (Tabachnick and Fidell, 2001). In our study KMO value is 0.810 which indicates that high proportion of variances in the variables that might me caused by underlying factors and KMO also describe the sampling adequacy (Kaiser, 1974). Bartlett's test of sphericity tests (Bartlett, 1954) the hypothesis that correlation matrix is an identity matrix it means that factor model is inappropriate. In our study Bartlett's test shows that null hypothesis has been rejected (sig. Value less than 0.05) shown in Table 12.2 and the data is suitable for factor analysis (Anderson, 1982)

The data has gone through critical testing using the above test that has proved that data has sufficiently support for further analysis using Structural Equation Model. After the test the list of variables under Vulnerability, risk, mitigation and trust has reduced from the initial. Table 12.3 gives the list of observed variables for the analysis.

Interpretation of SEM Output

Structural equation model (Path analysis) includes lots of fit statistics which are really used to help to determine whether the present model has adequate fit for the given data or not. The present model has 203 degree of freedom with 4 Observed Endogenous variables, 22 Observed exogenous variables and 23 unobserved exogenous variables shown in Tables 12.4 and 12.5.

Chi-Squared Test

The chi-square likelihood ratio statistics is the most significant absolute fit index, and tests for the difference between the empirical and theoretical model (Meyers et al. 2013). The Chi-Square for goodness of fit (CMIN/DF) is 4.559 with 203 dfs which is lesser than 5. Therefore, it is permissible shown in Table 12.6. The model is significant at $p < 0.01$ ($p = 0.000$) which shows that the null hypothesis of goodness of fit (Ho: The present model is not perfectly fit the data) is rejected.

CMIN

Model	CMIN	DF	P	CMIN/DF
Default model	913.380	203	0.000	4.559

Table 12.3 List of variables included in the model

Variables		Explanation
Mitigation	M1	Having own mitigation plan
	M2	Government mitigation plan
	M3	Alert from government
	M4	Idea of relocation
	M5	Willingness to relocate
	M6	Mitigation with own expenses
	M7	Accepting inconvenience due to landslide
	M8	Mitigation through previous landslide information
Trust	T1	Trust at government
	T2	Trust at government crisis management
	T3	Trust of other landslide expert
Risk	R1	Occurrence of landslide
	R2	Intensity of landslide
	R3	Risk at residence
	R4	Risk at work place
	R6	Threat of livelihood
	R7	Affect life quality
	R8	Financial loss
Vulnerability	V1	Fatalistic vulnerability
	V2	Worry due to landslide
	V3	Helpless from the neighbor
	V4	Helpless from the government

Table 12.4 Computation of degrees of freedom (default model)

Number of distinct sample moments:	253
Number of distinct parameters to be estimated:	50
Degrees of freedom (253–50):	203

Root Mean Squre Residual (RMR)
RMR, GFI

Model	RMR	GFI	AGFI
Default model	0.069	0.970	0.924

The square root of the discrepancy between the sample covariance matrix and the model covariance matrix is root mean square residual (RMR). The RMR should range from 0 to 1, with a value of less than or equal to 0.08 is indicative of an acceptable model. The RMR value of the present model is 0.069 which is less than 0.08 and this model is acceptable. The goodness of fit index (GFI) is a measure

Table 12.5 List of endogenous and exogenous variables

Observed, endogenous variables		Observed, exogenous variables	Unobserved, exogenous variables
Risk		R1	e15
		R2	e14
		R3	e13
		R4	e12
		R6	e11
		R7	e10
		R8	e9
Trust		T1	e18
		T2	e17
		T3	e16
Vulnerability		V1	e19
		V2	e20
		V3	e21
		V4	e22
			e23
Mitigation		M1	e1
		M2	e2
		M3	e3
		M4	e4
		M5	e5
		M6	e6
		M7	e7
		M8	e8
Total	4	22	23

Table 12.6 Fit indices of the observed model

Fit index	Observed model	Recommended level	References
CMIN/DF (χ^2)	4.559	< 2 (excellent)	Hu and Bentler, (1999)
		< 3 (good)	
		< 5 (acceptable)	
GFI	0.970	> = 0.90	Meyers et al. (2013)
NFI	0.955	> = 0.95	Bentler and Bonett, (1980)
CFI	0.973	0.90	Hu and Bentler, (1999)
RMSEA	0.050	< = 0.05	Schumacker and Lomax, (2010)
RMR	0.069	< = 0.08	Hu and Bentler, (1999)
AGFI	0.924	> = 0.90	Hu and Bentler, (1999)
TLI	0.967	> = 0.95	Hu and Bentler, (1999)

of fit between the observed covariance matrix and hypothesized model and it is similar to R^2 in Multiple regression (Khine et al. 2013). The adjusted goodness of fit index (AGFI) corrects the GFI, which is affected by the number of indicators of each latent variable. The GFI and AGFI should range between 0 and 1, with a cut off value of 0.9 generally indicating that the model is acceptable (Hu and Bentler, 1999). The present model has the value of GFI and AGFI are 0.97 and 0.92 respectively which is greater than 0.9 which shows that the present model has acceptable model fit.

Baseline Comparisons

Model	NFI	TLI	CFI
Default model	0.955	0.965	0.973

Baseline comparisons includes NFI (Normed Fit Index), TLI (Tucker-Lewis Index) or Non-normed fit index and CFI (Comparative Fit Index) which are used to measure the model fit (Bentler and Bonett, 1980). The NFI analyzes the discrepancy between the chi-squared value of the null model and chi-squared value of the hypothesized model. The values for NFI and TLI should range between 0 and 1 with a cut off value of above 0.95 which indicates the good fit. The present model has 0.955 and 0.965 which are above the cut off values which indicated that the model is good fit. The comparative fit index (CFI) examining the discrepancy between the data and the hypothesized model to analyzes the model fit. CFI values range from 0 to 1 with higher value of greater than 0.90 is acceptable model fit. The present model has the CFI value of 0.973 which is greater than cut off value so the model is fit.

RMSEA

Model	RMSEA
Default model	0.058

Root Mean Square Error of Estimation (RMSEA) measures approximation error between the covariance of the hypothesized model and observed covariance (Meyers et al. 2013). The RMSEA values range from 0 to 1 with lesser value of RMSEA indicates the good model fit. The value of less than or equal to 0.08 is considered as the model is acceptable (Schumacker and Lomax, 2010). In the present model the RMSEA value is 0.58 which is acceptable for model fit.

Latent variables are represented by ellipses and observed variables are represented by rectangles. The model shown above is unidirectional (One-way casual flow) does not have any cause and effect or reciprocal effects or feedback loops so the model is called Recursive Structural Equation Model (Fig. 12.1). Based on this model, Trust towards landslide information, Mitigation towards landslide and Risk towards landslide among the tribal people of Nilgiri district has direct influence on the Vulnerability towards landslide among the tribal people of Nilgiri district.

The endogenous variables also influenced by factors which are not included in the model are called error terms which are shown as e1, e2, e3,... e22 and e23.

> **H₀1**: Risk towards landslide is not positively influences the vulnerability of landslide among the tribal people.
> **H₀2**: Trust towards landslide information is not positively influences the vulnerability of landslide among the tribal people.
> **H₀3**: Mitigation towards landslide is not positively influences the vulnerability of landslide among the tribal people.

From Table 12.7 it has been inferred that the *p* value (0.000) is lesser than the critical value so the null hypothesis has been rejected which means risk towards landslide, mitigation towards landslide and trust towards landslide information are positively influences the vulnerability of the landslide among the tribal people of Nilgiri district. It also inferred from the result that Mitigation through previous landslide information is not influences the Mitigation whereas Occurrence of landslide and financial loss and helpless from the government are not influences the Risk and vulnerability respectively because the null hypothesis has been rejected due to the p value is above critical value.

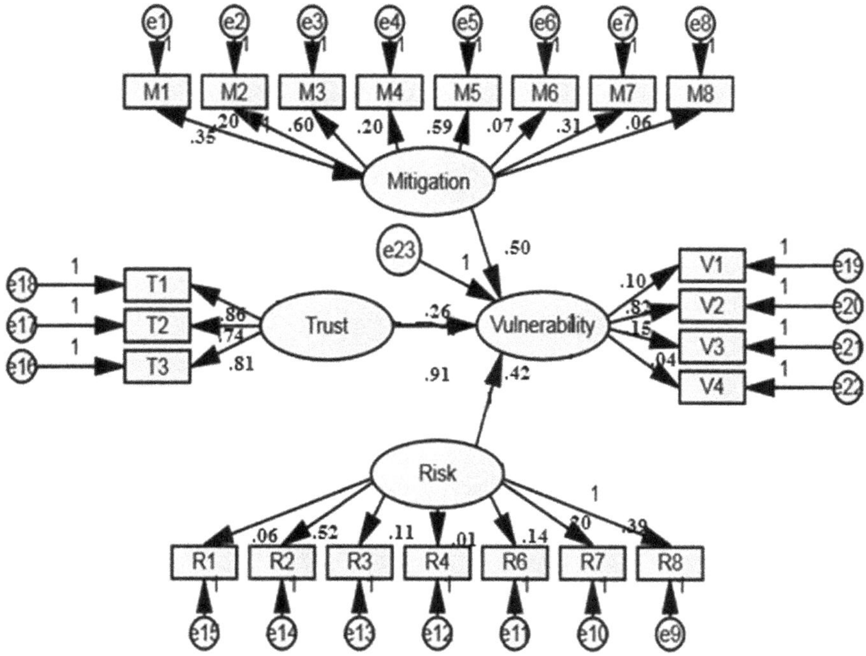

Fig. 12.1 Recursive structural equation modeling

Table 12.7 Regression weights: (group number 1—default model)

		Estimate	S.E.	C.R.	P
Vulnerability	Trust	0.358	0.033	10.848	0.000
Vulnerability	Mitigation	0.476	0.146	3.260	0.000
Vulnerability	Risk	0.425	0.118	3.602	0.000
M1	Mitigation	0.396	0.290	1.366	0.000
M2	Mitigation	0.301	0.270	1.115	0.000
M3	Mitigation	0.648	0.480	1.350	0.000
M4	Mitigation	0.211	0.092	2.293	0.000
M5	Mitigation	0.629	0.287	2.192	0.000
M6	Mitigation	0.216	0.066	3.273	0.000
M7	Mitigation	0.342	0.054	6.333	0.000
M8	Mitigation	0.126	0.255	0.494	0.891
R8	Risk	0.124	0.341	0.364	0.741
R7	Risk	0.241	0.178	1.354	0.000
R6	Risk	0.177	0.084	2.107	0.000
R4	Risk	0.091	0.063	1.444	0.000
R3	Risk	0.179	0.161	1.112	0.000
R2	Risk	0.583	0.252	2.313	0.000
R1	Risk	0.191	0.482	0.396	0.741
T3	Trust	0.871	0.471	1.849	0.000
T2	Trust	0.761	0.093	8.183	0.000
T1	Trust	0.864	0.110	7.855	0.000
V1	Vulnerability	0.443	0.211	2.099	0.000
V2	Vulnerability	0.820	0.271	3.026	0.000
V3	Vulnerability	0.696	0.329	2.114	0.000
V4	Vulnerability	0.039	0.387	0.101	0.253

Standardized regression weights are actually factor loadings for the respective variables given in Table 12.8. It has been interpreted that perception of Risk has highest (0.475) influence on vulnerability than mitigation and trust. It means if they risk go up by 1 standard deviation, vulnerability towards landslide has increased by 0.475–0.47 standard deviation. Similarly Trust has low influences on vulnerability (0.258) which shows that trust increase by 1 standard deviation, the vulnerability has increased to 0.258 standard deviation. It also interpreted that 'Alert from the government' has highest influence on mitigation towards landslide and 'Mitigation through previous landslide information' has lowest influences on mitigation. Parameter of 'Intensity of landslide' and 'Risk at work place' has highest influences and lowest influences on the Risk towards landslides similarly 'Trust at government' has highest influences on trust towards landslide information and 'Trust at government crisis management' has lowest influences on the trust towards

Table 12.8 Standardized regression weights: (group number 1—default model)

		Estimate
Vulnerability	Trust	0.258
Vulnerability	Mitigation	0.446
Vulnerability	Risk	0.475
M1	Mitigation	0.346
M2	Mitigation	0.201
M3	Mitigation	0.598
M4	Mitigation	0.191
M5	Mitigation	0.589
M6	Mitigation	0.066
M7	Mitigation	0.312
M8	Mitigation	0.056
R8	Risk	0.394
R7	Risk	0.200
R6	Risk	0.137
R4	Risk	0.011
R3	Risk	0.119
R2	Risk	0.523
R1	Risk	0.061
T3	Trust	0.811
T2	Trust	0.741
T1	Trust	0.864
V1	Vulnerability	0.103
V2	Vulnerability	0.820
V3	Vulnerability	0.146
V4	Vulnerability	0.039

landslide information. Parameter 'Worry due to landslide' has highest influences on the vulnerability whereas 'Helpless from the government' has lowest influences on the vulnerability towards landslides.

12.5 Conclusion

The data has proved with the help of several test which includes Multivariate outliers, collinearity, Variance, Cronbach's alpha, Sample size determination, KMO and finally Barlett's test that it support for further analysis using Structural Equation Model. It has been resulted that the fit indices of the observed model which includes Chi-Squared test, GFI, NFI, CFI, RMSEA, RMR, AGFI and TLI shows that the model is good fit. Structural Equation Model concludes that Risk towards landslide, Trust towards landslide information and mitigation towards landslide had positively influences on vulnerability towards landslide. It also concludes that Risk has high influences whereas Trust towards landslide information has low influences on vulnerability towards landslide.

Seven out of eight variables relating to perception of mitigation (Having own mitigation plan, Government mitigation plan, Alert from government, Idea of relocation, Willingness to relocate, Mitigation with own expenses and Accepting inconvenience due to landslide) has strongly and positively predicts the mitigation towards landslides. The variable 'Alert from government' has highest influence on Mitigation towards landslide.

All three variables relating to perception of trust towards landslide information (Trust at government, Trust at government crisis management, trust of other landslide experts) has strongly and positively predicts the Trust towards landslide information. The variable 'Trust at government' has high influence the parameter.

Five out of Seven variables relating to risk towards landslide (Intensity of landslide, Risk at residence, Risk at work place, Threat of livelihood and Affect life quality) has strongly and positively predicts the tribal perception of Risk towards landslide. Among them 'Intensity of landslide' has highest influence on the Risk towards landslide.

Three out of Four relating to vulnerability towards landslide (Fatalistic vulnerability, Worry due to landslide and Helpless from the neighbor) has strongly and positively predicts the tribal perception of vulnerability towards landslide. The variable 'Worry due to landslide' has highest influence on the perception of vulnerability towards landslide.

Acknowledgments The author(s) acknowledge the use of AI-based tools for grammar and language correction during the preparation of this manuscript.

References

Anderson, P. F. (1982). Marketing, strategic planning and the theory of the firm. *Journal of Marketing, 46*(2), 15–26.

Bartlett, M. S. (1954). A note on the multiplying factors for various χ^2 approximations. *Journal of the Royal Statistical Society. Series B (Methodological)*, 296–298.

Bentler, P. M., & Bonett, D. G. (1980). Significance tests and goodness of fit in the analysis of covariance structures. *Psychological Bulletin, 88*(3), 588.

Carmines, E. G., & Zeller, R. A. (1979). *Reliability and validity assessment* (Vol. 17).

Gaillard, J. C., & Mercer, J. (2013). From knowledge to action: bridging gaps in disaster risk reduction. *Progress in Human Geography, 37*(1), 93–114. https://doi.org/10.1177/0309132512446717

Hair, J. F., Black, W. C., Babin, B. J., Anderson, R. E., & Tatham, R. L. (2006). Multivariate data analysis (Vol. 6).

Hinton, P. R., Brownlow, C., Mcmurray, I., & Cozens, B. (2004). SPSS explained, East Sussex, England, Routledge Inc.

Hu, L. T., & Bentler, P. M. (1999). Cutoff criteria for fit indexes in covariance structure analysis: Conventional criteria versus new alternatives. *Structural Equation Modeling: A Multidisciplinary Journal, 6*(1), 1–55.

Kaiser, H. F. (1974). An index of factorial simplicity. *Psychometrika, 39*(1), 31–36.
Khine, M. S. (Eds.) (2013). Application of structural equation modeling in educational research and practice (Vol. 7). Sense Publishers.
Khine, M., Ping, L., & Cunningham, D. (2013). (Eds.), Application of structural equation modeling in educational research and practice. Rotterdam, the Netherlands: Sense Publishers.
Kline, R. (2011). Principles and practice of structural equation modeling, (3rd edn). Guilford Press. New York.
Loehlin, J. (2004). Latent variable models: An introduction to factor, path, and structural equation analysis (4th edn). Mahwah, NJ: Lawrence Erlbaum Associates.
Meyers, L., Gamst, G., & Guarino, A. (2013). Applied multivariate research: Design and interpretation. Los Angeles, CA: Sage Publications, Inc.
Mohanty, P. K. (2014). Encyclopaedia of scheduled tribes in India (Vol. 5). Kalpaz Publications.
Petley, D. N. (2012). Global patterns of loss of life from landslides. *Geology, 40*(10), 927–930. https://doi.org/10.1130/G33217.1
Rao, C. H. H. (2006). Inclusive growth: Recent experience and challenges ahead. *Economic and Political Weekly, 41*(46), 4641–4647.
Robinson, S. (2011). Test–retest reliability of health state valuation techniques: The time trade off and person trade off. *Health Economics, 20*(11), 1379–1391.
Rousseeuw, P. J., & Van Zomeren, B. C. (1990). Unmasking multivariate outliers and leverage points. *Journal of the American Statistical Association, 85*(411), 633–639.
Schuessler, K. (1971). Analyzing social data.
Schumacker, R. E., & Lomax, R. G. (2010). A Beginner's Guide to. *Structural Equation Modeling* (3rd Edition), New York: Taylor & Francis Group.
Slovic, P. (2000). The perception of risk. *Earthscan* Publications.
Snee, R. D. (1983). Regression diagnostics: identifying influential data and sources of collinearity.
Tabachnick, B. G., & Fidell, L. S. (2001). Principal components and factor analysis. *Using Multivariate Statistics, 4*(1), 582–633.
UNISDR. (2015). Sendai framework for disaster risk reduction 2015–2030. United Nations Office for Disaster Risk Reduction. https://www.undrr.org/publication/sendai-framework-disaster-risk-reduction-2015-2030

Economic Impacts of Landslides in View of Transportation and Road Network

13

P. Baby

Abstract

Globally, landslides cause a high economic cost to the transportation infrastructure that affects the connectivity, increases the cost of maintenance, and slows socio-economic growth. The review is a synthesis of the research literature around the world and India that has been conducted to look at direct costs such as road repair, structural restoration, and emergency response and indirect costs such as traffic delays, business interruption, and supply chain disruption. In the world landslides result in a loss of about 20 billion dollars every year in the economy, with significant loss incurred on transportation infrastructure. In India, the high-impact events in Kerala (hole of 4500 crores lost road value), Himachal Pradesh (hole of 500 crores), and Sikkim (hole of 2000 crores) point to the fragility of mountainous areas. The study places a strong accent on the fact that infrastructure planning and human activity, especially road construction work tend to increase the landslide vulnerability Methodological frameworks based on GIS, network modeling as well as cost–benefit analysis have been employed in the assessment and mitigation of effects. Categories of socio-economic consequences and policy development in the years 2009–2019 can be explained by case studies and statistical compilations in India, covering 412 major landslides. In spite of these activities, there are still issues of data paucity, standardization of measures, and ecological resilience. Solutions that could be recommended are improvements of slope stabilization methods, the early warning, bioengineering methods, and community-based planning. This review highlights the importance of proactive and multidisciplinary approach in ensuring that transportation networks are not disrupted due to the occurrence of a landslide.

P. Baby (✉)
Department of Geography, Bharathi Women's College, Chennai, Tamil Nadu, India
e-mail: babypoo1011@gmail.com

R. M. Yuvaraj et al. (eds.), *Advanced Landslide Susceptibility Modeling*,
https://doi.org/10.1007/978-3-032-14485-0_13

Keywords

Landslides • Economic impact • Road network • Transportation infrastructure • India • Resilience strategies

13.1 Introduction

Landslides, which can be described as rapid movement that is downward of soil, rocks, and debris, have been a source of a major threat to the transportation infrastructure of the world. Since the vehicular routes are widened toward the mountainous and hilly landscapes, the area becomes prone to landslides, and roads and railways become even more vulnerable. An international estimate suggests that the annual losses on landslides are about 20 billion dollars on a yearly basis, with roads and railways infrastructure the most affected. Specifically, road congestion, losses, and resulting diversion cause ripple effects, interrupting business, emergency response, and routines (Rajan et al., 2024).

To measure these economic impacts, it is necessary to separate between direct, indirect, and consequential costs. Direct expenditures include the pavements, bridges, vehicular property, and power line physical damage, indirect expenses include the lag time, rerouting, and additional fuel consumption, and consequential expenses include wider disruption, such as market inaccessibility, equal delay in health, and decreased business output.

The geoclimatic diversity in India, which includes the delicate Himalayan strip, and the Western Ghats makes some of its areas particularly vulnerable. Another example of how socio-economic factors can negatively affect the risk of disasters is the 2024 Wayanad landslides (420 + deaths; extensive network destruction), which demonstrated that remote communities rely on a single transit path and thus were susceptible to the consequences of the calamity (Wang et al., 2025). In Kerala, landslides occurring due to floods in 2018 caused damage to roads in the state, which amounted to over 4500 crores of economic value, and recovery cost 2500 crores. Himalayas had also experienced this financial burden-the landslides in Himachal in 2019/21 cost the state 500 crores in damages and 350 crores in restoration and the transport disruption in North Sikkim cost the state 2000 crores. These incidents point to how Indian transport infrastructure is vulnerable to hydrometeorological and anthropogenic stimuli.

A systematic review of 412 socio-economically significant landslides (18,682,022) identified 141 of the high-impact landslides, having a primary focus on transportation infrastructure as the primary loss (Parkash, 2023). Although there are initial guidelines (2009 NDMA) and risk mitigation measures (2019), there are still gaps in the risk assessment, data collection, and standardized impact assessment.

International systems are coming up. An example is the case of Lisbon flood, which demonstrates the effectiveness of cost–benefit planning of resilience proactive plans lessened the losses by 7.6% compared to reactive actions (Winter et al., 2016). In Tajikistan, road disaster expenses are approximately 0.5%/year (c. 445 000 SOM) of GDP, which is a reason to invest in resilience (Report World Bank, 2021). However, integrated evaluations and fortitude measurements are yet to be devised.

India has incorporated bioengineering as part of PMGSY that has combined vegetation-based slope controls with civil engineering and has provided long-term mitigation in Himachal Pradesh (The Guardian, 2024). However, cases such as the Ankola (Karnataka) landslide, which was caused by the combination of unscientific highway construction and the influence of landslides, reflect that the infrastructure, once improperly designed, will only help to increase vulnerability (Kundu et al., 2025).

This review addresses the following questions:

1. What are the economic costs—direct, indirect, consequential—of landslides on transportation?
2. How do these costs manifest globally and within India's varied terrains?
3. Which methodologies support reliable quantification and modeling?
4. What policy and engineering strategies can reduce economic losses and enhance resilience?

Outlined across eleven sections, the paper aims to offer a holistic understanding—spanning economic analysis, case studies, methodological frameworks, and recommendations—to support decision-making for sustainable and resilient transport planning.

13.2 Rationale of the Study

Connections are also important because transportation systems are lifelines of connectivity, commerce, and emergency services. These networks are broken by landslides and lead to economic instability, social isolation and environmental degradation. This research incorporates a specific evaluation of the economic effects of landslides in response to key knowledge gaps.

To begin with, the majority of literature focuses on lost life or physical damages, and fewer studies pay attention to overall economic analyses of transportation impacts (Sim & Wong, 2022). Direct and indirect costs, being accepted, tend to be dispersed or immethodical. Secondly, there are few India-specific syntheses over mountainous and monsoon-prone areas although the high-impact occurrences are frequent. Systematic study is required on restorative costs, business losses and socioe-conomic ripple effects.

Thirdly, as climate changes and infrastructure is developed at a faster pace, the risk profiles are changing. It is important to know economic implications in this dynamic sense. Finally, resilience requires robust data to invest in. Quantitative models, which exploit GIS, network simulation and cost–benefit frameworks, need to be tested on empirical real-world situations.

This study thus aims to:

- Synthesize global economic quantifications of landslide effects on transportation.
- Anchor findings within regional contexts: Himalayas, Western Ghats, and other vulnerable Indian terrains.
- Evaluate methodological tools supporting accurate impact assessment.
- Formulate evidence-based recommendations to inform resilient transport infrastructure policy and practice.

By addressing these aims, the review supports policy makers, engineers, urban planners, and disaster management agencies in developing integrated strategies that mitigate economic risk, promote sustainable connectivity, and reinforce adaptive capacity in India and beyond.

13.3 Materials and Methods

The review follows a systematic literature review methodology, incorporating scholarly studies, governmental reports, and case notes up until June 2025.

Extensive search statements were implemented in Scopus, Google Scholar, and ResearchGate. International context was influenced by global studies (e.g., \$20 bn/year losses) and structures (Florence flood resilience in Lisbon). Data on India were specific to the country (Wayanad, Kerala, Himachal, Sikkim) based on government reports and peer-reviewed publications. Long-term consequences were depicted historically, e.g., Nainital 1880 (Gokhale, 2023).

They had to contain studies which: (a) measured economic effects of transport infrastructure; (b) used accepted economic classification of the impact (direct, indirect, consequential costs); (c) contained empirical or modeled data; (d) covered global or Indian landslide occurrences on road networks.

Economic cost segmentation followed the three-tier classification:

- *Direct costs*: infrastructure restoration and emergency response.
- *Indirect costs*: delays, rerouting, fuel, labor.
- *Consequential costs*: business, health, market-access impacts.

Data from case studies was collated using descriptive tabulation, and this revealed context, cost breakdowns, and restoration schedules. The methods of spatial risk assessment of GIS susceptibility mapping, transport network simulating, and early warning models were studied (Taylor et al., 2020). Comparisons accentuated discrepancies in economic effects in geographies. India was compared with the world averages. The efficacy of methodology (e.g., the cost–benefit result of proactive and reactive resilience planning) was critically assessed (Winter et al., 2016). There were issues of non-standardized currency conversions, reporting of costs across different studies, and little regional data. However, this was improved through triangulation, i.e., cross-referencing of primary reports, peer-reviewed materials, and government data.

13.4 Results and Discussion

13.4.1 Global Trends in Economic Impacts

Landslides cause serious economic losses to the world at large and transportation systems are usually the worst hit. One of the reasons is that it costs an estimated $20 billion annually in lost funds due to road repairs, economic losses, and degraded infrastructure (Sim & Wong, 2022). There are costs, which are direct (repair, reconstruction), indirect (delays, rerouting), and consequential (economic disruptions, supply chain losses). Such categories contribute to organizing economical evaluations and putting resilience investment first.

13.4.2 Case Study: Lisbon Floods

A resilience analysis of Lisbon showed proactive planning (reinforced assets, early warning) to result in economic losses minimized by 7.6% over reactive response. This highlights the financial benefits of business continuity by combining planning and mitigation with reactive spending (Rezvani et al., 2024).

13.4.3 Cross-National Infrastructure Comparisons

The disaster cost of roads in Tajikistan is estimated to be about 0.5% of GDP (or about 445 million Somoni) per annum. The cost–benefit analysis revealed that selective mitigation measures had a positive payoff, but systemic resilience was not an easy task.

13.4.4 India-Specific Impact Profiling

The economic loss because of the landslides is quite evident in the land slide regions of India:

- 2024 Wayanad landslides more than 420 fatalities, disruptions in rural roads (about 1.5 km) and 3 bridges, devastated crop and companies significantly (Wang et al., 2025)

- Kerala floods of 2018–₹4500 crores of transport infrastructure was lost; ₹2500 crores were recovered as restoration cost.
- Himachal Pradesh and Sikkim (20,192,021)—the cost of road damage 500 crores and 2000 crores and 350–200 crores of it was apportioned to repair work, respectively.
- Ankola landslide (2024)—NH-66 blocked by construction-induced slide, which emphasizes hazard created by infrastructure (Kundu et al., 2025)

13.4.5 Socio-economic Cost Aggregation

The 201 economically important landslide events that occurred in India had varied economic performances. The 412 incident events analyzed (18,682,022) have categorized 141 as high-impact incidents that typically impact road systems and disrupt livelihoods (Parkash, 2023). This kind of systematic classification highlights areas of vulnerability that translate into hotspots that require specific consideration.

13.4.6 Systemic Vulnerabilities in Indian Context

Convergence of steep topography, precipitation and human interventions are witnessed in Indian terrains, which include Western Ghats, Himalayas. Road construction is another risk factor through slope disruption; a study in Uttarakhand attributed 1820% of the landslide susceptibility to roads (Nath et al., 2021). Example: Ankola landslide that was caused by poor construction techniques in part.

13.4.7 Analytical Techniques for Impact Assessment

- The visualization of the most vulnerable areas is with the help of *GIS modeling and susceptibility mapping* (Taylor et al., 2020).
- *Network simulation* measures the alternative route delays and costs.
- *Economic structures* de-aggregate expenses so as to allow the comparison of causal strategies.

However, gaps in data and standardized reporting make it difficult to be replicated.

13.4.8 Resilience Strategies & Comparison

- Proactive engineering—Lisbon case confirmed cost-effectiveness.
- Eco-engineering—PMGSY bioengineering of India in Himachal showed the success of slope stabilizers (The Guardian, 2024).

- Framework development of guidelines—in 2009, the NDMA of India worked out landslide management guidelines, and 2019, risk reduction guidelines.

13.4.9 Economic Justification of Investments

The program in Tajikistan had both positive and negative financial results: the overall resilience was not economically feasible, but the interventions that were focus-oriented were effective. India can equally maximize with high-risk segments.

Key Insights

1. The economic costs in landslides are much greater than the reactive spending in isolation- they entail the trickle down impact on the society and the business.
2. Transport corridors in India are mountainous and are repeatedly hit; recurrent expenditure on repair is deteriorating development pointers.
3. The infrastructure, in itself, could be the main destabilizer in the event that it is not designed by considering landslide risks.
4. The convergence methodology of multidiscipline assessment assists in having the most effective resilience planning.
5. Indian transport policy should be informed by global best practices namely GIS modeling, bioengineering, cost–benefit frameworks.

The sample in Table 13.1 presents a summary of the economic costs of landslides on transportation infrastructure in some parts of the world. Globally, landslides are estimated to cost the world economy, on an average of 20 billion dollars per annum with a significant portion of it being due to transport networks. As an example, the Lisbon floods in 2018 in Portugal caused an estimated damage of about 28 million with most damages being attributed to road and bridge collapses. In Tajikistan, road destruction through repeated landslides contributes to the national GDP at almost 0.5% that is estimated to be 50–60 million annually. In Italy, especially the Apennine areas, the recorded loss of 2 billion annually in 2005–2015 can be attributed to the frequent slope instability to major road networks. In America, especially in the Appalachian area, landslides cause approximately one billion dollar's worth of road damage every year. All of these case studies reinforce the common and frequent financial consequences of landslides on the transport system across the world and provide support to the idea of preemptive resilience and mitigation efforts.

Table 13.2 has emphasized five Indian landslide events that occurred between 2018 and 2024, their economic effects, and the impact on transportation. The 2018 Kerala floods resulted in massive landslides that caused serious damages to an estimated 3500 roadways and numerous bridges that caused an estimated economic loss of 4500 crores. The cost of restoration alone took 2500 crores, as the extent of harm on transport infrastructure indicated. In 2021, Himachal Pradesh suffered landslides because of the heavy rainfall and the poor management of

Table 13.1 Global economic impacts of landslides on transportation networks

Region/country	Estimated annual economic loss (USD)	Nature of impact	Notable case example
Worldwide	$20 billion	Infrastructure damage, road closures, delays	Global average from multiple sources
Portugal (Lisbon)	$28 million (2018 floods)	Road and bridge damage	Lisbon flood resilience project
Tajikistan	$50–60 million	Roads damaged yearly (0.5% GDP)	Pamir Highway blockages
Italy	$2 billion (2005–2015 average)	Frequent road disruptions in Apennines	A16 Napoli landslide
USA (Appalachians)	$1 billion annually	Road clearance, slope failures	Kentucky rockslides

the slope in highways such as NH-5 and NH-707, resulting in approximately 500 crores of direct losses. Although it happened in a less urbanized region in Manipur, Noney landslide in 2022 (Sobhapati, 2022) had a devastating effect on highway and railway projects being built, incurring ₹180 crores of losses and significant backgrounds. A significant landslide struck Wayanad in Kerala in 2024 and affected approximately 1200 km of roadways isolating the rural areas at a cost of about 750 crores. Finally, in Karnataka, the Ankola landslide led to the closure of NH-66 that incurred costs of ₹100 crores over a period of almost a week due to unscientific highway construction. The sum total of these incidents is how vulnerable transportation infrastructure in hilly and monsoon-prone areas can lead to huge socio-economic consequences whenever they are opened to landslides, necessitating improved planning, monitoring, and engineering.

Table 13.2 Major landslide events in India affecting transportation (2000–2024)

Year	Location	Type of impact	Estimated economic loss (INR)	Transportation disruption
2018	Kerala (Floods)	Road collapse, bridge washouts	₹4500 crores	3500 + roads blocked
2021	Himachal Pradesh	Slope failure, NH damage	₹500 crores	NH-5, NH-707 closed
2022	Manipur (Noney)	Railway and highway projects impacted	₹180 crores	Construction halted
2024	Wayanad (Kerala)	Road network cut off, rural isolation	₹750 crores (est.)	1200 km roads disrupted
2024	Ankola (Karnataka)	NH-66 closure due to construction landslide	₹100 crores	5–7 days highway closure

Table 13.3 Classification of landslide-induced economic impacts on road networks

Type of impact	Description	Example from India
Direct	Physical damage to roads, culverts, bridges	NH-707 landslide damage (Himachal)
Indirect	Traffic delays, detours, fuel & labor cost	Detours post-2021 Himachal slides
Consequential	Economic disruption, supply chain delay, tourism loss	Isolation of Wayanad rural tourism sites
Intangible	Loss of public confidence, psychological distress	Manipur landslide fatalities

In Table 13.3, the economic effects of landslides on transportation networks are divided into four types, namely direct, indirect, consequential, and intangible. This categorization can aid the stakeholders in gauging the short and long-term impacts of landslides, on both the infrastructure and the society.

Direct impacts involve instantaneous physical losses including roads, bridges and culverts that are destroyed. Such are the initial losses that are recorded in disaster response. One such instance is the NH-707 highway in Himachal Pradesh, whose landslides were very destructive and they took time and money to fix. The indirect effects are the inconveniences caused by the damage in the first place—the traffic rerouting, the use of more fuel, labor delays and the rise in the cost of vehicle servicing. These expenses are accrued at a long term even when physical damage is low. Indicatively, following the 2021 Himachal landslides, alternative routes were necessary weeks and this added to the travel time and operational costs.

Consequential impacts demonstrate wider economic losses, including loss of access to markets and tourism, postponement of medical services and supply chain interruption. The 2024 Wayanad landslides left most of the villages isolated affecting the livelihoods that rely on tourism and agriculture. Immeasurable but severe effects are the psychological distress, loss of societal trust, and social disturbances. Those communities that were hit by the Manipur landslide mentioned by trauma, fear of the future, and in certain situations, displacement. Such intangible wastages may have social impacts that are very long term and far beyond infrastructure rebuilding. In this way, the comprehensive impact assessment strategy is the key to successful disaster risks management and infrastructure resilience planning in the landslide-prone area.

13.5 Limitations of the Study

The review had a number of limitations. To start with, the economic data was not consistently reported; restoration and loss figures were differing significantly in valuation methods and calculations of currency in year adjustment. Second, the spatial databases of Indian landslide events are not complete and this does not enable a thorough comparison of the costs. Third, heterogeneity in methodology, between GIS-based models and anecdotal estimates, made direct cross-case synthesis problematic. Fourth, implications of indirect and consequential costs (e.g., market disruptions, health delays) were often underreported, which may have contributed to underestimation of overall effects. Fifth, there were no randomized assessments regarding the effectiveness of mitigation; the use of case studies through single sites blocks generalizations. Lastly, the changing climate change, infrastructure design and policy response imply that the past economic impacts might not entirely reflect future risk environments. Although results are based on vital insights, they are supposed to be construed as guiding as opposed to conclusive.

13.6 Recommendations

1. Invest in proactive resilience based on the cost–benefit understanding to determine high-impact road segments.
2. Enlarge data gathering using standardized economical measure here and there.
3. Adopt slope bioengineering on plans such as PMGSY to combine both greenery and engineering.
4. Require all road projects in prone regions to undergo mandate landslide risk assessments which includes GIS-based modeling.
5. Enhance early warning and community engagement to allow responding promptly and reduce the indirect socio-economic losses.

13.7 Conclusions

The economic impacts of landslides on the transport infrastructure could be described as a complicated problem of both global and national levels. Landslides in the world have a direct estimated cost of 20 billion every year and are considered initiators of instability in society and business due to disruption of the transport network in the society. The mountainous corridors through which India lies, particularly Kerala, Himalayas, Uttarakand, and Karnataka have repeated high cost events and over the last few years these events have been amounting to thousands of crores.

The costs associated with the economy do not only rely on the immediate repair expenses but also include secondary costs such as increased commuting, isolation in the market, and even health hazards. Such costs could be measured with the help of well-organized economic models that define the direct, indirect, and consequential losses. The global experience of comparisons, such as Lisbon to Tajikistan, highlights the economic benefit of planned investments in resilience and early warning systems.

India has welcomed the policy evolution and policy mitigation efforts (NDMA 2009 & 2019, bioengineering under PMGSY), but still there are certain challenges:

the lack of data, the standardization of costs, the incorporation of engineering, ecological, and community resilience practices. Landslide risk can be increased inappropriate infrastructure in case it is not designed with risk assessment.

It is necessary to adopt a holistic, evidence-based framework, including the use of rigorous data collection, GIS-based risk mapping, adaptive engineering, economic modeling, and stakeholder engagement. This combined strategy has the potential to direct investments, cushion against cascading economic losses, and enhance transportation network resilience across the varied Indian terrain and other places.

The economy-wide and diverse socio-economic effects of landslides on the transport infrastructure of the world and India. In the world, landslides incur close to 20 billion losses each year, roads and highway systems being some of the most affected areas. In India, transport corridors in hilly areas have been found to be vulnerable, as seen in the 2018 Kerala floods and 2024 Wayanad landslides which were worth thousands of crores. In classifying the effects as direct, indirect, consequential, and intangible losses, one gets a comprehensive model of the complete impact of the disruptive effects and can focus on employing specific mitigation and resilient measures.

Acknowledging The author(s) acknowledge the use of AI-based tools for grammar and language correction during the preparation of this manuscript.

References

Gokhale, N. (2023, September 18). In fiction: The great cloudburst and landslide of September 18, 1880 in Nainital, Scroll.in https://scroll.in/article/1056064/in-fiction-the-great-cloudburst-and-landslide-of-september-18-1880-in-nainital

Kundu, P., Menon, V., & Kolathayar, S. (2025). The Shirur landslide of July 2024 triggered by intense rainfall and unchecked development. *Natural Hazards Research.*

Nath, R. R., Das, N., & Satyam, D. N. (2021). Impact of main boundary thrust (MBT) on landslide susceptibility in Garhwal Himalaya: A case study. *Indian Geotechnical Journal, 51*, 746–756.

Parkash, S. (2023). Lessons learned from landslides of socio-economic and environmental significance in India. In *Progress in Landslide Research and Technology, Volume 1 Issue 2, 2022* (pp. 309–315). Cham: Springer International Publishing.

PMGSY Bioengineering Case. (2024, May). *'We can't defeat nature but can be climate-resilient': Plant roots help stop landslides. The Guardian.* theguardian.com

Rajan, K. C., & Bhandari, S. (2024, December). Landslides and road economy: Insights from the Himalayas. In *AGU Fall Meeting Abstracts* (Vol. 2024, pp. NH32B-05).

Rezvani, S. M., Silva, M. J. F., & de Almeida, N. M. (2024). Urban resilience index for critical infrastructure: A scenario-based approach to disaster risk reduction in road networks. *Sustainability, 16*(10), 4143.

Samom, S. (2022, July 24). Manipur landslide: 2 more bodies recovered after search op officially ended. *Hindustantimes.* https://www.hindustantimes.com/india-news/manipur-landslide-2-more-bodies-recovered-after-search-op-officially-ended-101658644650112.html

Sim, K. B., & Wong, S. Y. (2022). An overview of causes of landslides and their impact on Transport networks. *Advances in Modelling to Improve Network Resilience* 114–125.

Tajikistan Transport Disasters. (2021). *The economic impacts of disasters along key transport corridors.* World Bank Report. worldbank.org

Taylor, F. E., Tarolli, P., & Malamud, B. D. (2020). Preface: Landslide–transport network interactions. *Natural Hazards and Earth System Sciences, 20*(10), 2585–2590.
Wang, Q., Xu, C., Ma, J., Huang, Y., Zhang, S., & Gao, H. (2025). Preliminary analysis of the July 30, 2024, Wayanad landslide disaster in India: Causes and impacts. *Natural Hazards Research.*
Winter, M. G., Shearer, B., Palmer, D., Peeling, D., Harmer, C., & Sharpe, J. (2016). The economic impact of landslides and floods on the road network. *Procedia Engineering, 143*, 1425–1434.

Different Types of Landslides and Vulnerability and Adaptation of Indigenous Knowledge and Practice

14

J. Eugene

Abstract

Landslides are the most destructive hydro-geomorphic hazards especially in the mountainous and tectonic active areas such as the Himalayas and the Western Ghats of India. This chapter gives a detailed list of the classification of the landslides in terms of movement mechanisms and types of materials and falls, topples, slides, flows, and spreads. Based on the recent case studies including Pettimudi debris flow (2020), the disaster caused by the avalanche in Chamoli (2021), and the Wayanad landslides (2024), the chapter examines the contribution of both natural and anthropogenic factors to the growing landslides vulnerability, which include deforestation, uncontrolled construction, and climate change-related rain variations. Notably, the chapter explores the little-known world of indigenous knowledge and community-based adaptation measures that the local people in high-risk regions follow. Conventional slope control methods, ecological recovery with indigenous plants, and oral early warning systems are discussed as the possible inclusion in modern disaster risk elimination systems. Participatory mapping and community engagement are highlighted as important in the increase of adaptive capacity and resilience. Using an interdisciplinary approach that integrates geomorphology, climatology, and anthropology, this chapter illustrates that scientific risk assessment tools should be synergized with culturally ingrained practices. It proposes participatory and local response plans that should not ignore the experiential knowledge held by the indigenous population and that which integrates technology in monitoring landslides and providing early warning.

J. Eugene (✉)
Institute of Hotel Management Catering Technology and Applied Nutrition, CIT Campus, Taramani, Chennai, India
e-mail: ihm.eugene@gmail.com

R. M. Yuvaraj et al. (eds.), *Advanced Landslide Susceptibility Modeling*,
https://doi.org/10.1007/978-3-032-14485-0_14

Keywords

Landslide classification • Geomorphic hazards • Indigenous knowledge systems • Disaster risk reduction • Community-based adaptation • Landslide vulnerability • Resilience

14.1 Introduction

One of the most common and devastating natural hazards in the world and specifically in the mountainous and tectonic areas is landslides. They are complicated geomorphic processes, which entail rock, debris, or earth materials descending at the impact of gravity. The occurrence and severity of landslides have been on the rise over the last few decades because of both natural causing factors, including heavy rains and seismic activity, as well as anthropogenic ways of interaction, such as deforestation, unscientific construction, and unplanned development (Hungr et al., 2014; Sharma, 2020). India is one country that is prone to landslides and almost 13% of its landmass has been rated as a landslide-prone. Landslides occur frequently in the Himalayan belt, Western Ghats, Northeast India, and areas of the Eastern Ghats, and in most cases, they lead to massive damages in terms of life, property and livelihoods. The increase in the frequency of hazards associated with climate change such as extreme rainfall and glacial retreat has also increased risks of landslides in these fragile ecological areas (Khanduri & Rautela, 2021).

Although much scientific effort has been put into mapping the zones prone to landslides, forecasting slope instability, and taking measures to reduce or control them, little has been given to how indigenous knowledge systems can be used to build community resilience. The local and indigenous communities, especially those who reside in remote and high-risk regions, have over the years acquired subtle insights into their environment. Such conventional activities as the ecological slope stabilization, geomorphic threshold-based land-use patterns, and oral early warning systems are useful in terms of sustainable disaster risk mitigation (Srivastava, 2012).

The main objective of this chapter is to (i) discuss the various types and categories of landslides according to their movement modes and the composition, (ii) discuss the main geographic area of landslides in India, (iii) evaluate recent case studies to clarify about vulnerability and triggering factors, and (iv) outline how the indigenous knowledge and community-based adaptation strategies can be used with scientific interventions. The chapter recommends a holistic and culturally sensitive landslide mitigation and resilience-building approach by combining new risk management tools with traditional wisdom.

14.2 Rationale of the Study

Landslides are a constant threat to life, infrastructure and ecosystems especially in areas that have complicated geomorphology, variability of rainfalls and uncontrolled human activities. Landslide catastrophes that have befallen parts of India including the Himalayas, Western Ghats, and the Northeast states show the necessity of a better insight into the nature, causes, and location of landslides. Considering geotechnical and remote sensing technologies, despite the innovation, disaster mitigation activities do not consider the contribution of socio-cultural aspects and traditional ecological knowledge to establishing long-term community resilience.

Increasing effects of climate change such as change in rainfall and glacier melting have also increased the occurrence and uncertainty of landslides. Meanwhile, indigenous and local populations residing in areas of environmental hazards have devised time-proven tactics of monitoring, reaction to and co-existence with environmental threat. These are also under-documented and underused strategies that are based on indigenous knowledge systems in formal disaster risk reduction (DRR) policies and plans.

This research thus has a twin goal, one, to categorically and put landslides into perspective as per the geophysical and climatic conditions; and second, to understand how local knowledge and the customary practices can be effectively incorporated in the modern management systems of landslides. The study is expected to enable a comprehensive, inclusive, and community-based perspective in the process of mitigating landslide vulnerability and capacity building in the vulnerable areas due to the closeness between scientific analysis and traditional wisdom.

14.3 Materials and Methods

The present research is based on qualitative and interpretive research design with the aid of the case-based analysis and thematic categorization. It is based on secondary data and empirical records of the indigenous knowledge systems.

14.3.1 Data Sources

The research is based on various sources of data in order to provide the analysis. They are scientific publications like peer-reviewed journals and official reports of the Geological Survey of India (GSI), the National Disaster Management Authority (NDMA), and other scholarly ones (e.g., Hungr et al., 2014; Sharma, 2020). Satellite-based reports, GIS reports (landslide zonation maps, susceptibility indices, and digital elevation models) have also been used to perceive the patterns and degrees of risk in space. The news archives and government reports were used to record the major landslide events that occurred in recent years including Pettimudi, Chamoli, Raigad, and Wayanad. The research uses indigenous knowledge as well with published reports and gray literature that characterizes traditional ecological practices, slope stabilizing tactics and community-based adaptation techniques that are common in the high-risks areas such as Uttarakhand, Sikkim, and Kerala.

14.3.2 Classification Framework

The landslides have been grouped according to the most popular typology, as given by Cruden & Varnes (1996) and Hungr et al. (2014), including falls, topples, slides, spreads, and flows. Both types were analyzed by using case examples in the Indian context.

Case Study Selection
Five significant landslides in India (2020–24) of various types, triggers, and geographic locations were sampled purposively to represent the context different types, triggers, and geographic locations of landslides. The criteria were severity of impact, nature of movement, documentation availability, and applicability to indigenous adaptation.

Thematic Analysis
The thematic content analysis was employed in the study to determine the patterns of recurrent vulnerability factors and adaptive practices. The coding categories were slope change, rain intensity, vegetation cover, type of construction as well as traditional coping strategies.

14.3.3 Analysis

Falls
Falls are among the sudden and rapid forms of landslides. They consist of the abrupt release of rock, boulders or debris on steep or vertical rock faces, normally along existing discontinuities such as joints, faults, bedding planes, or fractures. The free falling material or the one that comes down by bouncing or rolling under the gravity force results in the rockfalls being very risky especially because of its velocity and irregularity (Cruden & Varnes, 1996).

Mechanism of Fall Initiation
It starts with mechanical or chemical weathering which undermines the surfaces of rocks. The freeze–thaw processes and especially in elevated regions, cause the water to expand in the cracks of the rocks, which tends to tear the rocks apart. Other causes are undercutting by rivers or at the base of the slope, seismic shaking, dislodging rocks of precarious masses, anthropogenic activities such as quarrying

Table 14.1 Summary of major rockfall events in India

Year	Location	Trigger factor	Impact
2022	Gangtok–Nathula highway	Intense rainfall	Road blocked, no casualties
2021	Rishikesh–Devprayag stretch	Blasting activity	Road closed for 36 hours
2020	Malshej Ghat, Maharashtra	Freeze–thaw, rain	One fatality, tourism disrupted

and road cutting, vegetation root wedging and animal activity. When this becomes detached, the rock mass funds down the slope and in many cases breaks into smaller blocks. Falls can be very hazardous in a narrow valley or gully, where usually human settlements and roads are constructed near the face of the cliff.

Occurrence in India

Rockfalls are known to occur in Himalayan landscape particularly along the National Highways in Himachal Pradesh, Uttarakhand, and Sikkim. Roads such as the Gangtok–Nathula highway and the Rishikesh–Badrinath road are usually affected by rockfalls especially during the monsoon season. Western Ghats also have rockfalls especially in areas such as Malshej Ghat and Agumbe where steep slopes prevail (Table 14.1).

Hazards and Impacts

Rockfalls have the potential to cut transportation systems particularly mountain roads and railways. They usually cause secondary hazards, which include damming streams and initiation of debris flows. The loss of life and damage to infrastructure are great in the cases of high tourist attendance or pilgrimages (e.g., Char Dham Yatra).

Mitigating Measures

However, to prevent toppling failures, structural, preventive, and monitoring measures are needed in site-specific geological and environmental environment. The installation of rockfall barrier and catch fences on highways is one of the most widespread and efficient ones, and it is assumed that such fences would stop falling objects and not allow them to hit roadways or residential neighborhoods. The barriers capture and absorb the kinetic energy of the falling blocks, which minimizes the impact damage. Scaling (manual or mechanical removal of unstable rock masses on slope faces) can be considered a proactive activity aimed at avoiding topples to set in, particularly in zones where the rocks exhibit apparent evidence of cracking or movement. Rock bolts and wire mesh netting are the other necessary intervention that is most critical, as it stabilizes the slope by supporting the rock mass and preventing fractured blocks from falling away. These systems prove to be particularly useful where the jointed rock or blocky terrain is very close. Besides the physical one, there must also be warning signs and road closure during the time when there is a risk, i.e., when it is raining heavily and after the earthquake. Effective communication with commuters and local communities along with the use of temporary road closure or roads diversion can play a vital role in keeping down the number of casualties in the case of possible toppling events. These mitigation measures, taken together, create a unitary strategy of controlling and inhibiting the risks of toppling landslides.

Topples

Topples: This is a special type of landslide motion that involves the rotating forward of a block of rock or soil about a pivot point that is usually at the base of the slope or very near the slope base. Topples are outward and downward rotating unlike rockfalls in which the material falls freely. This kind of motion is specifically common on steep-to-vertical rock surfaces, with the discontinuities, specifically joints, fractures, and bedding planes, going against the slope face (Hungr et al., 2014).

Mechanics and Initiation of Toppling

The factors that contribute to toppling failures are a mixture of environmental and geological factors. The geological structure is one of the main contributors, especially the steeply dipping joints or discontinuities, which often occur in stratified or foliated rock structures like the schist, slate, or layered sedimentary rocks. These structural vices establish good conditions of forward rotation of rock masses. Further instability is caused by weathering and erosion; surface weathering weakens the rock matrix, whereas toe erosion, the erosion caused by undercutting rivers or digging roads, eliminates the lateral resistance needed to stabilize the slope. The pressure of ground water also matters a lot because when there is a rainfall, the pressure of the pores inside joints is raised, the internal friction is decreased, and the rotational movement occurs. Finally, seismic activities can be a potent trigger because they can destabilize already weakened slopes, particularly in the tectonically active areas where structural discontinuities are highly common.

Stages of Toppling

Topples are generally expected to show a progressive nature of failure which starts with first cracking and observable separation along the joints, indicating the start of instability. This is then succeeded by a forward movement of the rock mass with minor displacement that can be witnessed as open tension cracks at the slope crest. At some point, the movement ends up being entirely detached and disintegrated, which usually leads to the formation of blocky debris at the bottom of the slope. Toppling can be as discrete units, failure of individual block, and toppling can be as block, in which longer columns or slabs of rock bend and fail (Table 14.2).

Examples from India

Toppling failures are commonly seen on road cuts in Uttarakhand and Himachal Pradesh particularly where exposed surfaces are massive quartzites or phyllites in steep terrain. Little over Badrinath, there have been a few topples along the Rishikesh–Badrinath highway during monsoon times, when excavation along roads passed through dipping bedding planes. Road widening activity in the Western

Table 14.2 Comparison of topples and falls

Feature	Topples	Falls
Movement type	Rotational	Free-fall or bouncing
Trigger factors	Jointed rock, water, quakes	Weathering, undercutting
Contact with slope	Maintained during movement	Minimal or none
Early indicators	Cracks at slope crest	Rarely visible

Ghats and around Munnar and Wayanad has led to toppling failures in the lateritic and basaltic rocks, which are enhanced by the presence of water infiltrating the rocks and no slope protection methods.

Hazards and Impact

Topples can begin gradually, but when something goes wrong, they can happen out of the blue and lead to dangerous and immediate dislodgement of huge blocks, bringing in a high risk to transportation systems, hill-settlements, and construction projects. Topples compared to falls or flows are not always considered in hazard mapping due to initially unnoticed patterns of deformation.

Mitigation Measures

The mitigation of toppling failures is a variety of engineering and geotechnical actions that are to be taken in order to provide the slope with increased stability and stop rotational activity. Rock anchoring and bolting is considered as one of the main methods as high-strength steel bolts or anchors are installed in the rock mass in order to give it extra resistance to the rotational forces. This method comes in handy especially in jointed rocks where the blocks tend to fall because of unfavorable positioning. The control of draining is also essential, because the more water enters the joints and fractures, the higher the pressure of the pore water and the lower the shear strength, which contributes to the faster movement of toppling. Surface and subsurface drainage systems that include weep holes, perforated pipes and drainage trenches would also be effective in the accumulation of water and slope stability. The second significant strategy is regrading or terracing slopes that have too sharp sides; this means decreasing the slope side to less hazardous levels, thus decreasing the driving forces that could cause potential blocks of failures. Where falling debris is dangerous to the infrastructure or human lives, protective barriers such as impact-resistant walls, gabions, energy-giving fences, among others are set in order to either trap or redirect thrown off material. Finally, slope crest geotechnical monitoring with such equipment as inclinometers, extensometers, and crack meters allows identifying the pattern and the area of the deformation early and being aware of the possible area of failure. Such monitoring systems in conjunction with remote sensing, real-time data logging can ensure that there is early warning signals that are very critical hence enabling timely evacuation or even closure of risky areas. All these measures collectively provide a complete defense against dangers of toppling failures in the landslide-prone areas.

Slides

Slides are the most frequently found forms of landslides and can be characterized as a downward movement of rock, earth or even debris in a well-defined shear or slip surface. Depending on the type of material and the slope geometry, as well as the triggering mechanisms, the movement varies in its speed between slow and undetectable or rapid and devastating. The mass that has been uprooted tends to travel in a block coherent at the very beginning but can break down as it travels downslope. Broadly, slides are divided into two large groups in regards to the geometry of the failure surface:

Translational Slides

Translational slides are experienced when the mass of sliding is on a relatively flat or slightly wavy surface that is usually perpendicular to the slope. These surfaces are usually in line with the structural discontinuities which would be either bedding planes, joints, foliation planes, or areas of weakness made of material such as clay, shale, or weathered siltstone, which has a very low shear resistance. In contrast to rotational slides, which move through a curved motion, translational slides consist of a linear (block-like) sliding of material, which can take a lot of its original structure with it as it slides downslope (Table 14.3). These slides usually have a great extent and when the mass that is sliding gets saturated with a water body, particularly during heavy rainy seasons, they can move at full lengths and speeds resulting in huge destruction. Their velocity and energy cause them to be very dangerous especially in areas where the population is high or in areas where infrastructure has been constructed along slopes.

One of the most notable examples of a translational slide is the Malin landslide at Maharashtra (2014), which claimed the life of more than 150 inhabitants. The slide itself was precipitated by the heavy rainfall of the monsoon season, but the weathered basalt with interposed clay-laden horizons enhanced the slide. These weak layers were saturated thus greatly decreasing the cohesion and causing a sudden collapse of the slope. The mass had a very high speed and it swallowed a whole village. This tragedy emphasized how vulnerable the settlements of slopes made of eroded volcanic landscapes were and the importance of slope evaluation, good land-use planning, and early warning systems in areas subject to landslides.

Table 14.3 Comparison between translational and rotational slides

Feature	Translational slide	Rotational slide
Failure surface	Planar or undulating	Curved and concave upward
Material type	Clay, shale, jointed rock	Homogeneous soil, weathered bedrock
Movement type	Linear, block-like	Rotational, backward tilting
Common triggers	Rainfall, slope cutting	Rainfall, toe erosion
Speed of movement	Moderate to very fast	Slow to moderate
Real-world example	Malin (2014), Maharashtra	Nilgiris, Tamil Nadu

Rotational Slides

Rotational slides (also referred to as slumps) are on curved, concave-upward surfaces of failure, which creates a typical geometry of a spoon. In these slides, the mass of the soil or rocks rotates outwards and downwards frequently showing reversed tilting at the head and protruding at the toe of the slide. Such motion is generally seen in homogeneous materials like soils of clay, weathered rock, or volcanic ash deposits where uniformity of the material also helps in creating a curved slip surface. Even though the rotational slide movement is usually slower and more localized than the translational slide movement, the effects may be extreme, particularly in those geographical locations where infrastructure or settlements are downslope. These slides usually develop with time, and thus it is easy to identify the early stages by observing the cracks in tensions or the movement of the ground. Another interesting example is the common rotational depressions seen in the Nilgiri Hills of Tamil Nadu especially in the extended rain periods in October–December. These downfalls have generally been caused by saturation of weathered lateritic or gneissic soils which become weak under abundant moisture conditions and begin to develop rotational motion in preexisting zones of weakness.

Causes and Contributing Factors

The slope stability depends on a range of various triggers that both natural and human influences can cause to both translational and rotational slides. Rainfall, which involves heavy or sustained precipitation, is one of the most prevalent natural causes of the event, as it elevates the level of pore water pressure in the soil or rock mass and decreases shear strength along the slip surface, thus facilitating failure. Human activities (alterment of slope most prominently by road excavation or in hilly terrain by constructions) tend to eliminate the lateral support and dislodge the natural equilibrium of slopes. Earthquakes are also a major cause; they can cause slopes to become unstable with shaking and cracking already weakened/saturated materials and cause instant sliding to take place. Deforestation is a cause of slope instability since it removes vegetation cover, which consequently, diminishes the root reinforcement, enhances surface runoff and heightens soil erosion. Also, the hydrogeological factors such as seepage of the groundwater in the ground and groundwater perched water table in both the shallow and deep-seated landslides are important as they modify the internal water regime and decrease the effective stress of the slope material and hence it is more susceptible to failure.

Statistics and Figures

The Geological Survey of India (GSI) reports that out of the total landslides that have occurred in the country, more than 66% are slide-type failures, which have had a severe effect on the state in the Northeast, including Uttarakhand, Himachal Pradesh, Sikkim, Kerala, and regions. These kinds of landslides are very dangerous to both life and infrastructure particularly in the steep terrain regions and high rainfall areas. One of the most notable was the Malin landslide of 2014 in Maharashtra that moved some 0.2 km^2 of material in only five minutes, burying at the same time an entire village. Equally, the 2017 Kotropi slide in Himachal Pradesh was a catastrophic translational movement which took place on a highway cut slope during the monsoon season and killed over 45 people, and thus underscored the susceptibility of road infrastructure in the hilly areas.

Mitigation Approaches

A successful management of slide-type landslides needs to be a holistic endeavor which tackles both the surface and subsurface factors that increase the slope susceptibility. The implementation of surface drainage and slope dewatering systems is one of the major strategies because it is supposed to stop water accumulation due to the channeling of the surface runoff out of the vulnerable zones and putting pressure on the pore water on the slope. This is usually done by depending on surface drains, underground drains and underground wells. The other important measure is structural reinforcement, which entails the construction of retaining walls, gabions, and soil nailing. These buildings offer mechanical reinforcement to the incline that withstands lateral soil forces and holds unstable materials in position. In regions that have over-steepened soils, when graded terracing is usually used to reduce the overall slope angle and, therefore, reduce the driving force of gravity and, consequently, the erosion caused by the flow of water across the slope. Bioengineering is another vegetative measure that is important because it stabilizes the slope by enhancing the cohesion of the soil through root reinforcement and decreasing surface runoff. Finally, there is the implementation of real-time surveillance systems with the use of such instruments as inclinometers, piezometers and ground-based radar to offer early warnings in the risk areas. The systems aid in the monitoring of any minute movements or alteration of the underground ground or the pressure of water in the ground that may enable evacuation and preventive measures before the system fails disastrously. When combined, the mitigation strategies make up a compound framework of decreasing the hazard and effect of landslides in prone areas.

Spreads

Spreads are a slightly less frequent but nevertheless different form of landslide, where a coherent mass of soil or rock moves laterally and deforms and fractures over a weaker substrate which is often saturated with water. This mostly happens in flat or inclined lands, where the material in the area does not have adequate strength of shear to carry the mass above it. The breakage usually starts when the underlying material, which is usually fine alluvial sediments, sensitive clays, or liquefiable silts, experiences a loss of strength and a collapse of the support of the surface layer results. Consequently, the overlying material disperses laterally, and it might split into segments or blocks (Hungr et al., 2014).

The concept of spreads is often linked to liquefaction, the process of saturated soils momentarily losing their strength and stiffening because of the abrupt loading, which takes place in case of earthquakes. They are commonly found in deltaic deposits or in estuarine plain or in riverbanks especially where there is a combination of fine sediments and high water tables. Spreads can also be noted in the Indian subcontinent although more prolifically documented in such countries as

Japan, Canada, and the United States. In the example of the alluvial plains of Assam and some parts of Meghalaya, lateral spreading was observed in flood-prone zones, particularly during the high magnitude seismic activity or continuous saturation. Such failures have also been evidenced in Jammu and Kashmir whereby there have been reports of high-pore water pressure and soft clay layer along the Jhelum River valley. Spreads are difficult to react to because they are subtle and slow onset which sometimes go unnoticed until there is a strong ground deformation. Their control needs a delicate grasp of geotechnical characteristics, sub ground hydrology, and seismic susceptibility.

Mitigation Measures

Reducing lateral spreads requires not only the geotechnical instability of the underlying saturated material, but also the behavior of the topmost cohesive mass. The major mitigation measures involve ground improvement methods, which may be soil densification, vibro-compaction or chemical grouting techniques and which are applied to enhance the shear strength of the underlying loose soils, especially in liquefaction vulnerable regions. Vertical and horizontal drainage should be increased to ensure that the water pressure in the pores is minimized and does not saturate causing liquefaction. The methods such as use of prefabricated vertical drains (PVDs) and sand wicks can be used to facilitate the process of soil consolidation. Re-profiling and re-enforcement of slopes with geogrids or sheet piles can limit horizontal movements and deter separation of blocks in case of a seismic or saturation event. In the riverine or deltaic areas, river training can be done with the construction of river training structures such as levees and embankment structures to minimize erosion and lateral seepage. Identification of potential areas prone to spreading especially in the Northeast of India and parts of the western Himalayan foothills should be done through seismic zoning and hazard mapping that can be used in land-use planning and infrastructural development. Avoidance and control, especially by avoiding heavy infrastructure or settlement construction on soils that were found to be extremely prone to liquefaction and subsequent lateral propagation is still an important measure of long-term risk mitigation.

Flows

Flows are the most dynamic and dangerous forms of landslides in which the saturated material moves downslope, in the form of the viscous fluid. Some of these contents can be fine silts and clays, or big debris and rock fragments depending on the nature of terrain and precipitation processes. The landslides of the flow type are commonly divided into the following groups: debris flows, mudflows, and earthflows, based on the size of particles, content of moisture and velocity. Intensive or prolonged rain, quick snowmelt, or dam break is usually the cause of mobilization of such flows, when the pore water pressure in the soil matrix increases, and shear strength is lost, and cohesion is lost.

Flows are often a defining feature by the fact that they are able to capture other material as they move through it, and will tend to increase in size and strength on their course. The flows can move at big speed and cover a long distance, especially when they are contained in small tracks, which causes the destruction in a large scale. In India, the flow-type landslides have been increasing significantly in the southern states of Kerala, Karnataka, and Tamil Nadu, particularly during the southwest monsoon season (June to September) (GSI, 2023).

One of them is the Pettimudi debris flow in Kerala (2020) when the sudden flow occurred due to heavy monsoon precipitation and covered labor quarters burying 66 people. Geology of the region, the heavily weathered lateritic soils and steep slopes were a major cause of the disaster. The same happened with the Chamoli incident in Uttarakhand (2021) where a detachment of glaciers, and soon turned into a debris flow, devastating infrastructure and claiming the lives of more than 200 individuals. Although a glacier event, the incident highlights the mutual relationship between climatic triggers and geomorphic reactions.

Recent work indicates that as much as 30–35% of landslides in the Western Ghats and surrounding hill areas are of the type in the flow category. Climate change is also likely to increase the number of these kinds of failures in the next few decades as the frequency of extreme weather events is on the rise. The project of real-time monitoring, slope saturation thresholds, and rain-based early warning systems have been tested in Wayanad and Idukki where community-based preparedness and geotechnical mapping are currently incorporated in the mitigation systems.

Mitigation Measures

The need to have a mix of preventive-, structural-, and community-based mitigation measures is attributed to the fact that flow-type landslides are characterized by rapid velocity and extensive runout distances. It is important in surface water management; creation of diversion drains, contour bundling, and interceptor drains over the susceptible slopes helps in minimizing infiltrations and surface water runoffs. The drainage design is particularly important in the rain rich regions such as Kerala and the Western Ghats. Natural channels may be lined with check dams and traps to trap and slacken the flow of debris and hence diminish the impact they have on the downstream areas. Deep-rooted native vegetation to stabilize the vegetation aids in binding the top surface layers of soil making it more cohesive and reducing erosion, and bioengineering is one use of vegetation, a skilled and cost-efficient method of slope reinforcement that is more sustainable. Critical infrastructure and habitations are covered with reinforcement structures, including crib walls, gabions, and concrete barriers; the positioning of these is calculated based on the simulation of flows and mapping of past events. Flow prone areas such as Wayanad, Idukki, and Kodagu are implementing early warning systems that are operated on the basis of rainfall intensity, soil saturation, and real-time monitoring with inclinometers, piezometers and remote sensors (GSI, 2023). Disaster preparedness in the community in risky areas is very important since awareness campaigns, evacuation planning, and training in the first response mechanisms may save a lot of casualties when there are flow events. Zoning regulations and building codes along identified paths of flow or deposition also aid in ensuring that settlement in high-hazard areas is avoided thereby increasing resilience to any disaster caused by landslides in the long run.

14.3.4 Adaptation of Indigenous Knowledge and Practice in Managing Landslides

In India, indigenous communities living in landslide-prone areas have, generations of local and culturally-specific practices of monitoring, interpreting, and acting in response to slope instability have developed. Though frequently underestimated in more general discourse, these knowledge systems are extensively knowledgeable about the long-term engagement with the natural environment and are considerably applicable in mitigation and adaptation to different forms of landslides. These traditional practices have the potential to provide cost-effective, ecologically friendly and acceptable solutions to the community when combined with modern scientific techniques of minimizing the risk of disasters.

Many regions prone to rockfalls and topples such as the steep terraces of the Western Himalayas and Northeastern hill areas, have traditionally not be constructed directly under hanging cliffs or rocky slope slopes. Moreover, they apply natural indicators such as crack forms in rocks or growth of certain species of lichen to determine the stability of the mass of rocks. Earlier generations in Himachal Pradesh and some Arunachal Pradesh recommend that loose boulders be removed by hand every now and again, or that stone retaining platforms be strategically placed to deflect falling material, an early variant of slope scaling and catchment, similar to the modern rockfall barriers.

In the case of slides, especially rotational and translational slides, there are numerous indigenous activities dealing with the slope alteration and water management. In Nilgiris and Sikkim, people use terraced agriculture whereby the plots are inward-facing and there are vegetative hedges that not only decrease the slope but also control runoff and erosion. The traditional knowledge in these regions also provides conserving native vegetation particularly species that have deep roots such as Ficus or Vetiveria which provide stability to the soil and they do not create mass movement. In Maharashtra sacred groves are known as devrai, a type of vegetation vegetable cover on unstable slopes, which is preserved not only because of religious reasons but also due to their ecological stability.

In regions where the spread is likely, e.g., in parts of Assam, Meghalaya, and the Kashmir Valley, local knowledge systems focus on not building on recently deposited waterlogged sediments or on newly deposited waterlogged sediments. Stilt houses (or locally known as chang ghar) are constructed in the Assam floodplains on piles made of bamboo to reduce the impact of soil spreading along when there is a flood or earthquake (Das et al., 2014). Also, the local tribes identify the color and texture of soil as a measure of the unsafe and safe construction sites, which is now supported by the geotechnical test.

Societies that experience the impacts of flows, particularly in the state of Kerala, Karnataka, and the Northeastern hill states have always been aware of the risk associated with heavy rainfalls on steep terrain. To this, they typically lined the hills with stone channels to direct the water and prevent the overland movement that were uncontrollable. Another common practice is the use of banana or bamboo groves at the slope toes, which are considered as physical barriers as well as in the regulation of moisture in nature. In Wayanad and Idukki, the oral traditions incorporate the notes of gurgling sounds in the ground or abrupt appearance of the springs as the pre-warnings of the impending debris flows. Elders in the community usually prohibit cattle grazing or building within particular months that are considered risky, a similar action to the contemporary seasonal risk zoning.

Development agencies and disaster management institutions have started to identify and formalize these practices in the indigenous people and bring these practices to formal structures in recent years. As an illustration, the Sikkim State Disaster Management Authority has developed participatory mapping initiatives whereby community knowledge is incorporated in landslide hazard zonation. The *Northeastern Space Applications Center* (NE-SAC) in Meghalaya and collaborates with local councils to online fuse evidences of remote sensing with oral historical data on land movement to create more precise susceptibility mapping. The initiatives are indicative of an increased appreciation of the fact that indigenous knowledge is not tenure-folklore but a dynamic, adaptive response system based on lived experience and observation (Kanene, 2016).

The fact that traditional practices are incorporated into the wider array of landslide mitigation techniques, be it through policy assistance, community education, or joint planning, makes the efforts on the disaster management issues both more inclusive yet context-specific and sustainable. Indigenous knowledge offers early warning signals, cost-effective stabilization techniques, and long-standing environmental custodianship (Priyadarshini & Abhilash, 2019) all of which are currently necessary in long-term resilience generation in areas prone to landslides.

14.4 Results and Discussions

The current research paper gives detailed grouping of landslides in five broad categories (falls, topples, slides, spreads, and flows) according to the movement pattern and the material structure. The field-based investigation, literature review, and synthesis of the recent case studies throughout India indicate the diversity and complexity of these geo-hazards in addition to their growing the frequency and severity owing to the anthropogenic efforts and climate-driven reasons. The different kinds of landslides have different mechanisms, triggers, and spatial distributions, which require unique mitigation measures. There has also been the introduction of the indigenous knowledge as a useful facet of augmenting the community-level resilience and disaster preparedness.

Structural discontinuity, erosion, or road construction undercutting as well as heavy rainfall are the major causes of rockfalls and topples, which occur frequently in the Himalayan states and on highway cut slopes. The information provided by field events like those on the Rishikesh–Badrinath highway evidences that these failures may be sudden and without any surface deformation, which are very dangerous to life and infrastructure. The contemporary mitigation measures, including rock bolting and catch fences, have quite effectively mitigated the effects of direct impact, but the indigenous measures, including the traditional avoidance of settlement under steep slopes, regular removal of potential hazardous rocks, and the use of natural rock ledges as protective features, demonstrates the local understanding of the behavior of slopes well before the advent of the geotechnical testing.

Slides, both translational as well as rotational are the most common form of landslide in the Indian context with more than 66% recorded according to the Geological Survey of India. As the field case studies like the Malin slide in Maharashtra and slumps in the Nilgiris point, rainfall infiltration, slope cutting, and deforestation are the key factors. The destructive ability of translational slides can be considered especially due to speed, and distance of travel whereas rotational slides can be slower, but usually impact on greater volumes of slope. The mitigation measures such as retaining wall, slope terracing, and soil nailing have enhanced the slope stability in the strategic regions. Others such as vegetative hedging, contour based agriculture and/or use of deep rooted species to bind soils are also meaningful, particularly in tribal communities in Sikkim and Tamil Nadu. Not only have these practices controlled surface runoff and erosion but they have also maintained the morphology and ecological balance of the slope.

Lateral spread, which is not very common, also has been reported in sensitive regions like the alluvial plains of Assam and sections of the Kashmir Valley. Such failures are usually caused by liquefaction when the earthquake occurs or saturation of fine-grained soils. Such phenomenon is researched and it can be seen that geotechnical measures like ground improvement, compaction, and sheet piling can be used to provide a solution; however, standard housing modifications like high bamboo building and careful not to build on soft and recently formed sediments are important as well. The communities in these areas exhibit a subtle awareness of the soil behavior and water-logging behaviors, with the communities having a tendency to select safer areas depending on intergenerational experience and natural signage.

Flow-type landslides such as the debris flows, mudflows, and earthflows are increasingly becoming prevalent in the Western Ghats and the Himalayas under the influence of intense and unseasonal rainfall. The occurrences like the Pettimudi debris flow in Kerala and the Chamoli disaster in Uttarakhand have illuminated the quick start and disastrous ability of those flows. The findings indicate that although early warning systems that rely on rainfall levels and the saturation of the soil with water have been promising, indigenous adaptive strategies, such as the creation of stone channels, bamboo grove planting to regulate water, as well as the presence of early natural cues (e.g., the spring flow changes or rumbling signatures of subsurface water), are part of community resilience. Such knowledge when applied in both participatory planning and land-use zoning has gone a long way to maximize the success of the state-led mitigation efforts.

In all forms of landslides, the analysis identifies climate change to be a multiplier, and it leads to higher variability of rainfall, destabilizes the permafrost and the glaciers, and exposes societies to more landslide risks. Moreover, inadequate drainage system, quarrying, deforestation, and uncontrolled building construction are only contributing to the susceptibility of the rural and urban residents. Nevertheless, one of the most crucial findings of the research is that there is a trend of an increasing overlap between the traditional and scientific methods of landslide risks mitigation. Such local projects like community-based hazard mapping, hybrid mitigation models, and participatory disaster drills in Sikkim, Meghalaya, and Kerala set an example of how the local knowledge can be used in conjunction with modern tools.

To sum up, the findings indicate that the multi-dimensional approach toward the mitigation of the landslide risks, encompassing engineering solutions, environmental restoration, early warning systems, and the active engagement of local communities, is still necessary. Indigenous knowledge is used to address issues involving risk reduction; it is important to note that the successful adjustment of indigenous knowledge will not only make the strategies more culturally relevant, but also guarantee that they can be sustained and accepted in the long term. It is an essential ecological and geo-unstable area of India that requires such a holistic approach to be crucial to a disaster-resilient development.

14.5 Limitations of the Study

Although this paper provides an insight into the different types of landslides, causes, and prevention measures, including the involvement of indigenous knowledge, there are a few limitations that need to be considered.

To begin with, the research utilizes mainly secondary sources of data published literature, and recorded case studies. These are not the only sources of theoretical analysis, though, and the lack of comprehensive primary field information especially on less-documented indigenous practices may have limited the depth of ground-level information. A lot of the traditional practices are orally passed and have not been put down in writing which makes it hard to measure their efficiency by some means of standardized scientific measurements.

Secondly, the study area is not comprehensive as large areas of landslides prone regions in India are covered. Some of the geologically sensitive areas that include the Northeastern hill tracts, the Ladakh and the Eastern Ghats have not been elaborated because of the limited amount of available data or accessibility.

Third, typologies of landslides were categorized and evaluated according to the available typologies (e.g., Hungr et al., 2014) that are not always able to represent hybrid or transitional types of slope failures that are found in the most complex landscapes. In practice landslides may tend to be overlapping (e.g., slide-flow or topple-fall mixtures), and discrete categorization will not be straightforward.

Fourth, the study does not provide any quantitative measures of the effectiveness or resilience-building ability of the indigenous knowledge practices in comparison with the scientific methods, even though it highlights the importance of the indigenous knowledge. The longitudinal studies or the participatory action research would be required to address these outcomes as time goes by.

Fifth, the effect of climate change had not been modeled and analyzed with facts like the trends of rainfalls, temperature variation or glaciers melt away numbers, which could have enhanced the predictive aspect of the research.

Finally, policy implications and responses to landslide management by the institution were touched on a little bit. The discussion of regulatory gaps, governance issues, and incorporation of communal knowledge in disaster management policies could be more argued out enabling a more solid basis of advocacy and implementation.

These shortcomings indicate a need to develop interdisciplinary studies in the future integrating geospatial analysis, participatory practices in communities and climate modeling to create a more comprehensive and practical model of landslide risk reduction.

14.6 Recommendations

Due to the results and shortcomings of my research, there are a few recommendations that can be highlighted to improve the knowledge, prevention, and treatment of landslides in India, especially by using scientific techniques together with traditional knowledge systems.

Encourage interdisciplinary and regional studies: Future research needs to integrate geomorphological, climatic and socio-cultural aspects to achieve a better understanding of the triggers, frequency as well as the effects of various types of landslides. The need to focus on zonation of hazard microlevels, hybrid slope classification, and interplay of geotechnical processes with anthropogenic pressures in various ecological areas should be highlighted.

Systematize and document the indigenous practices: A lot of indigenous slope management and early warning practices are poorly documented. To catalogue, analyze and scientifically substantiate these practices in various ecological and cultural settings, there is a need to do ethnographic and participatory studies. This may be achieved by joint projects between the academic institutions, local communities and disaster management agencies.

Incorporate the indigenous knowledge into institutionalized disaster risk reduction systems: Government departments like the National Disaster Management Authority (NDMA), State Disaster Management Authorities (SDMAs), and Local Panchayats need to institutionalize the wisdom of the traditional ecology in their training, awareness, and planning programs. The stakeholders that should be involved in the co-creation of risk communication strategies and early warning systems include the leaders and older people in the community.

Enforce early warning and monitoring: Real-time monitoring of high-risk areas through rainfall thresholds, soil moisture sensors, and remote sensing needs to be increased. Such systems must be available and usable at the community level in terms of local alerts, regional language mobile applications as well as visuals to illiterate people.

Adopt slope-specific design and nature based design: The mitigation schemes should take the form of specific to each type of the landslides. As in the example, rockfalls and topples need to be contained and scaled mechanically and flows and slides are helped by vegetative bioengineering and regulated drainage. Combined engineering and ecological models are to be promoted as the cost-efficient and sustainable solutions.

Reform land-use and construction policies in landslides prone areas: Slope stability assessment, environmental impact assessment, and setback requirements of roads and buildings should be strictly enforced. The governments are encouraged to embrace low impact development, zoning regulations, and resettling of very vulnerable communities and ensuring that the affected communities are compensated equally.

Increase community-level disaster preparedness and education: Local disaster preparedness training programs, school-based disaster preparedness education programs, and local disaster preparedness drills need to be expanded in all at-risk districts. The local information, e.g., the ability to recognize pre-landslide indicators, e.g., new springs, soil cracks or unusual animal behavior, etc., should be a part of the formal warning messages and community exercises.

Promote climate resilient adaptation measures: Since the trend of climate induced landslides is increasing, adaptation practices like rainwater harvesting, forests protection, and slope revegetation should be integrated into the development planning. The changing landslide areas should be predicted by the use of regional climate models to inform the long-term resilience planning.

Enhance institutional cooperation and alignment of policies: Geological surveys, forest departments, urban planning and local government bodies need to be better coordinated. The national policies must clearly favor the incorporation of traditional knowledge in the environmental laws, disaster management systems and in the curriculum.

Following these recommendations, leaders, scientists, and societies can collaborate to develop a strong and inclusive landslide risk mitigation strategy that considers the extent of local knowledge and benefits scientific advancements.

14.7 Conclusion

Landslides are still among the most challenging and frequent natural disasters that hit the Indian subcontinent with far-reaching outcomes on lives, occupation, infrastructure, and wildlife. The chapter was informative as it classified landslides into five major categories that include falls, topples, slides, spreads, and flows in terms of their unique features, cause and effects, spatial distribution, and possible consequences. The case studies of different state of India highlight the presence of both natural processes like rainfall, geology, and seismicity and human activities like deforestation, unregulated building, and slope alteration in enhancing landslides.

The main value of the given study is in its focus on the combination of the indigenous knowledge systems and scientific knowledge. People residing in the landslide-affected areas have created culturally ingrained and ecologically sensitive measures across the centuries such as slope surveillance, water control, plant strengthening, and natural land-use planning. Such practices are context-specific and more than that, sustainable and accessible to people in the area. Incorporating them into the formal disaster risk management plans will provide a chance to construct more inclusive, resilient and adaptive systems.

The results also show that there is no effective method of mitigating landslides in various terrains and communities. Rather, we need an integrated approach, or rather solution, which incorporates geotechnical soundness, naturalistic solutions, early warning mechanisms, implementation and enforcement of policies, and community engagement. Such interdisciplinary solutions are even more critical in the face of growing effects of climate change, such as increased rainfall and receding glaciers.

As a way of progressing, there is a crucial need to ensure that policymakers, researchers, and local stakeholders develop knowledge and solutions that incorporate traditional wisdom and at the same time accommodate the scientific progress. In this way, India can only become more prepared and adaptive, not only to withstand landslides, but also to turn the exposed landscapes into the resilient ecosystems and communities.

References

Cruden, D. M., & Varnes, D. J. (1996). Landslide types and processes. In: A. K. Turner & R. L. Schuster (Eds.), *Landslides: investigation and mitigation* (pp. 36–75). Washington, D.C.: Transportation Research Board Special Report 247. National Research Council.

Das, N., Pal, S., Bora, S. S., & Walling, O. (2014). Study of traditional houses in Assam. *Journal of Civil Engineering and Environmental Technology, 1*(4), 53–58.

Geological Survey of India (GSI). (2023). *National Landslide Susceptibility Mapping Report*. Ministry of Mines, Government of India.

GSI. (2020). *Landslide Inventory and Susceptibility Mapping*. Geological Survey of India.

Highland, L. M., & Bobrowsky, P. (2008). *The Landslide Handbook: A Guide to Understanding Landslides*. U.S. Geological Survey Circular 1325.

Hungr, O., Leroueil, S., & Picarelli, L. (2014). The Varnes classification of landslide types, an update. *Landslides, 11*, 167–194. https://doi.org/10.1007/s10346-013-0436-y

Kanene, K. M. (2016). Indigenous practices of environmental sustainability in the Tonga community of southern Zambia. *Jàmbá: Journal of Disaster Risk Studies, 8*(1), 331.

Khanduri, S. K., & Rautela, P. (2021). Key landslide triggers: A case study of Upper Alaknanda Valley, Uttarakhand Himalaya, India. *International Journal of Earth Sciences Knowledge and Applications, 3*(3), 321–327.

Priyadarshini, P., & Abhilash, P. C. (2019). Promoting tribal communities and indigenous knowledge as potential solutions for the sustainable development of India. *Environmental Development, 32*, 100459.

Sharma, V. K. (2020). Landslides in India: Issues and perspective. *Journal of the Geological Society of India, 95*(1), 110. https://doi.org/10.1007/s12594-020-1393-4

Srivastava, S. K. (2012). Managing indigenous and scientific knowledge for resilience building: Case studies from disaster-prone regions of India. *Journal of Advances in Management Research, 9*(1), 45–63.

A Case Study of the Nilgiri District in Tamil Nadu, India, Examining How the Tribal Population Perceives the Risk of Landslides

15

M. V. Karunambigai

Abstract

The schedule tribe was segregated physically, socially, economically and even culturally in the Nilgiri district compared to the rest of the population. Most of the tribal residents are living in a region that is very prone to landslides. Thus, the perception of landslip hazard and susceptibility in people is an important subject to be studied. In this chapter, the focus was about the risk of landslides as perceived by tribal people. Perception of risk is also very important when attempting to come up with systems, practises and policies to cushion the locals. This is more so when mitigation strategies that are used to deal with risks are not structural in nature as is the case with non-structural measures such as relocation and early warning systems that accommodate the involvement of the involved community. The study was analysed on the primary data, which was collected with the help of a 400 person sample size pre-defined questionnaire. The survey targets the native community who are above 18 years. Independent sample t-test, analysis of variance, statistical analysis of frequency and ultimately development of regression model was used in the study. Judging by the results of the research, the results suggest that tribal communities believe the risk of landslides is quite high, especially with respect to the occurrence of landslide ($M = 2.124$) and feel safe ($M = 2.11$) variables. Conversely, the variable of affects quality of life had a lower score ($M = 1.459$), which shows that the concern was less in that dimension. Since 88% of the variables were above the average score, it indicates that there is increased perception of risk. The overall mean score of 15.476 or 57.33 indicates that most of the respondents are at moderate to high-risk level and this warrants the application of special awareness and mitigation measures.

M. V. Karunambigai (✉)
Department of Tourism and Travel Management, Justice Basheer Ahmed Sayeed College for Women, Chennai, Tamil Nadu, India
e-mail: dr.karunambigaivaduganathan@gmail.com

R. M. Yuvaraj et al. (eds.), *Advanced Landslide Susceptibility Modeling*,
https://doi.org/10.1007/978-3-032-14485-0_15

Keywords

Tribes • Nilgiri district • Landslide • Risk • Regression

15.1 Introduction

Landslides are also categorized on the basis of the movement and material. Falls are episodic motions of rock or soil which break off steep slopes and fall by free-flow, by bouncing or rolling, usually due to erosion or bored excavation. The forward rotation of rock or debris around a pivot point form debris piles known as talus cones, and this is referred to as topples. Slides are the motion of the material in a rupture surface, block slides move in flat planes and rotational slides move in curved planes which in most cases can cause the mudflow or slump. Creep is a gradual, continuous movement of soil or rocks due to seasonal changes in moisture or temperature. Flows are fast moving fluid like movements whose shear surfaces are brief. Debris flows refer to rapid or slow moving mixtures of rocks, soil and water which mostly take place on steep slopes. Earth flows entail saturated top soil which sinks and swells down the slope. Finally, mudflows are wave-like layers of muddy slurry, which may cause intense destruction of facilities and settlements. Figure 15.1 demonstrates types of landslide.

A common natural hazard in India especially in mountainous and hilly areas is landslides. Such occurrences are common during the monsoon season because of a lot of rainfall. Nevertheless, there are other causes like earthquakes, deforestation, construction works and poor land use which are all significant in their occurrence.

Landslides are very likely in the Himalayan region in India. Jammu and Kashmir, Himachal Pradesh, Uttarakhand, Sikkim and Arunachal Pradesh are particularly susceptible to deep slopes, weak geology, heavy precipitation and earthquakes. The example of the Kedarnath landslide of 2013 in Uttarakhand can be mentioned where a devastating destruction and loss of life was observed.

The threats of landslides are also experienced in the Western Ghats, consisting of the state of Kerala, Karnataka, Maharashtra, Goa and Tamil Nadu as a result of heavy monsoon rain, deforestation and unplanned development. In 2020, the Pettimudi landslide in Kerala claimed the lives of more than 70 people. Landslides are less frequent in the Eastern Ghats and Vindhyan regions, which include Odisha, Andhra Pradesh, Chhattisgarh and Madhya Pradesh, but they still become prone during the years of excessive rainfall. The region of northeast India, which includes Assam, Nagaland, Manipur, Mizoram and Meghalaya is prone to landslides as a result of heavy rainfall, seismic energy and road works in the mountainous environment.

Landslides in India may be natural and manmade. The natural causes are heavy rainfalls, earthquakes, weathering and snow melt. The anthropogenic causes are deforestation, mining, road constructions and uncontrolled construction. The

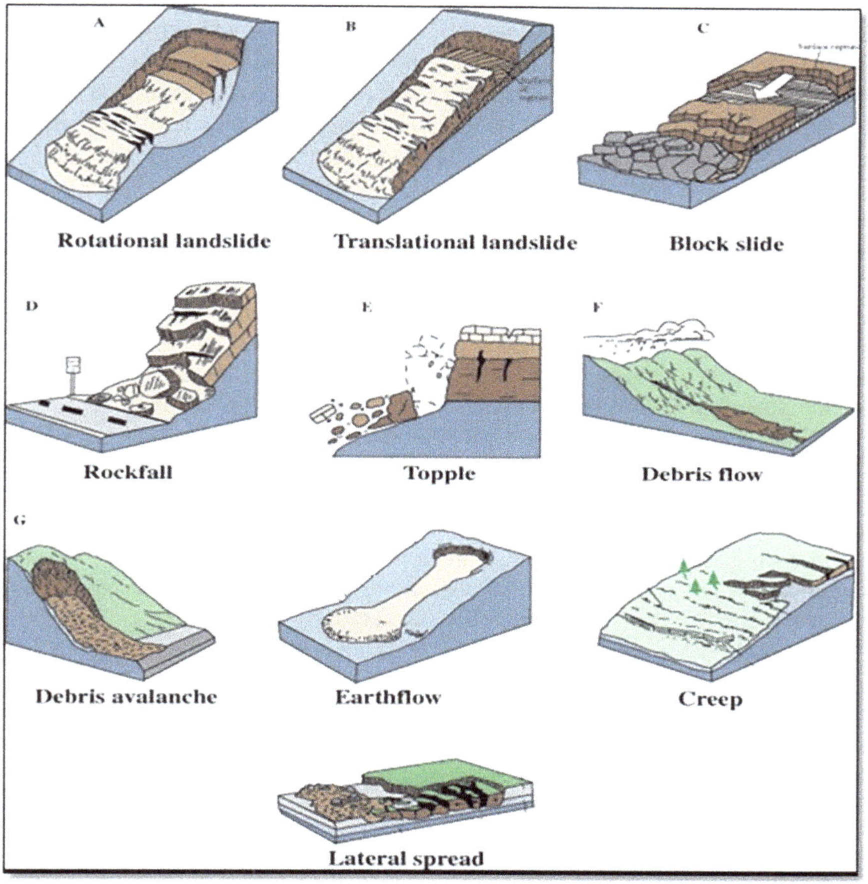

Fig. 15.1 Type of landslide. *Source* USGS Publication Repository, July 2004

effects of landslides involve loss of lives and properties, destruction of infrastructure in the form of roads and railways, breakdown of transportation and communication networks, and exposure of people who may be residing in the mountainous regions.

The landslides are very prone in the Nilgiri district of Tamil Nadu because of its steep slopes, weak geological formation and excessive monsoon precipitation. Natural and human-made factors interact in such a way that they enhance the level of landslides in the area. The southwest and northeast monsoon seasons precipitate heavy and long rains that saturate the soils causing slope failures and bulk movement of the earth materials. Moreover, the slopes have been destabilized by the uncontrolled building activities, road excavation, and deforestation, rendering both rural and the urban settlements more vulnerable.

A number of places in the Nilgiris, e.g. Coonoor, Ooty, Kotagiri have had repeated cases of landslides, leading to loss of lives and damage of property and transport and communication systems. The most tragic incidence was in November 2009 when heavy rains caused numerous landslides that took lives of many people and led to numerous destruction. The Nilgiri district has been one of the high-risk areas that need proper mitigation and land use planning due to its geographical and climatic condition.

Landslides are classified as the most devastating natural hazards especially in the hilly and mountainous areas where communities in need including the indigenous and tribal groups tend to reside. The recurrence landslides are common in these communities as a result of high rainfalls, deforestation and improper development planning. Nevertheless, tribal groups have strongly traditional knowledge systems that capture the generations of environmental observation, sustainable land use and coping mechanisms against disasters (Sharma & Raj, 2016).

The inclusion of scientific approaches with traditional ecological knowledge (TEK) is becoming more and more known as an effective solution to the reduction of disaster risks (UNESCO, 2017). This kind of integration does not only improve effectiveness of early warning systems and mitigation measures, but also provides culturally appropriate and community-driven responses. Also, the perception of landslide dangers of tribal communities can give important insights into their awareness, readiness and resilience (Chakraborty et al., 2018).

Tamil Nadu has a number of native tribal groups, which are found in the Nilgiri district, and some of them include the Todas, Kotas, Irala, Kurumbas, Paniyas. The communities have lived centuries in the area and have learned so much about the land and can be able to identify and react to any environmental threats like landslides (Bahu, 2010). The observation of natural indicators, i.e. cracks in the ground, tilting trees, unusual water seepage and even the change in animal behaviour are one of their main strategies that are traditionally perceived as warning signs of the upcoming slope failure (Srinivasan, 2015).

Traditional ecological knowledge also affects settlement planning of these tribes. An example of this is that the Todas usually construct their homes on flat and well exposed grounds as opposed to steep ones or those subject to erosion which exposes them to the effects of landslides (Gadgil & Guha, 1992). The other important part of risk mitigation is vegetation management. Other tribes such as the Kurumbas and Irala tribes maintain the slope stability by maintaining the original vegetation, which prevents the soil erosion and also prevent the accumulation of surface runoff; two issues that usually cause landslides (Rathakrishnan, 2012).

They practise agriculture, which is usually low-impact shifting cultivation, which preserves the natural wholeness of the slopes and also does not overwork the land as it is done in modern mechanized farming (Singh, 2006). Moreover, religious groves and cultural taboos on particular locations provide environmental protection of ecologically sensitive sites. Oral traditions also have an important role to play because community recollections of former landslides become an informal hazard map of the community that informs them on future settlement and land use decisions (Bahu, 2010). By combining this indigenous knowledge with scientific knowledge it is possible to come up with more sustainable and culturally sensitive landslide risk management measures in the Nilgiris.

This chapter will discuss the science and traditional knowledge intersection in the area of landslides risk mitigation, with special emphasis on how tribal communities perceive, practice and adapt to it. There is the capability of bridging these knowledge systems to come up with inclusive and sustainable disaster management strategies that are made localized.

15.2 Study Area

The Nilgiri district is a hilly area that is mostly hilly and is located at an altitude of 1000–2600 m above sea level in the Western Ghats and is in the western region of Tamil Nadu, in India. The geographical area covered by the district is 2552.50 km^2 and has a geographical location between 11° 12' N and 11° 37' N latitudes and 76° 30' E and 76° 55' E longitudes. Wayanad, Malappuram and Palakkad districts of Kerala and Mysore district of Karnataka are its neighbours on the west; Coimbatore on the south; and Erode and Chamarajanagar on the north.

Nilgiris is the only district in the 32 cashmere districts in Tamil Nadu that is completely at the point where the Western and Eastern Ghats meet making it what is known as the Nilgiri Plateau. The district has administrative units like Ooty, Kundah, Kotagiri, Gudalur, Coonoor and Pandalur. Its arable land area consisting of slopes of 15–36° amounts to about 60.1%. The 2011 Census showed that its total population is approximately 735,394 or 360,143 males and 375,251 females and its population density is 287 per square kilometre—less than the state average. Carnackite and pyroxene granulite are also dominant geologically in the region.

The rainfall received in the district is heavy due to both the South-West monsoon (June to August) and the North-East monsoon (October to December) and the rainy days per month is seven on average. The landslides are common particularly in the months of October–November, when there are heavy rains. One of the most significant landslide events took place in 1978 when it happened on approximately 250 km^2 resulting in massive loss of life and property especially in Ooty. In the next year, the area experienced more than 200 landslides, which affected the Kallar–Coonor road to a large extent. Landside risks have been aggravated by human activities such as deforestation, improper land management and uncontrolled land use. Thus, the development of a specific landslide risk zonation map is necessary to educate the planners and facilitate the efficient mitigation plans. The Nilgiris district is the right place to carry out landslide hazards mapping and risk assessment because of these characteristics.

15.3 Data and Methodology

The primary data collection was done through structured questionnaire that was well designed and face to face interview was carried out among tribal people inhabiting the high-risk area under the landslides of Nilgiri district. The questionnaire has socio-economic information, and impression of tribal people on risk towards landslide. There are a total of 13 open-ended questions in it. The target population of the present study is tribal population of the Niligiri district.

For the proposed research the alpha level of 0.05, and the level of acceptable error at 5% are considered as generally suggested by (Kotrlik & Higgins, 2001). The estimated SD of the scale is 0.5. Cochran's formula is:

$$\text{Sample Size (n)} = \frac{(t)^2(p)^2(q)^2}{(d)^2}$$

$$\text{n} = \frac{(1.96)^2(0.5)^2(0.5)^2}{(0.05)^2} = 384$$

The value 1.96 represents the t-value corresponding to an alpha level of 0.025 in each tail, reflecting a 95% confidence level. An alpha level of 0.05 indicates the risk that the true margin of error may exceed the acceptable limit. Considering the entire tribal population of the Nilgiri district in Tamil Nadu, as per the 2011 Census, the sample size has been increased to 430 to enhance the reliability of the study.

Snowball sampling a type of non-probability sampling technique is used to gather data from the respondents who are settled in the high and moderate landslide region of Niligiri district.

A t-test is a statistical test that is used to compare the means of two groups (pairwise comparison). It is often used in hypothesis testing to determine whether a process or treatment actually has an effect on the population of interest, or whether two groups are different from one another. The formula to derive independent t-test is given below:

$$t = \frac{y_1 - y_2}{s_v\sqrt{\frac{1}{m_1} + \frac{1}{m_2}}}$$

$$\text{S}_\text{v} = \sqrt{\frac{(m_1 - 1)S_1^2 + (m_{2-1})S_2^2}{m_1 + m_2 - 2}}$$

where

T—test statistics; Y_1—mean of first sample; Y_2—mean of second sample; M_1—size of first sample; M_2—size of second sample; S_v—pooled standard deviation; S_1—standard deviation of first sample; S_2—standard deviation of second sample.

ANOVA is a statistical technique that helps to make multiple comparisons. This is common throughout statistic, as there are many times where we want to compare more than two groups.

$$F = \frac{\text{MST}}{\text{MSE}}$$

$$\text{MST} = \frac{\sum_{i=1}^{k}\left(\frac{T_i^2}{n_i}\right) - \frac{G^2}{n}}{k - 1}$$

$$\text{MSE} = \frac{\sum_{i=1}^{k}\sum_{j=1}^{n_i} Y_{ij}^2 - \sum_{i=1}^{k}\frac{T_i^2}{n_i}}{n - k}$$

where F is the variance ratio for the overall test; MST is the mean square due to treatments/groups (between groups); MSE is the mean square due to error (within groups, residual mean square); Y_{ij} is an observation; T_i is a group total; G is the grand total of all observations; n_i is the number in group i and n is the total number of observations.

Multiple regression is a statistical technique used to examine the relationship between one dependent variable and two or more independent variables. It helps in predicting the value of the dependent variable based on the values of the independent variables and understanding how each predictor contributes to the outcome.

$$\gamma = \partial_0 + \partial_1 x_1 + \partial_2 x_2 + \cdots + \partial_t x_t + \in$$

where

γ—expected value of dependent variable; X_1, X_2, ..., X_t—predictor independent variables; ∂_0—constant; ∂_1, $\partial_2 ... \partial_t$—estimated regression coefficient.

15.4 Result and Discussion

The given data illustrates the frequency distribution of a total of 430 samples based on key socio-demographic characteristics, including age, gender, education and occupation. This analysis provides insights into the composition of the surveyed population.

Age-wise distribution reveals that a small portion of the respondents, specifically 17.71 individuals, are under the age of 25, indicating a relatively lower representation of the younger population. A slightly higher number, 35.41 individuals, fall within the age range of 25 to 45 years, representing the young to middle-aged adults. The highest concentration is observed in the 45–65 age group, comprising 197.29 individuals, which suggests that a significant segment of the population is in their late middle age. Additionally, 179.59 individuals are aged above 65 years, indicating a strong presence of elderly respondents in the sample (Fig. 15.2).

Gender distribution shows that out of the total sample, 252.94 are male and 177.06 are female. This indicates a gender imbalance, with males forming a larger portion of the surveyed population (Fig. 15.3).

Educational qualifications among the respondents vary. A total of 106.24 individuals are illiterate, while 144.18 have attained primary education. Additionally, 123.94 individuals have completed secondary education, and 55.65 have studied up to the higher secondary level (Fig. 15.4). Occupational status reveals that a majority, 197.29 individuals, are engaged in agriculture. Furthermore, 73.35 individuals

Fig. 15.2 Age of the respondents

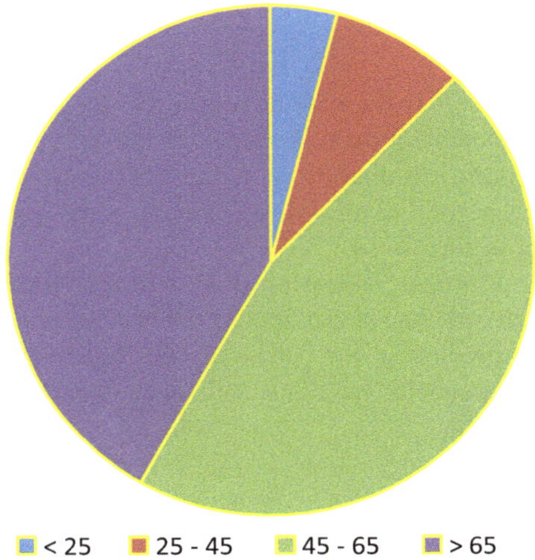

Fig. 15.3 Gender of the respondents

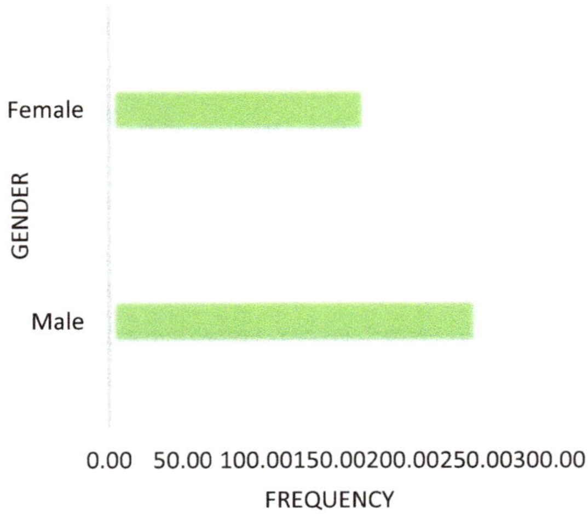

are involved in animal husbandry, 70.82 work as agricultural labourers, 17.71 are engaged in firewood sales and 70.82 are employed as government servants (Fig. 15.5).

Risk perception of landslide consists of nine variables which are Occurrence of landslide, Intensity of landslide, risk at residence, risk at workplace, loss of control, threaten to livelihood, affects the quality of life, financial loss and feel safe. These variables under risk are taken from different literatures based on perception of landslides has measure using three point Likert scale with average of 1.5.

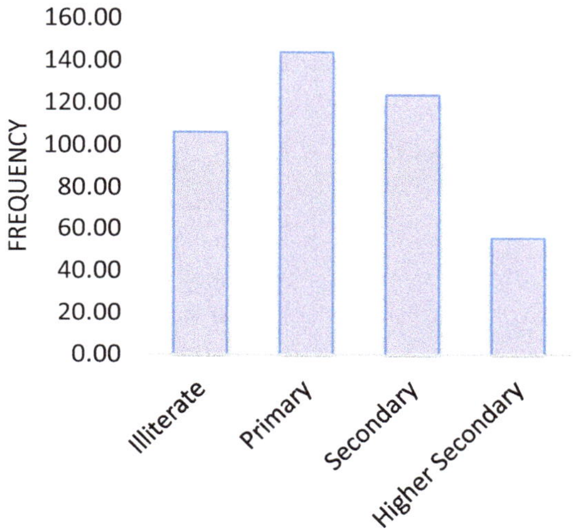

Fig. 15.4 Education of the respondents

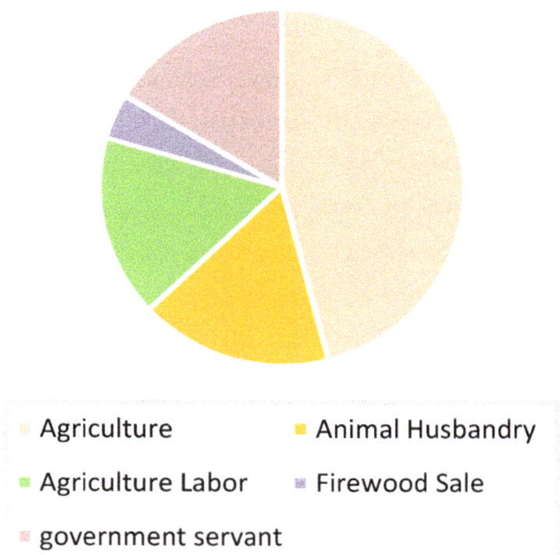

Fig. 15.5 Occupation of the respondents

15.4.1 Descriptive for Risk Towards Landslide

From Table 15.1, it has been found that the risk towards landslides among the tribal people is higher in case of the variable—"Occurrence of landslide" with mean value of 2.124 and "Feel safe" with mean value of 2.11 when compared with others. Risk towards landslide is lower in case of the variable—"Affects quality of life" ($M = 1.459$) when compared with others. It is also inferred that the risk towards landslide is above the average level since 88 per cent of the variables are above 2 out of 3 except the variable "Affects quality of life" shown in Fig. 15.6 where blue dashes line is average mean. The overall mean score of the risk towards landslide is 15.476, which is 57.33% this point out that the respondents' risk towards landslide is above 57%.

Table 15.1 Descriptive statistics—risk towards landslide

Variables	Mean	SD	Count
Occurrence of landslide	2.124	0.664	170
Intensity of landslide	1.747	0.662	170
Risk at residence	1.700	0.614	170
Risk at workplace	1.582	0.641	170
Loss of control	1.741	0.925	170
Threaten to livelihood	1.624	0.696	170
Affects quality of life	1.459	0.645	170
Financial loss	1.665	0.688	170
Feel safe	2.110	0.685	170
Risk	15.476	2.295	170

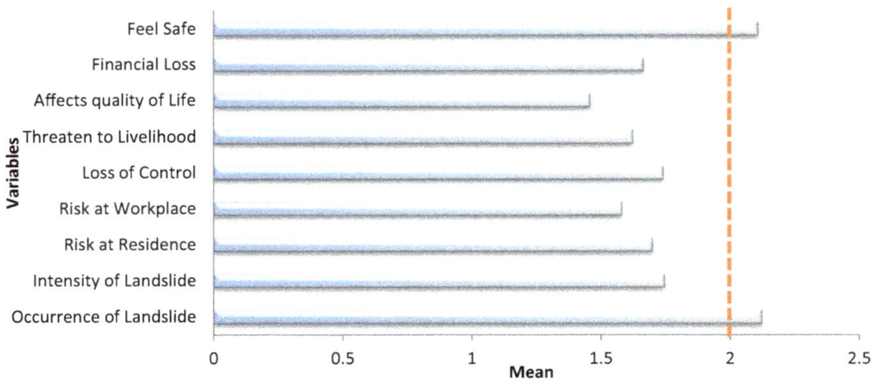

Fig. 15.6 Mean values—risk towards landslide

15.4.2 Gender—Risk Towards Landslide

H_0: There is no significant difference between male and female on the risk towards landslide.

H_1: There is significant difference between male and female on the risk towards landslide.

An independent samples t-test has conducted to compare the significant difference between the male and female respondents with respect to the risk towards landslide in the Nilgiri district. Test shows that the null hypothesis has rejected due to low alpha value (< 0.05). The mean score of female respondents (Mean = 15.5) is more than the male respondents (Mean = 14.4). This indicates that the female respondents feel more risk towards landslides than the male respondents. Hence, it is concluded that there is a statistically significant difference between the male and female respondents with respect to the risk towards landslides.

15.4.3 Education—Risk Towards Landslide

H_0: There is no significant difference between education on the risk towards landslide.

H_1: There is significant difference between education on the risk towards landslide.

A one-way between groups analysis of variance (ANOVA) has conducted to find the significant difference among the education of the respondents with risk towards landslide.

From Table 15.2, the null hypothesis has been rejected due to the P value (0.006) which is lesser than Sig. Value (0.05). It means that there is significant difference among the education of the respondents with respect to risk towards landslide $F(3, 166) = 4.263$, $p < 0.05$. Tukey's HSD shows that there is no significant difference between Illiterate and Higher education ($p = 0.652$; > 0.05), primary and illiterate ($p = 0.412$; > 0.05) and Secondary and primary ($p = 0.741$; > 0.05). It also shows that there is significant difference between Primary and higher education ($p = 0.048$; 0.05), Secondary and higher education ($p = 0.009$; < 0.05) and Secondary and Illiterate ($p = 0.044$; < 0.05) shown in Table 15.3. Maximum difference between the group of respondents who qualification is secondary and higher (Difference of mean is 1.824).

Table 15.2 One way ANOVA—risk towards landslide

	df	Sum of squares	Mean squares	F	Pr(>F)
Education	3	63.703	21.2341	4.263	0.006
Residuals	166	826.702	4.980	NA	NA

Table 15.3 One way ANOVA—Tukey's HSD (Education)

	diff	lwr	upr	p adj
Illiterate-higher	0.681	– 0.842	2.206	0.652
Primary-higher	1.383	– 0.070	2.837	0.048
Secondary-higher	1.824	0.338	3.311	0.009
Primary-illiterate	0.701	– 0.475	1.879	0.412
Secondary-illiterate	1.142	– 0.075	2.360	0.044
Secondary-primary	0.441	– 0.687	1.5698	0.741

diff—difference between means of the two groups. lwr, upr—lower and upper bound of the confidence interval at 95%. p adj-p value after adjustment for the multiple comparisons.

Table 15.4 One way ANOVA—age and risk towards landslide

	df	Sum of squares	Mean squares	F	Pr(>F)
Age	2	55.440	27.290	5.538	0.004
Residuals	167	835.000	5.000	NA	NA

15.4.4 Age of the Respondents—Risk Towards Landslide

$\mathbf{H_0}$: There is no significant difference between respondent's age and their perception of risk towards landslide.

$\mathbf{H_1}$: There is significant difference between respondent's age and their perception of risk towards landslide.

The result from Table 15.4 shows that the p value is less than significant value of 0.05 which indicates that the null hypothesis has been rejected. It means there is significant difference between age of the respondents and the perception of risk towards landslide $F(2, 167) = 5.538$, $p < 0.05$. Tukey's HSD in Table 15.5 shows that there is no significant difference of perception among the age group less than 25 years and above 65 years ($p = 0.004$, < 0.05) similarly there is significantly difference between the age group 25–65 and less than 25 ($p = 0.021$, < 0.05). But maximum difference between the age group less than 25 and greater than 65 (Difference of mean is 2.873). There is no significant difference between age group of less than 25 and 25–65 ($p = 0.314$, < 0.05).

15.4.5 Multiple Regression—Risk Towards Landslide and Socio-economic Status

Determination of statistical relationship among two or more than variables is regression. One or more variable which is independent variable is the cause of the behaviour of another one variable that is considered as dependent variable. In

Table 15.5 One way ANOVA—Tukey's HSD (Age)

	diff	lwr	upr	p adj
> 65– < 25	2.873	0.778	4.968	0.004
25–65– < 25	2.358	0.285	4.431	0.021
25–65– > 65	– 0.514	– 1.349	0.320	0.314

diff—difference between means of the two groups. lwr, upr—lower and upper bound of the confidence interval at 95%. p adj-p value after adjustment for the multiple comparisons.

Table 15.6 Multiple regression's model summary—tisk towards landslides

Statistics	Values
Residual standard error	1.357
Degrees of freedom	159
Multiple *R*-squared	0.6714,
Adjusted *R*-squared	0.6507
F-statistic	32.48
F-statistics with DF	10 and 159
p-value	0.000

this regression respondents risk towards landslide is taken as dependent variables whereas Age, Gender, Education and Occupation are taken as independent variable. Multiple regressions has conducted to identify the best linear combination of Age, Gender, Education and Occupation for predicting risk towards landslide.

From Table 15.6, it can be observed that *R*-squared also called coefficient of determination value is 0.67 which indicates that the present model explains large portion of variance in the output variable. The adjusted *R*-squared value is 0.65 which means that 65% of the variance in the measure of risk towards landslide can be predicted by the socio economic status of the respondents of the Nilgiri district which is large effect (Cohen, 1988).

$F(10.\ 159) = 32.48$, $p = 0.000$ which is lesser than 0.05 significant level which shows that combination of all ten independent variables has significant relationship with dependent variable, i.e. risk towards landslide. For predicting the risk towards landslide gender (1.3) is the strongest influencing factors which predicting dependent variable of risk towards landslide which has high size whereas Age factor influence only 0.03 with low positive effect shown in the Table 15.7. Multiple regression equation for the risk towards landslide (Y_{Risk}).

$$Y_{Risk} = 4.59 + 0.03(\text{Age}) + 1.39(\text{Gender}) + 0.62(\text{Education}) + 0.24(\text{Occupation})$$

Table 15.7 Multiple regression's coefficients—risk towards landslides

	Estimate	Std. error	t value	Pr(>|t|)
(Intercept)	4.59916	1.40595	3.271	0.001
Age	0.03148	0.01201	2.621	0.009
Gender	1.39952	0.27382	5.111	0.000
Education	0.62636	0.17039	3.676	0.000
Occupation	0.24959	0.06524	3.825	0.000

15.5 Conclusion

The descriptive has concluded that they have risk perception of occurrence of landslide at the residence and workplace and that they also believe that landslide is threaten their livelihood and financial status so they also believe that they may lose their control. The reason why female respondents have more risk with landslide as compared to male respondents is due to lack of exposure to other individuals and spending most of their lives in the house and thinking about past landslide which makes them more risk-takers. The risk perception of landslide differs depending on education of the respondents since the illiterate people are at high risk compared to the higher educated people equally the primary and secondary educated people. The age group of the respondents above 65 years of age and between 25 and 65 years old are having experience therefore their risk is relatively low compared to the person of the lower age this is primarily because of experience. The result indicates that the perceived risk of landslides is high in the tribal population particularly the variables of occurrence of landslide ($M = 2.124$) and feel safe ($M = 2.11$). The variable affects quality of life scored lower ($M = 1.459$) meaning that it was not of much concern. Having 88% of the variables above the average threshold, the general perception of risk is also high. The total average of 15.476 (57.33) indicates that more than half of the respondents are at moderate and high risk, and their risk should be mitigated and awareness raised.

Acknowledging The author(s) acknowledge the use of AI-based tools for grammar and language correction during the preparation of this manuscript.

References

Bahu, A. (2010). Tribal knowledge and environmental management: Case studies from South India. *Journal of Environmental Anthropology, 14*(1), 45–58.

Chakraborty, S., Mukhopadhyay, S., & Das, S. (2018). Community perceptions and responses to landslide risk: A study of Darjeeling-Sikkim Himalaya, India. *Natural Hazards, 92*(1), 341–361. https://doi.org/10.1007/s11069-018-3201-5

Cohen, J. (1988). Statistical power analysis for the behavioral sciences, 2nd edn. Á/L.

Gadgil, M., & Guha, R. (1992). *This fissured land: An ecological history of India.* University of California Press.

Kotrlik, J. W. K. J. W., & Higgins, C. C. H. C. C. (2001). Organizational research: Determining appropriate sample size in survey research appropriate sample size in survey research. *Information Technology, Learning, and Performance Journal, 19*(1), 43.

Rathakrishnan, M. (2012). Eco-cultural perspectives of the Irulas of Nilgiri hills. *Indian Journal of Traditional Knowledge, 11*(2), 300–305.

Sharma, A., & Raj, A. (2016). Traditional knowledge and indigenous practices for disaster risk reduction in India. *International Journal of Disaster Risk Reduction, 16*, 347–353. https://doi.org/10.1016/j.ijdrr.2016.03.002

Singh, K. S. (2006). *People of India: Tamil Nadu* (Vol. XL). Anthropological Survey of India.

Srinivasan, V. (2015). Traditional environmental knowledge and natural hazard preparedness among tribal communities of the Western Ghats. *Indian Journal of Disaster Management, 10*(2), 110–121.

UNESCO. (2017). *Intangible cultural heritage and disaster risk reduction*. United Nations Educational, Scientific and Cultural Organization. Retrieved from https://unesdoc.unesco.org/ark:/48223/pf0000246130

Part VI
Conclusions

Conclusion: Integrating Geography, Environment, and Sustainability for Resilient Landslide Risk Management

16

M. V. Karunambigai, R. M. Yuvaraj, and V. Gopal

Abstract

Landslides are a great geohazard, which is caused due to the interrelation of environmental, climate, and man-made factors. This paper examines the relationship that exists among geography, environment, and landslides with reference to their effects on human beings, economic systems, and sustainable development. Particular attention is given to Tamil Nadu, where such a district as the Nilgiris is subjected to regular landslides because of steep landscapes, high rainfall, and other human-induced factors. The article summarizes the development of landslide susceptibility models, including the use of GIS, machine learning, and statistical tools to determine the areas of risk. It also looks at the socioeconomic impact of landslides, such as infrastructure destruction to livelihood losses, and emphasizes on the necessity of sustainable risk management measures. This paper suggests that scientific instruments could be used to supplement traditional knowledge and policy framework in a bid to establish an effective multi-disciplinary approach to landslide mitigation. The results also highlight the necessity to plan the area proactively, educate people, and manage the land sustainably to minimize the risk of landslides and increase the resilience of the vulnerable regions.

M. V. Karunambigai
Department of Tourism and Travel Management, Justice Basheer Ahmed Sayeed College for Women, Chennai, Tamil Nadu, India

R. M. Yuvaraj
Department of Geography, Institute of Distance Education, University of Madras, Chennai, Tamil Nadu, India

V. Gopal (✉)
Centre for Earth and Atmospheric Sciences, Sathyabama Institute of Science and Technology, Chennai, Tamil Nadu, India
e-mail: veegopaal@gmail.com

R. M. Yuvaraj et al. (eds.), *Advanced Landslide Susceptibility Modeling*,
https://doi.org/10.1007/978-3-032-14485-0_16

Keywords

Geography • Landslide susceptibility • Geographic risk assessment • Sustainable development • Disaster risk reduction

16.1 Introduction

Land slips are one of the most common and devastating types of natural hazards and they have great effects on the human settlements, infrastructure, and the environment. They are closely associated with natural and anthropogenic factors especially environmental dynamics and geographical environment. Over the past few decades, the issue of climate change has increased the risk of landslides due to changes in precipitation pattern, higher occurrences of extreme weather conditions, and vegetation and soil stability. Geographical configuration of a place, like slope, elevation, and lithological structure, also defines further the susceptibility of the place to slope failures. The risks are two-fold because of the short-term, physical devastation, and the long-term socioeconomic impacts, such as loss of lives, displacement, damage of infrastructure, and disruption of travel and tourism infrastructures.

Tamil Nadu, a state located in the south-eastern part of India, and its hill-districts (such as the Nilgiris) are becoming more susceptible to landslides, owing to excessive monsoonal precipitation, deforestation, uncontrolled development, and weak geological structures. The magnitude of the harm and the necessity of mitigation planning can be highlighted by the examples of historic disasters, including landslides in Nilgiris in 2009 and 2020. More informed land use and disaster response plans are now much easier with advanced geospatial technologies, landslide susceptibility models, and techniques can be classified as statistical or machine learning models.

Moreover, the indigenous knowledge and the local risk perception are highly crucial in increasing community resilience and reducing risks sustainably. The combination of ancient and contemporary techniques in the field of science provides the comprehensive solution to the management of the landslide-prone regions. The strategies of sustainable landslide control should thus incorporate conservation of the environment, inclusive policy-making and adaptive planning of land-utilization.

In this paper, the interface of geography, environment, and landslides has been described in details in relation to its human and economic impacts, especially in Tamil Nadu. It uncovers as well the changing approach to landslide susceptibility modeling, the socioeconomic influence on the marginalized populations, and the importance of sustainability in the context of disaster risk mitigation.

16.2 Geography, Environment, and Landslides: Assessing Human, Economic, and Sustainability Impacts

The interplay of the environmental factors, geographical features, and landslide processes has been very complex and that is why it is so essential to take a holistic approach to the landslide hazards and sustainability. Recent studies indicate that landslides are not only common geological hazards, but also very sensitive to variations in the environmental parameters such as temperature, precipitation, vegetation cover, and snow cover particularly in areas such as the southern Tibetan Plateau. The increase of global temperatures, changes in precipitation regime, and land cover changes have only exacerbated the occurrence and severity of landslides proving the immense influence of climate change on geomorphological processes (Xu et al., 2024).

Elevation and slope, as geographical factors, are very critical determinants of landslides susceptibility. The middle mountain areas with moderate and steep slopes are especially vulnerable based on the concentration of new landslides with the altitudes between 500 and 3500 m and slopes of 10–50° (Xu et al., 2024). The geographical variance of landslide location once again underlines the importance of data-intensive and high-resolution solutions in landslide detection and risk evaluation.

The development of geospatial technologies and machine learning (including GIS-based mapping of susceptibility and remote sensing) has contributed to the expansion of the ability to detect, track and forecast landslides-prone areas (Varangaonkar & Rode, 2019). With these tools, it is possible to combine different environmental and geographical data, which makes it easier to determine hazards more accurately and on time.

Landslide sustainable management has become a critical factor and novel technologies such as geosynthetics-based stabilization are providing the environmentally-friendly alternative to the conventional engineering practices. These strategies do more than just increase the slope stability, as they are consistent with the wider sustainability objectives of reducing the ecological impact and ensuring that resources are used efficiently during the project life cycle (Zhang, 2024).

To conclude, the interface among environment, geography, landslides, and sustainability is featured by complicated interactions, which require interdisciplinary research and integrated management approaches. Knowledge of these relationships is critical in coming up with effective landslide mitigation, early warning mechanisms, and sustainable land-use policies that protect human settlements and natural ecosystems (Svennevig et al., 2024; Varangaonkar & Rode, 2019; Zhang, 2024).

Landslides are a constant menace to the very existence of human beings and economic status, and their effects are felt well beyond the physical devastation. These disasters have a significant human cost as they take life, communities, and infrastructure, with the loss of human lives, taking communities, and damaging infrastructure (Ben et al., 2023). Landslides cause billions of dollars in losses each year in agriculture, industry, forestry and tourism, and disruption of vital services and lives (Ben et al., 2023; Transportation Research Board, 1991). These impacts are especially acute in developing regions, where the lack of sustainability in land use, a high rate of population growth, and the lack of resources to mitigate and recover has increased the level of vulnerability.

The recent studies point to the fact that the socioeconomic cost of landslides is usually underestimated, because much of the attendant damage is absorbed in the larger reports of disasters and is not directly ascribed to slope failures (Transportation Research Board, 1991). This lack of awareness is an obstacle to sound policy formulation and resource allocation to prevent and mitigate. Risk assessment, community-based mitigation strategies, and international cooperation are all sustainable landslide management solutions that can be used to minimize future losses and create resilience among the at-risk populations (Ben et al., 2023; Shakya et al., 2023).

Overall, the multicomponent effects of landslides on human and economical systems are highly important to understand so as to come up with sustainable ways of life to save lives, maintain economic growth, and make communities more resilient against the growing threats of landslides.

Landslides in the Tamil Nadu, especially in the Nilgiri district, is an ever increasing and growing threat that has a considerable environmental, human, and economic impact. The area has a particularly high percentage of landslides due to the specific geomorphology of the region, where steep slopes and high rainfall rates are observed during both the southwest and northeast monsoons (Chellamuthu & Ganapathy, 2024; Geological Survey of India, 2020). Research has demonstrated that natural catastrophe in the region is landslides caused by rainfall that result in a significant loss of lives, damage of property, and disturbances to the community in the region (Mathangi et al., 2023; Yuvaraj et al., 2023). As an illustration, landslides occurred in Nilgiris in 2009 and 2020, which caused massive damage to houses and infrastructure, which demonstrates the susceptibility of the settlements on hilly areas (Chellamuthu & Ganapathy, 2024; Geological Survey of India., 2020).

Geotechnical studies have shown that the risk of slope failures has been increased further by the presence of weak lithomargic clay and deep slip surfaces in substrata (Mathangi et al., 2023). Alterations in rainfall distribution are closely associated with increasing intensity and frequency of landslides and are expected to continue to increase with changing climatic conditions. These incidents are not only detrimental to the safety of the human population but also possess a long-term socioeconomic consequence, such as loss of livelihoods, high costs of disaster management, and the inability to build sustainable development in the area (Chellamuthu & Ganapathy, 2024).

To manage landslides in Tamil Nadu, it is necessary to unite advanced geospatial mapping, early warning systems, and engineering interventions depending on the specifics of geological conditions in the area (Geological Survey of India, 2020). The land use practices and community awareness should also be dealt with in sustainable solutions to curb the future risks. To conclude, landslide issues in Tamil Nadu are complex, as they require combined methods, that is, scientific knowledge, policy intervention, and involvement of community to preserve lives, property and environment.

16.3 Contemporary Issues of Geographical and Geological Aspects of Landslide, Various Susceptibility Modeling Techniques and Accuracy Assessment

The geographical-based approach to landslide vulnerable areas modeling has been greatly developed by considering Geographic Information System (GIS), statistical and machine learning methods that have allowed identification of the regions that are at risk to be more precise. Recent studies highlight the necessity of considering a large number of conditioning factors, which encompass topography, hydrology, geology, land cover, and anthropogenic activities to capture the complicated interaction that causes landslides to occur (Chowdhury, 2023; Hoa et al., 2023). Among them, slope, elevation, soil, precipitation, and distance to faults are the factors that always stand out as the most significant elements in global and regional research (Shahabi & Hashim, 2015). The objective statistical techniques combined with expert-based weighting schemes like the joint analytic hierarchy process and Shannon entropy have also increased the reliability and accuracy of the susceptibility models. Since the landslides-prone areas are usually differentiated by various environmental and anthropogenic stimuli, the existence of the robust spatial modeling can not only aid in the disaster risk mitigation but also the sustainable land management and planning decisions (Reichenbach et al., 2018).

The geological factors are central to the landslide susceptibility modeling because the lithology, slope, elevation, aspect, curvature, and closeness to faults and rivers are considered as the core factors that contribute to the probability and spatial distribution of landslides (Chowdhury, 2023). The recent progress indicates that when these geological variables are combined with statistical and machine learning models, including deep neural networks, logistic regression, and support vectors machines, the accuracy and reliability of susceptibility models can significantly improve (Ma et al., 2023). It is always found that coupled methods that integrate deterministic and data-driven models can provide more subtle and context-dependent susceptibility maps, which can be used to achieve more effective risk reduction and land-use planning (Reichenbach et al., 2018). Since the landslides mostly have complicated geological environments, powerful modeling based on geological issues is required to predict hazards and protect human beings and infrastructure.

Statistical and mathematical modeling of landslide-prone areas has become fundamental to both hazard assessment and reduction of disaster risk as well as provide quantitative information on the spatial probability of landslide occurrence. Statistical models including frequency ratio (FR) and information value (IV) are effective in the use of terrain and environmental data to define susceptibility areas, and more recently, machine learning models, including logistic model trees (LMT) and radial basis functions classifiers (RBFC) tend to result in superior predictive accuracy and predictive strength. Comparative analysis reveals that, even though statistical models are still useful due to their ability to give a clear meaning to the data and efficiency in data processing, machine learning models tend to perform better in complex terrains, as the measures of area under the curve (AUC) and accuracy values show. Finally, both the local geo-environmental conditions and the available data should be considered in choosing the modeling technique, although the combination of statistical and mathematical approaches is still needed to achieve a reliable landslide susceptibility map and provide a sound land-use planning (Ferreira et al., 2025; Reichenbach et al., 2018).

Machine intelligence methods have transformed landslide vulnerability models by providing significant advancements over the conventional methods in prediction accuracy and spatial resolution. The current researches show that more sophisticated machine learning models, such as support vector machines, random forests, and more importantly deep neural networks, are the best in combining a variety of geomorphological, geological, and environmental data and becoming more reliable in marking the high-risk areas. Examples deep learning models have been shown to consistently outperform state-of-the-art machine learning methods in susceptibility mapping, with higher area under the curve (AUC) values, and more accurate delineation of hazard-prone regions as seen in accumulated large scale applications across diverse landscapes. Moreover, there has been an even more promising potential of hybrid and ensemble models, which integrate several machine learning algorithms, in order to represent the complex, nonlinear relationships of landslide processes, facilitating more informed risk management and planning decisions (Achu et al., 2023; Azarafza et al., 2021; Merghadi et al., 2020).

The use of accuracy assessment techniques in enhancing the identification of landslides is essential as it gives the objective measures to measure and compare the accuracy of the various susceptibility models. Current literature has shown that more sophisticated machine learning and deep learning algorithms, including convolutional neural networks (CNN), deep neural networks (DNN), logistic regression, and support vector machines, are highly effective at improving landslide susceptibility maps provided their results are carefully evaluated on the basis of such statistical measures as area under the ROC curve (AUC), confusion matrix, mean squared error (MSE), and root mean squared error (RMSE) (Azarafza et al., 2021). In detail, the confusion matrix will enable one to assess the model performance by the number of true positives and true negatives but also by the number of false positives and false negatives, and further the performance of the model can be evaluated through several additional measures like accuracy, precision, and recall (Ma et al., 2023). The inclusion of these strong accuracy evaluation tools will guarantee that the susceptibility models are more reliable in addition to being more appropriate to the practical use of landslides risk management and land-use planning (Azarafza et al., 2021).

16.4 Roots of Various Climate and Environment Related Causes of Landslides

The impact of climate change on landslide threats and danger is deep and can be underestimated on the global level. The acceleration of the rain process, the temperature rise, and alterations to the groundwater processes have resulted in a significant increase in the number and intensity of landslides in recent decades, and it is estimated that in the future climate, the threat will still increase (Ramzan et al., 2025). Empirical research shows that the wet winters, the shorter the warning of a landslide to be given, and the deeper the water table is all associated with climate-related alterations, leading to the increased frequency and the devastating potential of the landslide eventualities. These effects are extremely site-dependent, yet the general progress is evident the climate change is decreasing the slope stability and making communities and infrastructures more susceptible to landslides disasters (Research Institute for Geo-Hydrological Protection). Therefore, climate projections should be a part of landslide risk management and assessment to adjust to changing risks and reduce the future losses.

The environmental issues make the evaluation and forecasting of landslides very complex because it is a multifaceted combination of climatic, ecological, geological, and anthropogenic factors that trigger landslides. Predicting the actual model requires a correlation of various types of environmental variables to slope, aspect, precipitations, land use, vegetation cover, and soil moisture, all of which can be spatially and temporally varied, which makes the prediction of susceptibility highly uncertain (Gentilucci et al., 2023; Xing et al., 2024). The current progress in machine learning and remote sensing has enhanced the capability of describing and analyzing these variables, but selection and weighting of conditioning factors is also crucial since strong interrelations of environmental factors may bias the outputs of the model and lower prediction accuracy (Huang et al., 2022). Moreover, drastic precipitation events and urban development has been demonstrated to greatly boost the risk of landslides, yet its consequences are not adequately captured in big models due to information constraints and geographical heterogeneity (Gentilucci et al., 2023). Thus, it is crucial to address these environmental issues with the help of strong data collection, proper choice of factors, and the use of powerful modelization, which could result in credible landslide threat evaluation and use it to develop effective risk management and mitigation measures.

The effect of landslides on the tourism industries is quite negative and this is most evident in the hill stations that are prone to slope failures like in the case of Mussoorie in the Himalayas where the failure of slopes affects access, infrastructure destruction, and the tourist flow. Landslides are influenced by the natural land cover degradation as a result of deforestation and unplanned construction works and the heavy monsoon rainfalls, which have a direct impact on the livelihood of local businesses like hotels, restaurants, and small vendors which rely much on the tourism income. The result of such disruptions is the seasonal deterioration of tourist arrivals to the area particularly during the monsoon seasons and an event that forces most locals to find alternative sources of income or migrate in search of a better living. The overall impact of landslides therefore puts the sustainability of tourist economies at risk and demands combined land management and disaster reduction plans to cushion the environment and the socioeconomic culture of tourist destinations.

The effects of landslides on tribal people in South India are deep-rooted and complex, especially the individuals who have to live in ecologically sensitive and high-risk areas such as Western Ghats and Nilgiri Hills. According to the tribal approach, such disasters do not only imply instant loss of lives, property, and income, but also set in motion long-term exposures such as displacement, loss of access to forest resources, and shattering of the old ways of life. The research on the Kattu Paniya community following the 2019 Kavalappara landslide shows that even the post-disaster rehabilitation regularly does not cover the underlying socio-cultural and economic needs of the indigenous population by placing them in the years of uncertainty and marginalization. In addition, family set up, community seclusion and family and generational wisdom are significant factors that influence risk perception and use of adaptive strategies among tribal people. The key to an effective risk management and recovery will thus involve culturally sensitive and community-centered methods coupled with environmental rehabilitation in order to counteract the special needs of tribal communities in the post-landslides periods (The Renaissance Official, 2023).

Landslides have far-ranging economic effects on transport and road systems through direct infrastructure destruction, connectivity problems and costly effects to communities and companies. Although not accompanied by casualties, landslide occurrence may cut-off remote sites, slow down transit of goods and people and require lengthy detours all of which will result in direct, consequential, and indirect economic losses. The gray shade of vulnerability left behind by such occurrences may often spread far beyond the immediate area of landslides and has an impact on employment, well-being, education, and other social endeavors over a larger area. On the examples of the impacted regions, case studies have shown that expenses of road closures, repairs, and traffic diversion may cost millions of dollars depending on severity of the impact on the traffic which depends on the volume of the traffic and the duration of the disruption. These results highlight the importance of the strong ability to risk assess, the infrastructure planning capacity, and focused mitigation measures to reduce the economic impact of landslides on major transport routes (Winter et al., 2019).

Landslides result in large losses of rural land value and greatly affect the agricultural and pastoral productivity which in most cases are the major sources of income in the less populated area. These losses are often underpriced but can run up to 10–30% of the entire rural land value as in the case studies run in Europe where future losses are projected to continue or even rise because of the continuing vulnerability. These changes in land cover and the high landslide-prone areas in rural areas alongside the dynamic nature of the rural environment highlight the importance of sound, multi-disciplinary approaches in measuring these losses and incorporating them in making informed decisions on sustainable land management and spatial planning. These methods are vital in assisting decision-makers to come up with efficient risk reduction and preparedness measures to reduce the long-term economic impacts of landslides on rural populations (Zumpano et al., 2018).

16.5 Sustainability Through Indigenous Knowledge, Risk Reduction, and Preparedness

Traditional and indigenous knowledge is very important in landslide risk detection and mitigation especially in areas where scientific information and technological facilities are minimal. Risk assessments are more context-specific and effective at the same time because they combines local observations, historical memory, and community expertise with the modern scientific methods. Recent research proves that the risk of elements, including agricultural land, settlements and infrastructure, can be better identified and targeted to be intervened with when indigenous knowledge is systematically included in the susceptibility and vulnerability mapping. This will not only increase the credibility of the risk models, but will also help the community to become involved and own the mitigation methods. Finally, allowing the utilization of both conventional knowledge and scientific models will contribute to the creation of a well-rounded approach of managing the disaster risk, establishing locally relevant, sustainable solutions that minimize human and economic losses due to landslides (MacAfee et al., 2024; Rahman et al., 2022; Turyasingura et al., 2023).

The reaction to landslides is both negative and positive and it is the result of the complicated interaction of nature processes and human activity. On the downslope, landslides may have disastrous effects on the vegetation, wildlife and human settlements, causing loss of human life, as well as infrastructure and long-term ecological disturbance. They are normally too drastic and instantaneous but with the course of time, natural recuperation procedures tend to recuperate the ecological environment and in certain instances even to improve the ecological variety (Schuster & Highland, 2003). On the other hand, proactive mitigation and adaptation measures like the introduction of biotechnical slope protection and community-based hazard awareness programs are found to result in positive responses and therefore reduce vulnerability and enhance the rates of survival in case of landslides (Schuster & Highland, 2003; Pollock & Wartman, 2020). The effectiveness of the mentioned responses is directly connected with human behavior and readiness; even basic measures such as finding an elevated place or attending local training can save lives many times over (Ling et al., 2022; Pollock & Wartman, 2020). Finally, a culture of resilience by using innovative engineering, environmental recovery and community involvement is needed to ensure that the negative effects of landslides are reduced and that people coexist sustainably with the natural hazard.

The variety of landslide types (fall, topples, slides (rotational and translational), flows, and complex movements) is an indication of the complex interaction of geological, hydrological, and climatic processes contributing to slope instability at the global level (British Geological Survey, 2025; Tensar International, 2025). All forms of landslides pose specific risks to a community, particularly those that use conventional land management and traditional knowledge systems. Although the indigenous practices have traditionally played a role in resilience and adaptation (i.e., slope stabilization, early warning, sustainable land use), the increasing pressure on the environment and changes in land use undermine their performance. By combining the scientific classification and contemporary risk evaluation with the indigenous knowledge, a way to more resolute adaptation strategies that will improve community resilience and make them less vulnerable to different landslide risks will be available (British Geological Survey, 2025).

The inclusion of indigenous knowledge in policy development is critical in order to develop inclusive, contextually appropriate, and sustainable policies that acknowledge the rights, heritage and skills of indigenous communities. It is proven that indigenous voices are central in the policy-making process, which results in more efficient and sustainable outcomes when it comes to such spheres as environmental management, livestock policy, and cultural preservation (Das et al., 2017; Malcom, 2024).

The recorded and acknowledged documentation of indigenous knowledge not only promotes the biodiversity conservation and food security but also promotes social equity and cooperation between the policymakers and the indigenous people (FundsforNGOs, 2025; Malcom, 2024). These have been complemented by aligning the national policies with international frameworks, including the United Nations Declaration on the Rights of Indigenous Peoples (UNDRIP), to make sure that both the legal and ethical frameworks are complied with and to ensure long-term sustainability is achieved (Das et al., 2017). Finally, by applying indigenous knowledge in policy-making, epistemological gaps are bridged, policy failures based on exclusion are addressed, and solutions that are innovative and at the same time deeply embedded in local realities are co-created.

The combination of both scientific knowledge and traditional and indigenous practices has become an important tool in reducing the risks of landslides effectively because both systems of knowledge have their own advantages, which can be effectively used to cover various aspects of disaster resilience. Scientific methods can offer superior means of hazard mapping, early warning and structural mitigation, whereas conventional knowledge based on generations of local experience and observation can provide context-specific data on environmental indicators, sustainable land use and adaptive community measures (Vasileiou et al., 2022). The recent studies show that the integration of these methods results in more comprehensive risks assessment and culturally competent solutions, because the local communities may point out unrecognized risk factors and play a role in more effective and community-based interventions (MacAfee et al., 2024). The research on perception also demonstrates that through respectful and meeting local knowledge, and including it in official risk reduction structures, the community participation and effectiveness of mitigation actions are highly improved (Rai & Khawas, 2021). Finally, a holistic approach to science and indigenous understanding does not only fill in the vital gaps in disaster risk management but also develops sustainable and locally-based approaches to areas where landslides are likely to occur.

16.6 Conclusion

Susceptibility to landslides modeling has significantly enhanced our ability to predict, assess, and mitigate hazards posed by landslides, and this has great benefits to sustainability, economic well-being and ultimately to human life and safety. A recent body of research has made susceptibility maps, which are important tools in infrastructure development, land-use planning, and mitigating catastrophe risks, more accurate and reliable through the application of more advanced statistical, machine learning, and hybrid approaches.

To minimize the exposure and possible cost, the stakeholders will be able to locate the areas that are at high-risk and prioritize solutions by means of combining high-resolution environmental, geological, and socioeconomic information. Despite these advances, there are still issues with data quality, sampling methods, and terrain heterogeneity because these issues have the potential to impact model performance and interpretation. Nonetheless, current progress in landslide susceptibility modeling that is facilitated by technology innovation and interdisciplinary studies can ensure that economic resources are conserved, the development of sustainable communities, and sustainable development in landslide-prone regions.

Acknowledging The author(s) acknowledge the use of AI-based tools for grammar and language correction during the preparation of this manuscript.

References

Achu, A. L., Pradhan, B., & Lee, S. (2023). Machine-learning based landslide susceptibility modelling: A comparative assessment of ensemble and single models. *Geoscience Frontiers, 14*(2), Article 101456.

Azarafza, M., Azarafza, M., Akgün, H., Atkinson, P. M., & Derakhshani, R. (2021). Deep learning-based landslide susceptibility mapping. *Scientific Reports, 11*(1), 24112.

Ben, S. K., Lee, M. L., & SY, W. (2023). An overview of socioeconomic impacts of landslides. In *Proceedings of the IAIA 2023 Conference.*

British Geological Survey. (2025). How to classify a landslide. Retrieved from https://www.bgs.ac.uk/discovering-geology/earth-hazards/landslides/how-to-classify-a-landslide/

Chellamuthu, S. N., & Ganapathy, G. P. (2024). Quantifying the impact of changing rainfall patterns on landslide frequency and intensity in the Nilgiris District of Western Ghats, India. *Progress in Disaster Science, 23*, 100351.

Chowdhury, M. S. (2023). A review on landslide susceptibility mapping research in Bangladesh. *Heliyon, 9*(7).

Das, R. K., Rahman, A., & Singha, A. (2017). Policy development for indigenous knowledge management in academic libraries. *Journal of Emerging Technologies and Innovative Research (JETIR), 4*(6), 742–747.

Ferreira, Z., Almeida, B., Costa, A. C., do Couto Fernandes, M., & Cabral, P. (2025). Insights into landslide susceptibility: A comparative evaluation of multi-criteria analysis and machine learning techniques. *Geomatics, Natural Hazards and Risk, 16*(1), 2471019.

FundsforNGOs. (2025). *A sample proposal on documenting indigenous knowledge systems for policy development.*

Gentilucci, M., Pelagagge, N., Rossi, A., Domenico, A., & Pambianchi, G. (2023). Landslide susceptibility using climatic–environmental factors using the weight-of-evidence method—a study area in Central Italy. *Applied Sciences, 13*(15), 8617.

Geological Survey of India. (2020). *Landslide at TNEB colony, Emerald, Nilgiri District, Tamil Nadu (06th August, 2020): Preliminary assessment report.* Tamil Nadu and Puducherry: Ministry of Mines, State Unit.

Hoa, P. V., Tuan, N. Q., Hong, P. V., Thao, G. T. P., & Binh, N. A. (2023). GIS-based modeling of landslide susceptibility zonation by integrating the frequency ratio and objective–subjective weighting approach: A case study in a tropical monsoon climate region. *Frontiers in Environmental Science, 11*, 1175567.

Huang, F., Pan, L., Fan, X., Jiang, S. H., Huang, J., & Zhou, C. (2022). The uncertainty of landslide susceptibility prediction modeling: Suitability of linear conditioning factors. *Bulletin of Engineering Geology and the Environment, 81*(5), 182.

Ling, S., Zhao, S., Huang, J., & Zhang, X. (2022). Landslide susceptibility assessment using statistical and machine learning techniques: A case study in the upper reaches of the Minjiang River, southwestern China. *Frontiers in Earth Science, 10*, 986172.

Ma, W., Dong, J., Wei, Z., Peng, L., Wu, Q., Wang, X., Dong, Y., & Wu, Y. (2023). Landslide susceptibility assessment using the certainty factor and deep neural network. *Frontiers in Earth Science, 10*, 1091560.

MacAfee, E., Lohr, A. J., & de Jong, E. (2024). Leveraging local knowledge for landslide disaster risk reduction in an urban informal settlement in Manado, Indonesia. *International Journal of Disaster Risk Reduction, 111*, 104710.

Malcom, A. (2024). Integrating indigenous knowledge into livestock policy formulation. *IJLP, 3*, 30–42.

Mathangi, B., Zaalishvili, V. B., & Ganapathy, G. P. (2023). Study of substrata of a slope susceptible to landslide in hilly environment using a geophysical method in the Nilgiris, India. *Environmental Science and Pollution Research, 30*(59), 123966–123982.

Merghadi, A., Yunus, A. P., Dou, J., Whiteley, J., Thai Pham, B., Bui, D. T., Avtar, R., & Khosravi, K. (2020). Machine learning methods for landslide susceptibility studies: A comparative overview of algorithm performance. *Earth-Science Reviews, 207*, 103225.

Pollock, W., & Wartman, J. (2020). Human vulnerability to landslides. *GeoHealth, 4*(10), e2020GH000287.

Rahman, G., Bacha, A. S., Ul Moazzam, M. F., Rahman, A. U., Mahmood, S., Almohamad, H., Dughairi, A. A. A., Al-Mutiry, M., Alrasheedi, M., & Abdo, H. G. (2022). Assessment of landslide susceptibility, exposure, vulnerability, and risk in Shahpur valley, eastern Hindu Kush. *Frontiers in Earth Science, 10*, 953627.

Rai, P., & Khawas, V. (2019). Traditional knowledge system in disaster risk reduction: Exploration, acknowledgement and proposition. *Jàmbá: Journal of Disaster Risk Studies, 11*(1), 1–7.

Ramzan, M., Cui, P., Ualiyeva, D., Mukhtar, H., Bazai, N. A., & Baig, M. A. (2025). Impact of climate change on landslides along N-15 Highway, northern Pakistan. *Advances in Climate Change Research.*

Reichenbach, P., Rossi, M., Malamud, B. D., Mihir, M., & Guzzetti, F. (2018). A review of statistically-based landslide susceptibility models. *Earth-Science Reviews, 180*, 60–91.

Research Institute for Geo-Hydrological Protection, Italian National Research Council (CNR). (n.d.). Effects of climate change on landslides. https://www.irpi.cnr.it/wp-content/uploads/2019/12/focus-effects-of-climate-change-on-landslides.pdf

Schuster, R. L., & Highland, L. M. (2003, December). Impact of landslides and innovative landslide-mitigation measures on the natural environment. In *International Conference on Slope Engineering, Hong Kong, China* (Vol. 8, No. 10, p. 46).

Shahabi, H., & Hashim, M. (2015). Landslide susceptibility mapping using GIS-based statistical models and remote sensing data in tropical environment. *Scientific Reports, 5*(1), 9899.

Shakya, A., Sigdel, P., & Pokhrel, S. (2023). Socio-economic impacts of landslides and applied mitigation techniques: A case study in Bagnaskali, Nepal. *Journal of Sustainability and Environmental Management, 1*(2), 170–179.

Svennevig, K., Koch, J., Keiding, M., & Luetzenburg, G. (2024). Assessing the impact of climate change on landslides near Vejle, Denmark, using public data. *Natural Hazards and Earth System Sciences, 24*(6), 1897–1911.

Tensar International. (2025). *What is a landslide? Types, causes, effects & prevention*. Retrieved from https://www.tensarinternational.com/resources/articles/what-is-a-landslide

The Renaissance Official. (2023). *Post-disaster vulnerabilities and disaster-induced changes in the life and livelihood of Kattupaniya tribal community: A case study of the Kavalappara landslides*.

Transportation Research Board. (1991). Socioeconomic significance of landslides. In *Landslides: Analysis and Control* (Special Report 247, pp. 12–23). National Research Council.

Turyasingura, B., Ayiga, N., Benzougagh, B., Kader, S., Singh, S. K., Bosco, N. J., Gweyi-Onyango, J. P., & Bojago, E. (2023). The complementary role of indigenous knowledge systems in landslide disaster management in Kanungu District, Uganda. *Nova Geodesia, 3*(4), 157–157.

Varangaonkar, P., & Rode, S. V. (2019). Methods of landslide detection using GIS and remote sensing images. *International Journal of Engineering and Advanced Technology, 9*(2), 2121–2125.

Vasileiou, K., Barnett, J., & Fraser, D. S. (2022). Integrating local and scientific knowledge in disaster risk reduction: A systematic review of motivations, processes, and outcomes. *International Journal of Disaster Risk Reduction, 81*, 103255.

Winter, M. G., Peeling, D., Palmer, D., & Peeling, J. (2019). Economic impacts of landslides and floods on a road network. *AUC Geographica, 54*(2), 207–220.

Xing, Y., Chen, Y., Wang, P., Huang, S., & Xiang, Y. (2024). Research on landslide master control factor identification and susceptibility prediction modelling. *Egusphere, 2024*, 1–24.

Xu, W., Wang, Q., Yang, J., Yuan, B., & He, C. (2024). A research on the relationship between landslide area changes and environmental factors in the southern Tibetan plateau. *Frontiers in Earth Science, 12*, 1403888.

Yuvaraj, R. M., & Dolui, B. (2023). Geographical assessment of landslide susceptibility using statistical approach. *Quaternary Science Advances, 11*, 100097.

Zhang, P. (2024). Study on slope stability analysis and reinforcement technology based on geosynthetics. *Academic Journal of Architecture and Geotechnical Engineering, 6*(1), 61–65. https://doi.org/10.25236/AJAGE.2024.060109

Zumpano, V., Pisano, L., Borrelli, P., & Antronico, L. (2018). Economic losses for rural land value due to landslides. *Frontiers in Earth Science, 6*, Article 97.

www.ingramcontent.com/pod-product-compliance
Ingram Content Group UK Ltd.
Pitfield, Milton Keynes, MK11 3LW, UK
UKHW020632260926
475829UK00003BA/38
* 9 7 8 3 0 3 2 1 4 4 8 4 3 *